Sanitation for Foodservice Workers

Third Edition

Sanitation for Foodservice Workers

THIRD EDITION

Treva M. Richardson
Wade R. Nicodemus

VAN NOSTRAND REINHOLD
New York

Library of Congress Catalog Card Number 81-2144

ISBN 0-8436-2205-9

Printed in the United States of America

Designed by Jack Schwartz

Van Nostrand Reinhold
115 Fifth Avenue
New York, New York 10003

Van Nostrand Reinhold International Company Limited
11 New Fetter Lane
London EC4P 4EE, England

Van Nostrand Reinhold
480 La Trobe Street
Melbourne, Victoria 3000, Australia

Nelson Canada
1120 Birchmount Road
Scarborough, Ontario M1K 5G4, Canada

16 15 14 13 12 11 10 9 8 7 6 5

Library of Congress Cataloging in Publication Data

Richardson, Treva M.
Sanitation for foodservice workers.

Bibliography
Includes index.
Food service-Sanitation. 2. Food handling.
I. Nicodemus, Wade K. II. Title
TX943.R5 1981 363.7'296 81-2144
ISBN 0-8436-2205-9 AACR2

Contents

Preface to the Third Edition

Sanitation for Foodservice Workers is designed to educate persons who have an interest in the field of food sanitation. Foodservice management and personnel, students, housewives—any person who has a desire to learn of the origins and consequences of foodborne illness and the measures that are used to prevent it will profit from this book. The book also provides information about and reasons for the regulations and procedures necessary to maintain a successful sanitation program in a foodservice establishment.

Such education must include more than the passing on of information, however. Following correct procedures, rules, and proper personal habits must become a way of life. It is hoped that the many case histories presented in this book will aid in establishing this philosophy and make a lasting impression on the reader.

The foodservice industry today is a complex business. Eating habits and distributing procedures have changed immensely since World War II. It is not uncommon for a finished product to have had many contributors; thus, each ingredient possesses the possibility of contamination. The finished product may be shipped thousands of miles from the plant of the final processor. This multiple handling of food in all phases and the long distance movement from manufacturer to consumer requires extreme care and no small amount of regulation to protect the person who buys the food. The consumer depends upon the manufacturer for his health and well being. Unfortunately, many foods can be easily contaminated and many can also readily support the growth of those bacteria and molds that can cause foodborne illness.

The foodservice/lodging industry is America's largest employer and one of the largest dollar-volume industries. Over 160 million meals per day are served in this nation's food and beverage establishments. In 1979, the restaurant industry reported sales of approximately 102 billion dollars. It is essential that every individual connected with the manufacture, transportation, preparation, and serving of food is well acquainted with the measures that are necessary to protect the health of the consuming public. That is what this book is all about.

Fortunate indeed are those foodservice managers whose employees have been trained in sanitary procedures, are knowledgeable as to why such practices are necessary, and possess a fund of interesting, clearly understood food sanitation facts applicable to their specific work routines. It has been said that the successful presentation of a modern sanitation course to foodservice personnel represents a greater benefit to management than almost any other food industry development of recent years.

The growth of the training movement in the United States is interesting. It was begun in 1938 by two state health departments. Soon, other states and communities were experimenting with training programs for "food handlers" as they were termed at that time. During World War II the training was intensified as a means of reducing the number of foodborne disease outbreaks. By 1946 the attendance of foodservice personnel at health department training sessions was required by many community food sanitation ordinances.

Sanitation-conscious food industry leaders cooperated with health agencies in selecting material essential for industry use. However, health departments with limited staffs often found it difficult to carry on continuous, effective programs. Acceptable industry-developed courses made their appearance, taught by industry-employed sanitarians, dietitians, etc., or by supervisory personnel.

By the late 1970s, many state health agencies had once again assumed the major role in training foodservice personnel, due to action by a number of state legislative bodies. These assemblies passed laws requiring persons who contact food during selling, such as meat cutters, or who prepare food commercially, to be trained and certified in sanitary procedures. This appears to be the trend at the present time.

Sanitation for Foodservice Workers is divided into six main sections. The first three chapters contain essential information about bacteria, molds, yeasts, and other organisms that cause foodborne diseases. The material is presented in an easily readable format and sets the tone for reader understanding of the dangers of microorganisms and the reasons for sanitation. The physical characteristics and growth patterns of bacteria are stressed, along with the importance of the physical environment. The major foodborne communicable diseases are also discussed.

Chapters four through eight present the histories and major characteristics of the most common and feared food poisoning organisms in the United States. The major emphasis is placed on *Staphylococcus aureus, Clostridium perfringens, Clostridium botulinum,* and the Salmonella. Many case histories are utilized to portray the effect of these bacteria on unfortunate human consumers. These chapters will show the reader where the bacteria live in nature and which foods are most likely to contain them.

Pest control is a vital part of any sanitation program. Insect pests such as cockroaches and flies and rodent pests such as rats and mice carry diseases, spoil stored food, and, if noticed, can destroy customer confidence. Methods of detection and control are therefore thoroughly discussed with a heavy emphasis on proper maintenance of the physical environment and the safe storage of food.

The fourth section is devoted to effective dishwashing and sanitizing procedures. Different types of dishwashing machines and how they do their cleaning are discussed. The reader is also acquainted with the makeup of detergents and sanitizers and how they do their job. Emphasis is placed upon reasons for the use of these agents and protective measures that should be taken to ensure that patrons receive table service that is both clean and safe.

Food protection is related not only to sanitary practices, but also to the ability of the personnel to maintain a sanitary establishment. The physical characteristics of the plant are important in this respect. If, for example, the floors, counters, and walls are not made of easily cleanable materials and the lighting is inadequate, bacteria can accumulate and spread to food and utensils. It is important that the reader is acquainted with this facet of sanitation in addition to procedures such as clean hands, prompt refrigeration, and proper holding temperatures, which are designed to prevent contamination of food.

The final section consists of two chapters designed to give the reader a historical perspective of the development of some of the more interesting and controversial legal philosophies which envelop the foodservice industry, such as the implied warranty of merchantibility and the "foreign or natural" concept of substances that sometimes make their appearance in food. These chapters are not essential to an understanding of foodservice sanitation but can be enjoyably read and discussed. They are documented with many case histories and courtroom decisions.

Acknowledgments

This book, like many others, could not have been written without the aid of several key persons. I would like to acknowledge a debt of gratitude to Catherine Bradley, Librarian, and Eleanor Egizii, Assistant Librarian at the Illinois Supreme Court Library. Chapters 15 and 16 could not have been written without their invaluable library assistance. My special thanks also to Genevieve Hanover, Mason City Librarian, whose diligence in securing reprints, many from overseas, was most appreciated.

I would also like to thank my staff of young typists, Tina Rakestraw and Cheryle Otten, who, in addition to handling my classroom paperwork also found time to type much of this book. Jennie Jones and Tami Wilcoxson were also valuable contributors on the typing team. Special plaudits are due to Trisha Anderson, whose proofreading probably preserved my sanity.

And last, but certainly not least, I am grateful for the patience of my lovely wife, Lois, and my three children-at-home, Mike, Patti, and Joann who suffered through much clutter and many lonely evenings in order to get this revised edition to press.

Wade R. Nicodemus

Sanitation for Foodservice Workers

Third Edition

1 Bacteria

I. Bacteria
  A. Where Found
  B. Classification
    1. History
    2. Scientific Names
II. Characteristics
  A. Physical Characteristics
  B. Staining
  C. Metabolic Tests
  D. Serotypes
  E. Vegetative Cells
III. Use
IV. Influence of Environment
  A. Temperature
  B. pH
  C. Spores
V. Toxins
  A. Exotoxins
  B. Endotoxins
VI. Reproduction
  A. Generation Time
  B. Exponential Growth
  C. Growth Curve
VII. Summary

BACTERIA

It is mid-August, 1942, in a large military camp in the South. A division of troops has arrived for a training bivouac. The temporary encampment is soon erected and training begins. Within a week, a case of dysentery appears among the troops, and by September 1st, a full-scale epidemic has begun that will rage for the entire month. By the time the epidemic runs its course, a healthy division of 15,000 men will have been crippled by 1,557 cases of dysentery. The culprit will be revealed as a type of bacterium, *Shigella paradysenteriae*, transmitted to the soldiers through their food in the mess hall.

In June and July of 1963, 1,190 cases of food poisoning resulted from another type of bacterium that had made its home in spray-dried milk, which was used in preparing food in school canteens in England. This culprit was *Staphylococcus aureus*.

In 1953, in Sweden, different bacteria, called *Salmonella typhimurium*, penetrated a slaughterhouse and multiplied prodigiously. The resultant slaughter occurred, not only in the slaughterhouse, but in the thousands of homes that experienced 8,845 cases of food poisoning and 90 deaths. The actual number of cases probably exceeded 10,000. The salmonellae reached the population through infected meat.

Food poisoning, with its accompanying nausea and diarrhea is a very unpleasant experience, even in the comfort of one's own home. One hundred eighty-eight cases of food poisoning within the small confines and limited restroom facilities of a railway train is a disaster. Just such an event did occur on a train in 1959. The type of bacterium involved was *Clostridium perfringens*. It reached its human destination through roast turkey dressing.

In 1970, in Sioux City, Iowa, bacteria somehow lodged on a restaurant meat slicer. Before they were discovered and killed, they caused 250 cases of food poisoning through the various meats, such as roast beef, that were sliced by this equipment. The offending bacteria in this case were *Salmonella enteritidis*.

The five outbreaks of foodborne illness just discussed involved about 12,000 people, in whose lives these bacteria played a temporary but dominating role. What are bacteria? What roles do they play in our lives?

Where Found

The great majority of bacteria are either harmless or helpful to us. Some cause disease, but most live out their lives without ever having contact with people.

Bacteria can be found anywhere and everywhere. They have been found in the air, in lakes and streams, mud, hot springs, soil, snow, and in and on all forms of life, both living and dead.

Most bacteria are saprophytes. They obtain their needed nourishment from dead or decaying forms of life. In meeting their own nutritional needs, they cause decay by decomposing the substances they inhabit. They are generally beneficial, therefore, except when they cause decay in those meats, vegetables, and fruits that we use as our own food.

Some species of bacteria are parasites. They obtain their life requirements from the tissues of living things and frequently cause disease. Disease-producing forms are called pathogens. Some diseases caused by pathogenic bacteria are:

Disease	Bacterial Agent
Tetanus	*Clostridium tetani*
Diphtheria	*Corynebacterium diphtheriae*
Food poisoning	*Staphylococcus aureus, Clostridium botulinum, Salmonella enteritidis*
Dysentery	*Shigella dysenteriae*
Typhoid fever	*Salmonella typhi*
Whooping cough	*Hemophilus pertusis*

Disease	Bacterial Agent
Tuberculosis	*Mycobacteria tuberculosis*
Cholera	*Vibrio cholera*
Scarlet fever	*Streptococcus pyogenes*

Classification

Bacteria have traditionally been considered microscopic members of the plant kingdom. They are, in fact, very old members of the plant kingdom. The ancestors of present-day bacteria were among the earliest forms of life on this planet and probably existed even before animal life. They were undoubtedly an important food source for the first tiny animal life.

History. The first observer of the microscopic world to which bacteria belong was Antony Van Leeuwenhoek (1632–1723), a Dutchman whose hobby was making lenses. He used these lenses in constructing primitive but ingenious microscopes for his own personal use. While observing stagnant rain water, saliva, plant leaves, cow dung, and many other substances, he discovered the world of bacteria, protozoa (one-celled animals), and many other microscopic forms of life.

It was not until many years later, about 200 in fact, that the Frenchman Louis Pasteur (1822–1885) related bacteria to "diseases" (spoiling) of wine and beer, and the German Robert Koch (1843–1910) discovered that anthrax (a disease of domestic animals that is sometimes transmitted to man) was caused by a specific type of bacterium.

Scientific Names. All forms of life have scientific names. This is necessary because the same plant or animal may exist in different parts of the world, yet have different local names. The bird called the house sparrow in England is a good example. This same bird occurs in many countries but under the following names:

Denmark	Graaspurv
England	House Sparrow
Germany	Haussperling
Holland	Musch
Portugal	Pardal
Spain	Gorrion
Sweden	Hussparf

The bird is also a resident of the United States where it is called the English Sparrow.

It is sometimes necessary for scientists in different countries to discuss various forms of life with regard to control, preservation, use, and research. All forms of life therefore must have, in addition to their common or local names, a worldwide name. This name is called its scientific name. The scientific names of five food-poisoning bacteria were mentioned earlier in this chapter.

Since all forms of life have scientific names, it will be helpful if we take some time to understand the classification system that is used. There are generally considered to be two major groups of living things called the plant kingdom and the animal kingdom. For our example, we will classify man, a member of the animal kingdom.

There are many kinds of animals, such as, worms, crayfish, clams, jellyfish, snakes, cows, elephants, and, of course, birds and bees. Therefore, animals are placed in certain groups. These groups are called *phyla* (singular: *phylum*) and all of the animals within a phylum have at least one major characteristic in common. Man is placed in the phylum of the chordates because he has a nerve cord (spinal cord) running the length of his back. This is the major characteristic of chordates.

There are, however, many different kinds of chordates, that is many different kinds of animals with spinal cords. Fish, frogs, snakes, birds, mice, and man all have spinal cords. The phylum of chordates is therefore divided into a number of classes, one of which is called mammals. Mammals are warm-blooded chordates with hair and nails. This description fits man, even though some of us may be a bit short on hair. Man, therefore, belongs to the class of mammals.

There are many kinds of mammals. There are gnawing mammals (mice), flying mammals (bats), hoofed mammals (horses), and among other kinds, upright mammals. This group, or order of mammals, that includes man is called the primates.

The order of primates is divided into a number of families. One of the families is the family of man. It includes all of the types of man who have trod the face of this earth over the past two million years or so. The family of man includes several groups or *genera* (singular: *genus*) of man. Present-day man belongs to the genus *Homo* of the family of man.

A genus contains one or more species or distinct entities. A genus and species name constitute the scientific name of a life form. Man's scientific name is *Homo sapiens;* Neanderthal man's is *Homo neanderthalensis.*

The following summary of the classification scheme shows each group and what it contains.

Group	Contains
Phylum	Classes
Class	Orders
Order	Families
Family	Genera
Genus	Species

There is some lack of agreement at the present time as to which phylum bacteria belong in. They are a very old and primitive form of life. Bacteria are, however, a class of one-celled living things. This class contains ten orders of bacteria. Each order contains families, each family consists of several genera, and each genus contains species. Although there are, in all, about 1,600 species of bacteria, only a small percentage of these cause disease.

At the beginning of this chapter, five outbreaks of food poisoning were discussed, each caused by a different species of bacteria. Two of these, *Salmonella typhimurium* and *Salmonella enteritidis,* belong to the same genus. Species in different genera are not as closely related to each other as species within a genus. The names of bacteria give us information about relationships and tell us exactly which individuals we are discussing.

CHARACTERISTICS

Physical Characteristics

Bacteria are very small. In fact, they are microscopic. Their size is measured in units called microns. One micron is .000025 (one twenty-five thousandths) of an inch, or, expressing it another way, there are about 25,000 microns in one inch. Most bacteria are about 0.5–2.0 microns in width and 0.5–10 microns in length. Many, however, vary from these sizes.

You would naturally expect such a small object to weigh very little. This is certainly true. A single bacterium may weigh as little as 4/10,000,000,000,000 (four ten-trillionths) of a gram (there are 28 grams in one ounce).

Bacteria tend to occur in one of three basic shapes: the coccus (plural: cocci), which is sphere shaped; the rod or bacillus (plural: bacilli), which is cylinder shaped; and the spirillum (plural: spirilla), which is spiral or corkscrew shaped (Fig. 1-1). Some species of coccus-shaped bacteria occur as grapelike clusters of cells and are called staphylococci (Fig. 1-2). Other cocci occur in chains and are called streptococci (Fig. 1-3).

Staining

Since bacteria are extremely small, they must be treated with a stain to be observed. The acceptance or rejection of a chemical stain can reveal information about the chemical and physical makeup of the bacteria being studied.

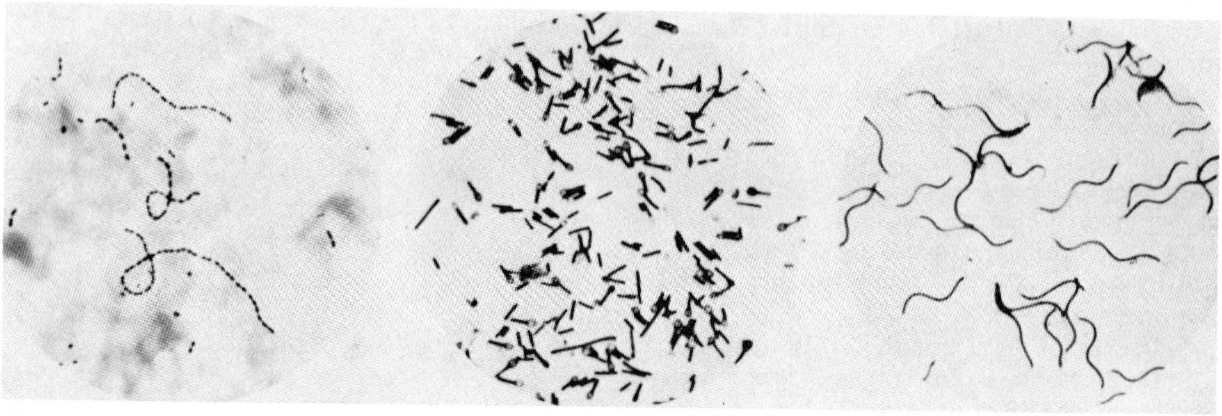

Figure 1-1 The three cell shapes that occur among bacteria: (a) coccus; (b) rod; (c) spiral. From Stanier/Doudoroff/Adelberg, *The Microbial World*, © 1970, 3rd edition, p. 120. Reprinted by permission of Prentice-Hall, Inc., Englewood Cliffs, New Jersey.

A commonly used staining procedure is the Gram stain, developed in 1884 by a Danish physician, H. C. J. Gram (1853–1938). The cells are stained with purple dye. A decolorizing agent is then added. Some bacteria retain the dye in spite of the decolorizing agent. These bacteria are called gram-positive. Bacteria that become decolorized are called gram-negative. Gram-negative bacteria will then usually accept a red dye. As a result of this difference in color, gram-positive and gram-negative bacteria can be distinguished easily from each other with a microscope.

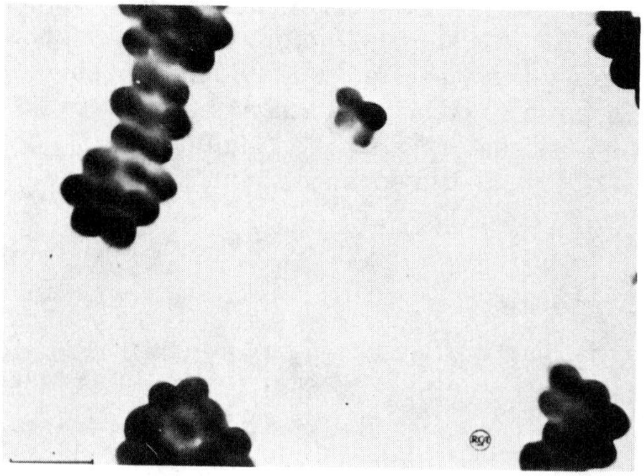

Figure 1-2 Staphylococci

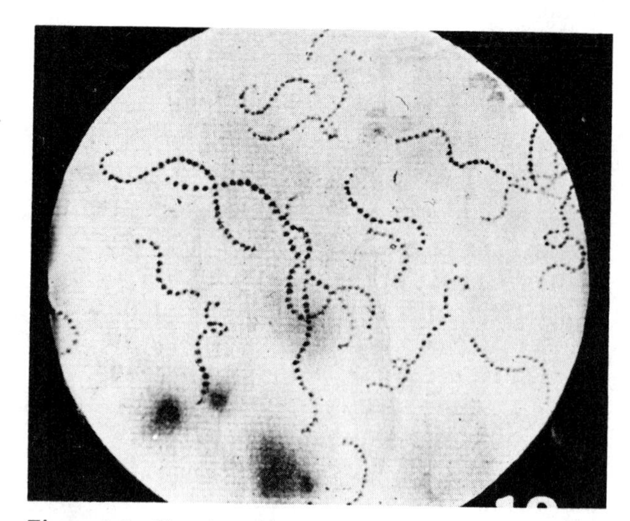

Figure 1-3 Streptococci

These two types of bacteria differ in several important respects. Gram-positive bacteria, for example, are much more sensitive to penicillin than gram-negative bacteria. The two types of bacteria tend to produce different types of toxins (chemicals liberated by some bacteria that are harmful to the organism harboring the bacteria). Gram-positive bacteria tend to be less sensitive to freezing than gram-negative bacteria. Gram-positive rods are more sensitive to radiation than gram-negative rods.

Metabolic Tests

We have seen that living things are usually classified according to structures that they possess (mammals have hair and nails, for example). This method of classification has limited application with bacteria, however. Since they consist of only one cell, there are a limited number of structures available to observe. Therefore, other characteristics must be used in conjunction with physical traits. These characteristics reflect what the different species of bacteria can do. The tests for these characteristics are called biochemical or metabolic tests.

In attempting to identify a type of bacterium that is suspected of causing a foodborne illness, we must first raise many of them and observe their physical traits. Then we proceed to biochemical tests.

Our initial task is to find a medium on which these bacteria will thrive. Not all bacteria can grow or grow well on the same medium. Different bacteria have different nutritional needs depending on their chemical makeup. Once we have discovered the best growth medium for our bacteria, we can eliminate as possibilities all other species that do not grow well on this medium.

As our original bacteria reproduce from small numbers to millions they tend to form a colony, a visible mass on the culture medium. We can now note certain physical characteristics of this colony. Do these bacteria form a large or small colony? What color is it? What is its shape? Is it smooth or rough, flat or domed in appearance? These are physical traits and very helpful to us in eliminating more possibilities.

At the same time we can determine whether these bacteria require oxygen or not. If they use oxygen from the air they are called aerobic bacteria. If they cannot grow in the presence of air they are called anaerobic bacteria. *Clostridium botulinum*, which causes botulism, is anaerobic; *Salmonella newport*, which is one of several salmonellae that cause salmonellosis food poisoning, is aerobic. We can also determine at what temperature our bacteria grow best (optimum temperature). We can apply Gram's stain to a sample from a colony and determine whether they are gram-positive or gram-negative. The food-poisoning strains of *Staphylococcus aureus* are gram-positive whereas the salmonellae are gram-negative.

It is now desirable, and probably necessary, to subject the bacteria to biochemical tests to determine which species we have. Samples can be added to separate media containing various carbohydrates, such as, lactose, glucose, sucrose, or mannitol, to find out if they will ferment any of these carbohydrates. This ability also varies among bacteria. We can also test for the ability to use different proteins or amino acids. The chemical, indole, sometimes results from the breakdown of a particular amino acid. Indole is routinely tested for, since only some species produce it.

Different bacteria also produce gases typical for the species. Some of the possibilities are hydrogen sulfide, carbon dioxide, hydrogen, methane, ammonia, and nitrogen. Other tests will determine whether or not our bacterium can digest gelatin, coagulated serum, or the casein in milk. It may also digest various starches and fats. Among other much-used tests are a determination of whether it can use sodium nitrate or sodium nitrite and whether it can decolorize the dyes methylene blue and litmus.

These are some of the tests that can be used to identify a type of bacterium, if they are properly done. However, they must be carried out by skilled personnel in a properly equipped laboratory. The tests also take time. A week may pass between the food-poisoning episode and the identification of the culprit. This creates obvious handicaps in dealing with a food-poisoning situation.

Serotypes

Many living things contain chemicals, mostly certain proteins, which stimulate certain cells of the host they have entered to produce chemicals called antibodies. These chemicals in invading bacteria, viruses, pollen, and so forth, which cause the antibody formation, are called antigens. It is the antigens in pollen, bee stings, and various foods that cause hay fever and other allergies. Every antigen causes the formation of a specific antibody. That antibody will react specifically with the antigen that caused its formation, and with no other.

Many bacteria can be classified or identified by the antigens they possess. They are often used to distinguish different strains of a species. For example, at one time all pneumococci (pneumonia-causing, coccus-type bacteria) were thought to be identical. Then it was discovered that certain complex carbohydrates that occurred in their outer capsule fell into some 75 or more distinct antigenic groups. These groups are called serotypes.

Antigenic differences have been found among the streptococci, clostridia, shigella, and salmonellae. These groups all contain food poisoning organisms. Through the use of serologic procedures, bacteria can be described in terms of antigens in greater detail than is possible by other means.

Vegetative Cells

The form of bacterium encountered most frequently in the environment is the vegetative cell. The vegetative cell is the active, growing, and reproducing form of bacterial cell. These cells are not very resistant to heat. Most vegetative cells are killed by exposure to a temperature of 160°F. (71°C) for several minutes and a temperature of 127°F. (53°C) is all that is required to prevent multiplication of the vegetative cells of any of the food-poisoning bacteria.

Vegetative cells are much more resistant to cold than to heat. Although all food-poisoning types cease multiplication by 37°F. (3°C), many of their vegetative cells survive not only the freezing process, but months of freezer storage as well.

USE

Many species of bacteria have been utilized by man in the manufacture of various products. Some are used to produce foods, such as, cheese, pickles, and sauerkraut; others produce antibiotics and vitamins. Bacteria are used to produce acetone and butanol, two important industrial chemicals.

Sewage treatment relies partly on bacterial decomposition. Bacteria are also utilized in curing tobacco. Nitrogen-fixing bacteria live in the root nodules of legume plants, such as, clover, alfalfa, and peas. These bacteria convert atmospheric nitrogen into chemicals needed for plant growth. Some of the uses of bacteria are shown in the following list.

Type of Bacteria	Product Made
Acetic acid bacteria	Vinegar
Lactic acid bacteria	Cheese and butter
Lactobacillus delbrueckii	Commercial lactic acid
Leuconostoc mesenteroides	Dextran (a substitute for blood plasma)

Type of Bacteria	Product Made
Butyric acid bacteria	Acetone and butanol (industrial chemicals)
Bacillus subtilis	Enzymes useful in the textile, beer, and tanning industries
Streptomyces olivaccus	Vitamin B_{12}
Streptomyces griseus	Streptomycin
Nitrogen-fixing bacteria	Soil chemicals needed for plant growth

INFLUENCE OF ENVIRONMENT

Temperature

The life of a bacterial vegetative cell is a tenuous one. It is totally at the mercy of its environment. A variety of factors, some physical, some chemical, regulate or affect its life and growth.

Temperature affects the life and growth of all living things. The temperature range within which life can exist extends from about 23° to 194°F. (−5° to 90°C). The lower limit exists because living things are composed largely of water and water solutions have freezing points near 32°F. (0°C). The upper temperature limits are set by the melting points of various essential fatty compounds in the cells and by the stability at high temperatures of vital proteins and nucleic acids. Within this range of temperature, every living organism has its own survival range and its own optimum growth range governed by its particular chemical makeup.

With respect to temperature range, bacteria fall into one of three groups. Those bacteria that grow best below 68°F. (20°C) are called psychrophiles. Bacteria that grow best above 113°F. (45°C) are called thermophiles. Mesophiles grow best between these two temperatures.

The bacteria involved in foodborne illnesses are mesophiles. High and low temperature extremes either prevent their growth or kill them. Therefore, thorough cooking, effective reheating, and prompt refrigeration of foods are the keys to successful prevention of foodborne illnesses.

For each species of bacteria there is a temperature at which it is most active and at which growth and reproduction occur most rapidly. This is called the optimum growth temperature. It is the temperature at which the major chemical systems of the bacteria function most efficiently. These systems are disrupted by temperatures that are too high or too low. The optimum growth temperatures of the common food-poisoning bacteria follow.

Species	Optimum Growth Temperature	
Clostridium botulinum, types A & B	95°F.	35°C
Clostridium botulinum, type E	86°F.	30°C
Clostridium perfringens	115°F.	46°C
Staphylococcus aureus	95°F.	35°C
Salmonella species	99°F.	37°C

The highest temperature at which a species is capable of growth is called its maximum growth temperature; the lowest temperature at which it will grow is its minimum growth temperature. As the temperature rises above the maximum growth temperature, the vegetative cells die rather quickly. At temperatures below the minimum growth temperature, the bacteria do not grow or multiply, but neither are they killed. *Staphylococcus aureus*, a very common food-poisoning species, survives freezing very well.

High temperatures are much more injurious to bacteria, therefore, than low temperatures. High temperatures kill. Low temperatures only prevent or slow growth and multiplication. Bacteria are still present and will grow and multiply if the food in which they exist is allowed to warm to their minimum growth temperature or higher. Foodborne illness can then result, with disastrous consequences for a foodservice establishment.

Cold temperatures, particularly subfreezing temperatures, are sometimes aided by storage time. It has been determined (Straka, 1969) that for many bacteria, the number killed increases as storage time at subfreezing temperatures increases.

(at 0°F.) Days of Storage	Percent of Initial Population Killed
1	22
5	38
11	44
19	53

As you can see from Table 1-1, the full range of temperatures that encompasses the growth range of the most prevalent food-poisoning bacteria extends from 38°F. to 122°F. (3 to 50°C). Food should be allowed to exist for only short periods of time within this range.

pH

The pH of a medium denotes whether it is acidic or basic (alkaline). The pH scale of measurement extends from 0 to 14 (see the following scale). A pH of 7 represents a neutral environment, neither acidic nor basic in nature. If the pH of the medium falls between 0 and 7 it is acidic, with the acid characteristics becoming greater as the pH numbers become lower. A pH of 6 is ten times more acidic than a pH of 7. A pH of 5 is ten times more acidic than a pH of 6. If the medium has a pH of 8, it is basic. A pH of 8 is ten times more basic than a pH of 7.

The pH scale

pH	
1	
2	
3	Acid
4	
5	
6	
7	Neutral
8	
9	
10	
11	Basic
12	
13	
14	

Most bacteria, including the food-poisoning bacteria, grow well at pH values between 6.0 and 9.0. Beamer and Tanner studied the heat resistance of *Staphylococcus aureus* in fruit juices whose pH was between 2.85 and 4.2 and found that heat resistance depends on pH. It was lowest in gooseberry juice (pH 2.85) and

Table 1-1
Growth range of common food-poisoning bacteria.

Species	Growth Range Minimum	Growth Range Maximum
Staphylococcus aureus	44°F. (7°C)	114°F. (46°C)
Clostridium perfringens	55°F. (13°C)	122°F. (50°C)
Clostridium botulinum	38°F. (3°C)	118°F. (48°C)
Salmonella species	42°F. (6°C)	114°F. (46°C)

highest in tomato juice (pH 4.2). In other words, as the pH became more acidic and farther away from the desirable growing conditions, this species of bacteria became less resistant to high temperature. This rule applies, generally, among the bacteria. White studied the effect of pH on the heat resistance of *Streptococcus fecalis* and found that it was lowest at the lowest pH tested (pH 2.8) and highest at the highest pH tested (pH 6.6).

Salmonella are completely inhibited in foods with a pH less than 4.5. Although many foods will support the growth and toxin production of the staphylococci, foods with a pH below 5.0 inhibit toxin formation (Bergdoll, 1969).

Figure 1-4 shows the growth of *Clostridium perfringens*, a common food-poisoning species, at different pH conditions. As you can see, this species exhibits growth only between a range of pH 5.5–8.0. Most foods fall within this pH range. Other pH levels in foods are:

> Low acid foods: pH above 5.3—meat, milk, poultry, seafood, peas, corn, lima beans.
>
> Medium acid foods: pH 4.3–5.3—spinach, asparagus, beets, pumpkin.
>
> Acid foods: pH 4.2 and below—berries, sauerkraut, tomato juice.

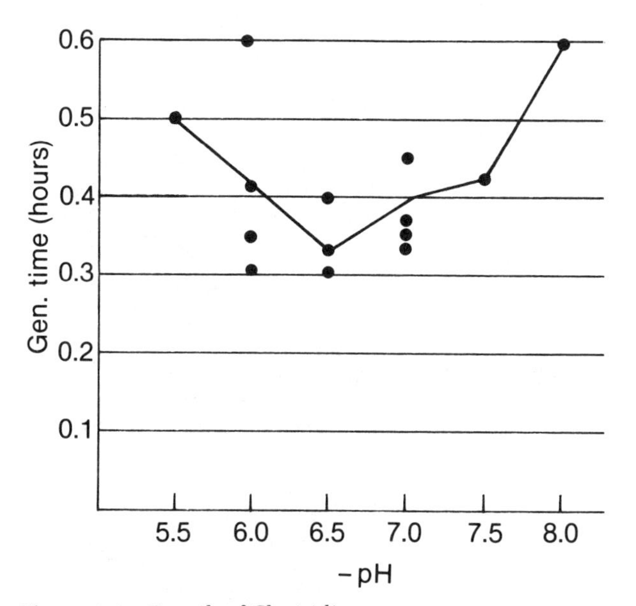

Figure 1-4 Growth of *Clostridium perfringens* at different pH conditions. The generation time is the time that it takes for a colony of bacteria to double in number. A short generation time indicates rapid growth.

All bacteria have a maximum and minimum pH value, above or below which they cannot grow (Table 1-2). The pH of the food also affects the growth rate of bacteria and their resistance to heat. The closer the pH approaches the maximum or minimum value for any species of bacteria, the slower they grow and the less resistant they are to heat.

Table 1-2
Maximum and minimum pH values for some food-poisoning bacteria.

Species	pH Value Minimum	pH Value Maximum
Clostridium botulinum	4.7	8.9
Clostridium perfringens	5.0	9.0
Staphylococcus aureus	3.0	9.0
Salmonella species	3.4	9.0

Spores

Certain of the rod-shaped bacteria are capable of producing a highly resistant, inactive cell called an endospore (Fig. 1-5). Each rod produces one spore. Spore forming, or sporulation as it is properly called, is not a type of reproduction since each vegetative cell produces only one spore. There is, therefore, no increase in numbers due to the sporulation process.

Spore formation is a survival mechanism, and a very good one. A spore can survive conditions that are unfavorable for the vegetative cell since it resists heat, freezing, drying, and most sanitizing chemicals. Since a spore is inactive it requires no food. Its ability to with-

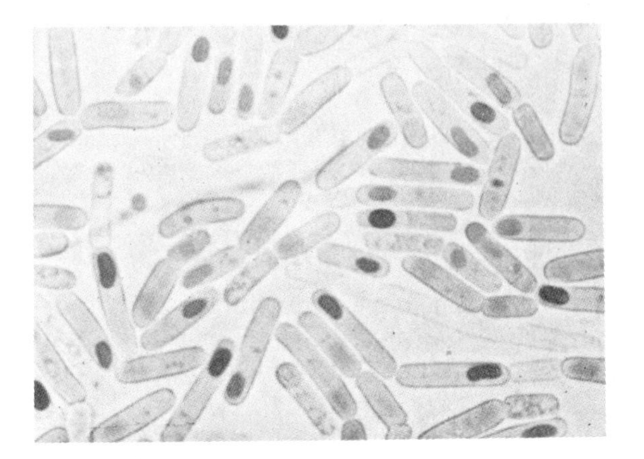

Figure 1-5 Developing endospores in a bacillus. From R. G. Wyckoff and A. L. ter Louw, "Some Ultraviolet Photomicrographs of B. subtilis," *J. Exp. Med.* (1931) 54:3, p. 451.

stand destructive agents is probably unequaled by any other form of life.

Spore formation can also provide for the spread of the organism. Since a spore is very light it can be carried through the air to a new medium. When conditions are proper for bacterial growth, that is, food, moisture, proper pH, and temperature, it will germinate into a vegetative cell and start a new colony or growth.

In this way the organism can survive long periods of adverse conditions, if necessary. Spores of the anthrax bacillus (a disease of cattle that can be transmitted to man), for example, have been found capable of germination after a lapse of as much as 60 years. Spores have been recovered and grown from canned goods that were 115 years old. As you can see, the spore form of a bacterium is very difficult to kill. Its major property, as far as the food industry is concerned, is its ability to withstand heat. This property is called heat resistance.

Although spores of some of the thermophilic bacteria may require up to several minutes of 179°–194°F. (80°–90°C) temperatures to kill them, the vegetative cells of most bacteria are killed within the temperature range of 131°–149°F. (55°–65°C). In other words, vegetative cells are rather easily killed by proper cooking heat. Spores, on the other hand, while varying in heat resistance according to species, are far more resistant than their corresponding vegetative cells. For example, the vegetative cells of *Clostridium botulinum* and those of the spoilage-causing, flat-sour bacteria are easily killed. The spores of *Clostridium botulinum*, however, can survive boiling temperatures for three hours or more and the spores of the flat-sour bacteria can withstand boiling temperatures for well over 12 hours.

Among the factors affecting the survival of spores in a food medium are:

1. Time/temperature relationships. The time that it takes to kill the spores existing in a food depends on the temperature used. Temperature, understandably, is the key tool at our disposal to make our foods safe. The higher the temperature, the less time required to destroy spores that may be present. The same is true in the case of vegetative cells, of course, but the temperature must be considerably higher to kill spores. As an example of the importance of temperature, note that at 212°F. (100°C) it takes about 1200 minutes (20 hours) to kill a large number of flat-sour bacteria spores; at 230°F. (110°C) the time drops to 190 minutes or about 3 hours; at 248°F. (120°C) the time required is 19 minutes.
2. The number of spores present. In general, the greater the number of spores that are present, the longer it takes to kill them. This is probably due to a protective effect, as the innermost spores in clumps are protected by the outer ones from the high temperature.
3. The organism. Different species of bacteria vary in their heat resistance. Resistance will vary even among different types and strains of the same species. The heat resistance of spores of the type E strain of *Clostridium botulinum* is only about one-thousandth that of types A and B. The spores of different strains of *Clostridium perfringens*, another common food-poisoning species, also vary considerably in their heat resistance.

TOXINS

Exotoxins

Since bacteria are living organisms, chemical reactions occur within them just as they do within all forms of life, both plant and animal. These chemical reactions are called metabolism. The metabolism of some bacteria results in the production of chemicals that damage or kill plant or animal cells. These chemicals are called toxins. Exotoxins are proteins that tend to be unstable when heated. Until recent years, it was believed that these toxins were liberated into the medium by the growing bacteria, hence the term exotoxin (*exo-* meaning outside the body). Evidence now indicates that these toxins are liberated upon the death and decomposition of the bacteria. Thus, if food is the medium, an exotoxin can cause disease or sickness even if the toxin-producing bacteria are no longer alive. Staphylococcus and botulism food poisoning result in just this manner. The toxin in the food causes the resulting sickness.

Some exotoxin-producing bacteria are capable of existing in the human body. In these cases, the exotoxins produced act directly on the human body. Resulting diseases include diphtheria, cholera, tetanus, and gas gangrene.

Endotoxins

Bacterial endotoxins tend to be heat stable, or opposite in this respect to exotoxins. They are also produced during the growth of some bacteria and released upon their decomposition. However, rather than being protein in nature, the endotoxins are molecules composed of a fat-carbohydrate-protein complex. Only gram-negative bacteria produce endotoxins. Among the noted producers of endotoxins are members of the genera *Salmonella*, *Shigella*, and *Escherichia*, all of which contain species that cause foodborne illness.

When the symptoms caused by a toxin are chiefly those of stomach or intestinal upset, the toxin is referred to as an enterotoxin (*enteron*—intestine). The food-poisoning bacteria best known for formation of enterotoxins are *Staphylococcus aureus*, *Clostridium perfringens*, *Bacillus cereus*, and many species of the genus *Salmonella*.

If the toxin produced affects the nervous system, it is called a neurotoxin (*neuron*—nerve). *Clostridium botulinum* produces a neurotoxin.

REPRODUCTION

Generation Time

The only form of bacterial cell capable of reproduction is the vegetative cell. The spore form can germinate into a vegetative cell but only the vegetative cell can undergo reproduction.

Reproduction in a vegetative cell takes place when the cell simply divides into two cells (Fig. 1-6). The two cells then grow and mature. The time required for this process is called the generation time. On a large scale, the generation time usually refers to the period of

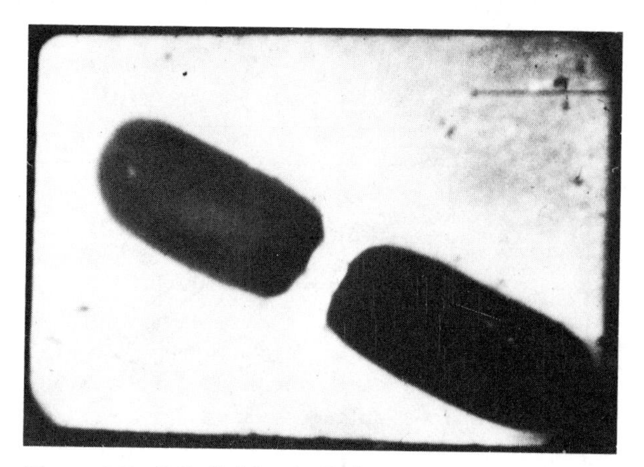

Figure 1-6 Cell division in *Escherichia coli*.

time it takes for a colony of bacteria to double in size (Table 1-3).

Table 1-3
Increase in numbers of a single bacterium with a generation time of 20 minutes.

Time (Min.)	Number of Bacteria
0	1
20	2
40	4
60	8
80	16
100	32
120	64
140	128

Each species of bacteria has its own generation time dependent on environmental conditions, such as, temperature, moisture, pH, and type of food. If environmental conditions are favorable, the generation time for some bacteria, such as *Clostridium perfringens,* can be as short as nine minutes. This results in a large increase in numbers in a short period of time. (Compare this with Table 1-2 where bacterial growth increased in numbers 64 times within a two-hour period.)

Exponential Growth

When a developing colony of bacteria grows, the increase in numbers of the bacteria follows a predictable and explosive pattern. Since the number of cells doubles each generation, the total population increases as the exponent of 2. This type of growth is called exponential growth.

Growth Curve

This rate of growth cannot continue unabated, however. Eventually a decrease in food and increase in wastes will slow the growth and finally cause a decrease in numbers. The population growth and decline in a colony of bacteria follows a demonstrated pattern that is graphed in Figure 1-7. This graph is called a growth curve and it refers to the life span of a typical bacterial colony.

In Figure 1-7, A to B on the growth curve is called the lag phase. It occurs frequently as the bacteria adjust to their medium. During this time there is little or no multiplication, and

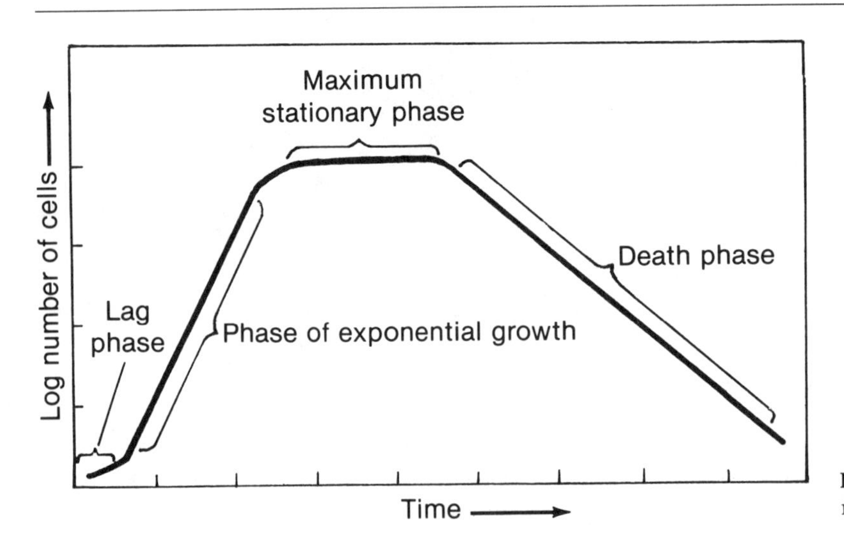

Figure 1-7 Growth curve of a bacterial colony.

occasionally a temporary decrease in numbers occurs.

C to D on the growth curve is called the exponential growth phase. This is the phase of explosive population growth. Each species of life has its own characteristic generation time under various environmental conditions. This segment of the curve represents the maximum reproduction potential of the organism for that particular environment. The cells produced during this phase are all viable cells. No spores are produced since the environment is suitable for growth and reproduction.

The exponential phase is the phase from which food must be protected. During this phase, bacterial populations can reach tens or hundreds of millions, and even billions, of bacteria per gram of food. If this phase occurs in human food, and if that food is ingested, foodborne illness usually results.

The length of the growth phase depends on the amount of food present. In most foods, this phase will last for hours because of the large mass of food available (Table 1-4).

Table 1-4
Increase in numbers of one bacterium per gram of food over a nine-hour period of time (comparable to overnight holding before refrigerating or reheating). A generation time of twenty minutes is assumed.

Hours	Number of Bacteria Per Gram
1	8
2	64
3	512
4	4,096
5	32,768
6	262,144
7	2,107,152
8	67,108,864
9	134,217,728

In time, as the competition for food and space increases, the multiplication rate slows and levels off. An essential nutrient may be insufficient. This is shown as E to F on the growth graph and is called the maximum stationary phase. Here, the reproductive rate is balanced by the death rate and the total number of bacteria remains constant. Finally, as nutrients become depleted and wastes increase, the death rate increases and the number of bacteria decreases (G to H). Eventually, the colony dies.

SUMMARY

Bacteria are microscopic one-celled forms of life. They occur in three basic shapes: coccus or sphere shaped, bacillus or rod shaped, and spirillum or corkscrew shaped. Some of the coccus and bacillus forms cause foodborne illness.

Since bacteria are extremely small, they must be treated with a stain in order to observe individuals. The acceptance or rejection of a stain can also reveal information about the chemical and physical makeup of the bacteria being studied. Bacteria can also be subjected to a number of metabolic tests to determine their characteristics. Some bacteria can also be identified through the antigens that they possess.

The great majority of bacteria are either harmless or helpful to us. Some cause disease, but most species have no direct relevance to man. They occur worldwide and can be found in or on virtually anything that exists, both living and dead.

Most bacteria are saprophytes; the remainder are either parasites or are photosynthetic. Parasitic forms attack other forms of life and cause disease. Many species of bacteria have been utilized by man in the manufacture of various products.

Bacteria are greatly influenced by their environment. All species have a temperature range, pH, and food type in which they thrive best. Some bacteria require oxygen; others cannot thrive in its presence.

Some bacteria produce spores during their growth cycle. Spores are highly resistant to heat, cold, drying, and most chemicals. Spores can be easily spread from one place to another. They can then germinate into vegetative cells.

Two types of toxins can be produced by bacteria. Exotoxins are proteins that tend to be unstable when heated. Endotoxins are molecules composed of a fat-carbohydrate-protein complex. They tend to be stable when heated.

When the symptoms caused by a toxin are chiefly those of stomach or intestinal upset, the toxin is referred to as an enterotoxin. If the toxin produced affects the nervous system, it is called a neurotoxin.

The only form of bacterial cell capable of reproduction is the vegetative cell. Reproduction takes place by cell division. The time that it takes for a colony to double in size is known as its generation time.

The growth of a bacterial colony follows a predictable pattern known as the growth pattern. This pattern consists of lag, growth, stationary, and death phases.

Review Questions

1. Why is it necessary for bacteria to have scientific names?

2. What is the difference between a genus and a species?

3. Name and describe the basic shapes of bacteria.

4. What is the difference between staphylococcus and streptococcus bacteria?

5. Why is it frequently necessary to stain bacteria?

6. What is the difference in behavior between gram-positive and gram-negative bacteria during the Gram stain?

7. What physical characteristics of a bacterial colony are helpful to note?

8. What are metabolic tests?

9. List several metabolic tests that can be utilized in identifying bacteria.

10. What is the difference between aerobic and anaerobic bacteria?

11. What is meant by the optimum growth temperature of a bacterium?

12. *Salmonella newport* and *Salmonella enteritides* both cause food poisoning. They are different serotypes. What does this term mean?

13. Where can bacteria be found?

14. What is the difference between parasitic and saprophytic bacteria?

15. In what ways are bacteria useful to man?

16. In what ways are bacteria harmful to man?

17. What is the difference between a vegetative cell and a spore? Which is more resistant to high temperatures?

18. Describe the three types of bacteria with regard to temperature range.

19. a. What is meant by the maximum growth temperature of a bacterium? b. What is meant by the minimum growth temperature? c. If you are in the foodservice business, why would it be helpful to know these temperatures?

20. Which is most effective in killing food-poisoning bacteria, a high temperature or low temperature?

21. a. What is meant by the pH of a food? b. List several foods whose pH renders them susceptible to attack by food-poisoning bacteria.

22. a. What is the purpose of spore formation in bacteria? b. Under what condition are spores formed? c. How do spores compare with vegetative cells in their ability to withstand heat?

23. a. What are toxins? b. Compare exotoxins and endotoxins. c. Name two food-poisoning bacteria that produce exotoxins. d. What are enterotoxins? e. What is a neurotoxin? Which major food-poisoning bacterium produces a neurotoxin?

References

Anderson, P. H. R. and D. M. Stone. (1955). "Staphylococcal Food Poisoning Associated with Spray-Dried Milk." *J. Hyg.* 53:387–397.

Berg, R. W. and W. E. Sandine. (1970). "Activation of Bacterial Spores. A Review." *Jour. Milk Food Technol.* 33:435–441.

Bergdoll, M. S. (1969). "Bacterial Toxins in Food." *Food Technol.* 23:532–533.

Brock, T. D. (1970). *Biology of Microorganisms.* Prentice-Hall, Inc. Englewood Cliffs, New Jersey.

Frobisher, M. (1968). *Fundamentals of Microbiology.* 8th ed. W. B. Saunders Co. Philadelphia, Pennsylvania.

Hart, C. J., W. W. Sherwood, and E. Wilson. (1960). "A Food Poisoning Outbreak Aboard a Common Carrier." *Pub. Hlth. Rep.* 75:527–531.

Jordan, M. C., K. E. Powell, T. E. Corothers, and R. J. Murray. (1973). "Salmonellosis Among Restaurant Patrons: The Incisive Role of a Meat Slicer." *A.J.P.H.* 63:982–985.

Kuhns, D. M. and T. G. Anderson. (1944). "A Fly-borne Bacillary Dysentery Epidemic in a Large Military Organization." *A.J.P.H.* 34:750–755.

Lundbeck. H., U. Plazikowski, and L. Silverstolpe. (1955). "The Swedish Salmonella Outbreak of 1953." *J. Appl. Bacteriol.* 18:535–548.

Stanier, R. Y., M. Doudoroff, and E. A. Adelberg. (1970). *The Microbial World.* Prentice-Hall, Inc. Englewood Cliffs, New Jersey.

Straka, R. P. and J. L. Stokes. (1959). "Metabolic Injury to Bacteria at Low Temperatures." *J. Bacteriol.* 78:181–185.

Thimann, K. V. (1963). *The Life of Bacteria.* 2nd ed. Macmillan Publishing Co., Inc. New York.

White, H. R. (1953). "The Heat Resistance of Streptococcus faecalis." *J. Gen. Microbiol.* 8:27–29.

I. Viruses
   A. Characteristics
   B. Reproduction
   C. History
   D. Symptoms
   E. Foods
II. Yeasts
   A. Characteristics
   B. Habitat
III. Molds
   A. Characteristics
   B. Aflatoxins
   C. Machinery Mold
IV. Protozoa
V. Roundworms
VI. Flatworms
   A. Tapeworms
   B. Chinese Liver Fluke
   C. Sheep Liver Fluke
VII. Summary

2 Other Organisms Causing Foodborne Illness

While it is true that most foodborne illnesses are caused by bacteria, there are other agents that can effect this type of illness. Viruses, fungi, protozoa, and worms are the major biological agents that occasionally play a role in foodborne illness.

VIRUSES

Viruses are the smallest of all known organisms. Unlike other forms of life, they are not cellular, nor do they reproduce as we normally think of that process. They differ radically from all cellular organisms in their structure, chemical composition, and development (Fig. 2-1).

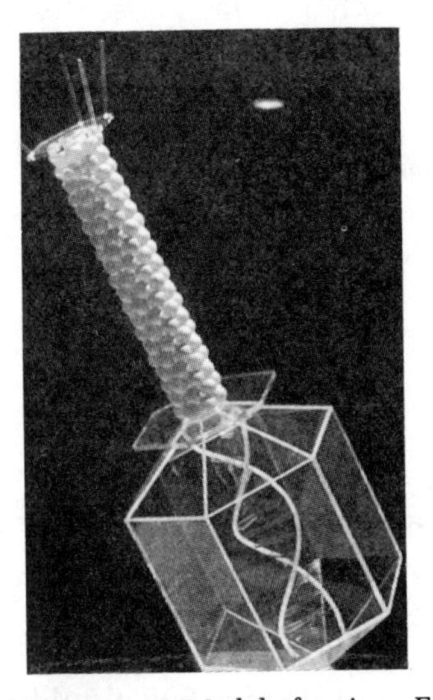

Figure 2-1 Model of a virus. From C. H. Heimler and C. D. Neal, *Principles of Science.* Charles E. Merrill Publ. Co.

Characteristics

Viruses can develop only within the cells of a host organism—they are parasites. Viruses are submicroscopic in size, detectable only with the electron microscope. They have a relatively simple chemical composition, consisting typically of a protein coat enclosing a single kind of nucleic acid, either RNA or DNA (Fig. 2-2).

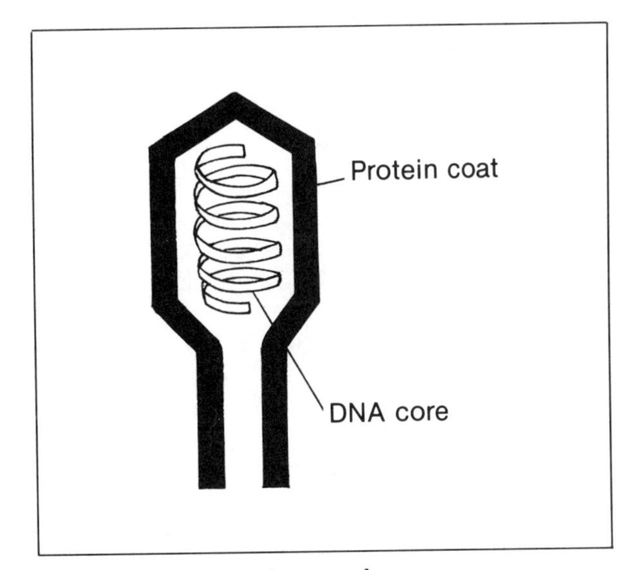

Figure 2-2 Diagram of a virus showing DNA core and protein coat. From C. H. Heimler and C. D. Neal, *Principles of Science.* Charles E. Merrill Publ. Co.

DNA is the chemical in chromosomes that determines hereditary traits. It does this by making a chemical, RNA, which directs the synthesis of other chemicals in the cell.

Reproduction

When a virus enters a cell, it causes the cell to make components of the virus rather than components of the cell. These components are then assembled into new virus particles, which escape from the cell and enter or infect other cells of the host.

The virus accomplishes all of this by attaching itself to a cell wall. When the core of the virus contacts the cell wall or cell membrane, it dissolves a small hole in it and the viral DNA enters the cell. This DNA contains the genetic information necessary to produce new virus particles. If the host cell is susceptible to the virus, its own protein-making machinery will be taken over by the intruder, and proteins and DNA needed for new virus particles will be manufactured. These chemicals are then assembled into new virus particles that escape from the cell, frequently causing the cell's death.

History

The existence of viruses has been known since the 1880s, even though they could not be seen by microscopes then in use. This was an active period in finding and identifying microorganisms. During this time, a cup of unglazed porcelain was devised to filter bacteria from broth cultures. When a broth culture containing bacteria was passed through this filter, all of the organisms were held back and the filtrate was clear and apparently sterile.

However, in 1892, D. J. Svanowsky showed that mosaic disease of tobacco leaves could be transmitted to healthy plants by the filtered juice from infected leaves, even though the filtered juice contained no microorganisms that could be seen under the microscope. He assumed that the infectious agent was a microorganism smaller than any known.

During the following 30 years, many serious diseases, some communicable, were found to be caused by these tiny agents. The one criterion that they all possessed was the ability to pass through pores in a filter that could retain all other known forms of life; thus, they became known as filterable viruses.

In 1935, Wendall Stanley was able to crystallize the tobacco mosaic virus. The crystals consisted mainly of protein. A few years later, the nucleic acid component was identified, in this case RNA.

Symptoms

In the case of human beings, the harm done by a virus infection depends primarily on the type of cell or organ that is attacked. The polio virus is commonly found in the human intestine. Here, it causes only mild symptoms that are usually passed off as "intestinal flu" or a "summer cold." Occasionally, the virus passes into the bloodstream, but without harmful effects. In rare instances, however, it invades nerve cells and damages or kills them, causing paralytic polio. There have been instances of this virus disease being transmitted through unpasteurized milk.

Infectious hepatitis outbreaks as well as polio outbreaks have been linked to food contaminated with viruses. In some instances, the foods were contaminated by carriers. In other instances, the foods, mainly oysters and clams, were harvested from water polluted with human sewage. Infectious hepatitis is the most frequent of the foodborne viral diseases.

Viruses also cause other human diseases, some of which are serious, some of which are merely annoying. Among these diseases are the common cold, small pox, yellow fever, influenza, mumps, measles, chicken pox, German measles, and rabies. Viruses also cause warts in people and cancerous tumors in some animals.

Foods

Foods have served as transmitting vehicles for viruses on many occasions. Viruses are parasites, however, reproducing only in the cells of a host organism. They do not grow and reproduce in food as bacteria do. If they are present in or on a food, they are simply carried unchanged into the body of the person consuming the food.

Foods become infected with viruses:

1. Through a virus infection in the animal that was the food source.
2. Through living in water contaminated with human sewage. Shellfish become contaminated with the virus of infectious hepatitis in this way.
3. Through preparation or handling by a human carrier.
4. Through droplet contamination from the respiratory tract of an infected person in some stage of a viral disease.

Although the resistance of viruses to heat is not nearly as great as that of spore-forming bacteria, such viruses as hepatitis and poliomyelitis can withstand temperatures approaching 212°F. (100°C). Other viruses, however, can be inactivated by heating at 149°F. (65°C) for one-half to one hour, or by a brief heating at 165°F. (72°C).

YEASTS

No other group of microorganisms has been more closely associated with the progress and well-being of the human race than the yeasts. A brewery and bakery that used yeast are known from 2000 B.C. in Thebes, Greece. The use of yeasts in brewing beer and wine and in baking bread has had worldwide occurrence.

Yeasts are desirable because they are used in the manufacture of some foods; they are undesirable because they can cause spoilage of other foods. Yeast fermentations are useful in the manufacture of such foods as beer, wine, bread, vinegar, and some types of cheese; they are undesirable when responsible for the spoilage of sauerkraut, dry fruits, fruit juices, syrups, molasses, jellies, honey, meat, wine, beer, and other foods, most of which have a high sugar content.

Characteristics

Yeasts are a group of microscopic, one-celled fungal organisms, usually spherical, ovoid, or elliptical in shape. Each lives as a separate, complete individual. They are nonmotile and colorless. Most yeasts grow best under aerobic conditions but one important group, which causes fermentation, can grow anaerobically. Yeasts possess no chlorophyll and so cannot carry on photosynthesis.

A typical yeast growth consists of small, oval cells that multiply by budding, a process in which some of the protoplasm bulges out of the cell wall (Fig. 2-3). The bud enlarges until it is almost equal in size to the mother cell. A cross wall is then formed between the two cells and they separate. Yeast cells also form spores (ascospores) as a method of reproduction. The number and appearance of the spores depends on the species of yeast. Yeasts grow faster and reproduce more rapidly than mycelial fungi, such as molds, and also bring about chemical changes more rapidly.

The temperature range for the growth of most yeasts is similar to that of molds, with the optimum temperature being around 77°–86°F. (25–30°C) and the upper limit at about 99°–117°F. (37–47°C). Some yeasts can grow at temperatures of 32°F. (0°C) or lower. Vegetative cells are usually killed by temperatures of 122°–136°F. (50–58°C) for 10–15 minutes; spores require 140°F. (60°C) for the same length of time. Yeasts grow best in an acidic environment of pH 4.0–4.5. They usually do not grow well in an alkaline medium. They also grow best under aerobic conditions, although fermentation types can grow anaerobically.

Habitat

Compared with other major groups of microorganisms, such as bacteria and molds, yeasts are rather few in number. There are only about 350 species.

Yeasts are widely distributed in nature. They commonly occur on grapes and other fruits and vegetables, which they help to decompose. They can be found in soil, dust, manure, and water. Flowers, insects, and honey are also frequent habitats. Some species live on the skin, in the mouth, or in the throat of human beings. They occasionally cause infections in or on these areas and in the respiratory system.

Yeasts occur in fresh and marine waters, particularly if pollution products such as sewage are present. Buck et al. (1977) reported on an examination of shellfish collected along the Connecticut shore of Long Island Sound.

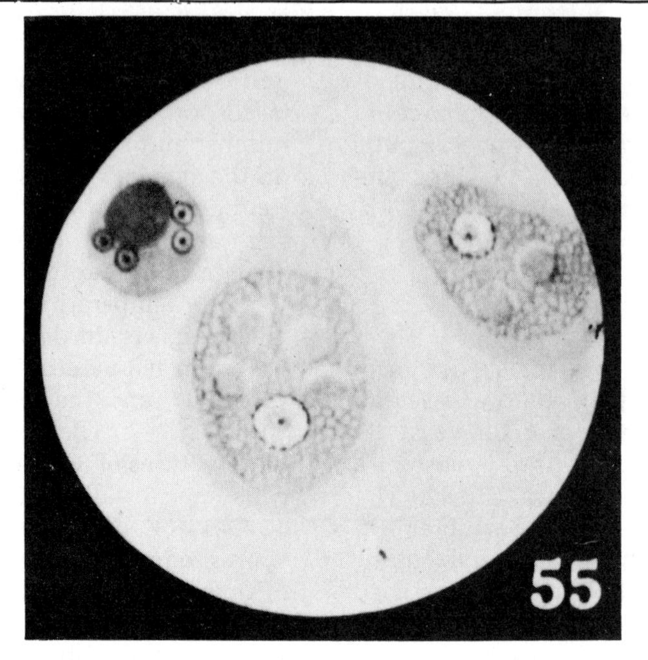

Figure 2-3 A yeast cell budding.

Shellfish taken were oysters, northern quahogs, and edible blue mussels. Human-associated yeasts were routinely recovered from all three species of shellfish. Even though some species of potentially pathogenic yeasts occurred with greater frequency than others, recovery was higher for all yeast species isolated from shellfish collected from polluted waters.

In general, sugars are the best energy food for yeasts. Foods with a high sugar content, such as fruits, jellies, honey, and syrup, are the most likely to be spoiled by these organisms, and therefore require prompt refrigeration. Bottled beverages, cottage cheese, and milk are sometimes attacked. Some yeasts use organic acids and alcohol for food and can be a problem in the brewing and pickling industries.

Carbon dioxide produced by bread yeasts, such as brewers' yeast, is responsible for the rising of bread. The alcohol produced during this fermentation reaction evaporates during baking. In the manufacture of wines and beer, the alcohol produced during fermentation of the mash or grapes is the desirable product.

Yeasts have also caused problems in the manufacture of mayonnaise and salad dressings. A case that shows the need for constant vigilance and sanitation occurred in a plant manufacturing these products. The company began the manufacture of mayonnaise and salad dressing on a small scale before World War II and expanded production considerably during the war. In addition to these products, pickles and pickle relish were produced. All operations were carried out in a single plant, and the same machinery was used in producing all of the various products. Early in 1947, a large number of spoiled jars of mayonnaise began to be returned to the plant. Their shelf life had been very short. Large numbers of yeast were isolated from the spoiled mayonnaise. Inspection of the plant revealed unsanitary conditions and machinery contaminated with yeast. It was necessary to close and disinfect the plant.

The defenses against yeasts are the same as for all microorganisms—prompt refrigeration, adequate heating, proper sanitizing of equipment and utensils, and personnel cleanliness. Yeasts are not known as food-poisoning agents but they do cause food spoilage and off flavors.

MOLDS

Molds are a type of fungus plant familiar to most people. Molds include most of the fuzzy, cottony, or powdery growths that appear on foods, such as, bread, tomatoes, peaches, strawberries, and jellies and jams. They may be green, blue, black, white, or another color. Many of them are useful to man, but others must be guarded against since they cause food spoilage. Their spores are as widespread as those of yeasts and bacteria, but they are able to grow in situations where bacteria cannot due to high acidity or low moisture content.

Characteristics

A growing mold consists of a mass of branching, intertwined, hairlike filaments called hyphae (Fig. 2-4). These hyphae may consist of elongated cells arranged end-to-end and separated by walls, or they may consist of cells with no crosswalls and freely flowing protoplasm. The entire mass of these hyphae is called the mycelium. Some of the hyphae penetrate into food and decompose it to obtain nutrients. This causes the food to spoil. Other hyphae project into the air and form reproductive structures that are characteristic for each type of mold. These structures contain spores and are responsible for the color and fuzzy appearance of the mold.

Most molds are saprophytes, obtaining their food from any environmental medium on which they germinate. The spores are produced in large numbers and are small, light, resistant to drying, and readily distributed by air currents. They differ in size, shape, color, and smoothness according to the species of mold.

Molds, like all fungus plants, lack true roots, stems, and leaves. Since they lack leaves, they also lack chlorophyll and, thus, cannot make their own food. Their food must come from their growth medium. Like the yeasts, they grow well during the night. In general, most molds require a lower moisture content in their growth medium than do yeasts and bacteria.

The optimal temperature for the growth of most molds is about 77°–86°F. (25°–30°C). Some can grow at higher temperatures, particularly those forms that are parasitic on warm-blooded animals. A number of molds also grow well at colder temperatures, a few growing at temperatures as low as 14°F. (−10°C). Mold growth on refrigerated foods certainly occurs. Species of *Aspergillus*, a common food-spoiling genus, are noted for their ability to grow at low temperatures.

Most molds, both vegetative and spore forms, are killed in five to ten minutes by moist heat at a temperature of 140°F. (60°C). Some species, however, are considerably more heat resistant. Spores tend to be more heat resistant than the body of the mold (mycelium) and usually require a 10°–20°F. (6–11°C) higher temperature for their destruction in a given time. Many species of *Aspergillus*, and some of *Penicillium* and *Mucor*, three food-spoiling genera, are somewhat heat resistant.

Although moist heat of 140°F. (60°C) is lethal to most mold spores, dry heat is a different matter. Spores tend to be rather resistant to dry heat and a temperature of 250°F. (121°C) for as long as 30 minutes leaves the spores of some species unaffected. Mold spores are also fairly resistant to freezing.

Those molds that grow on food and cause spoilage are aerobic, that is, they require oxy-

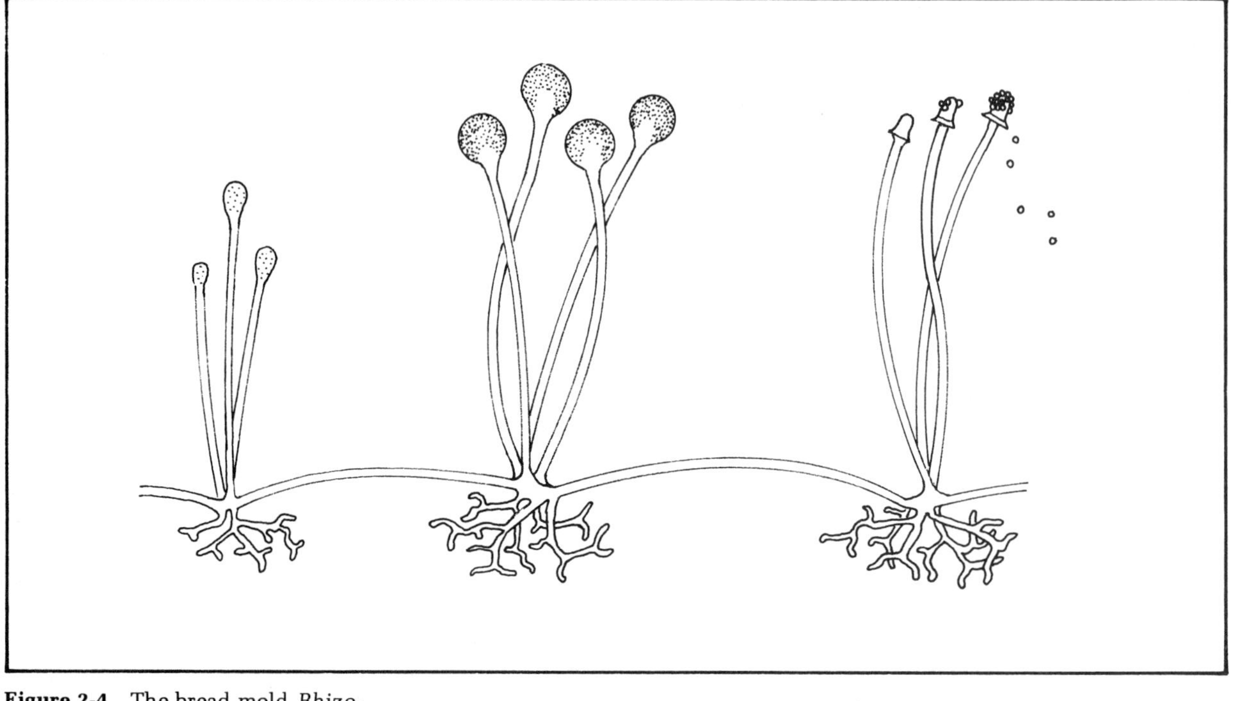

Figure 2-4 The bread mold, *Rhizopus*, showing horizontal hyphae and projecting reproductive structures containing spores. From Stanier/Doudoroff/Adelberg, *The Microbial World*, © 1970, 3rd edition, p. 148. Reprinted by permission of Prentice-Hall, Inc., Englewood Cliffs, New Jersey.

gen for their growth processes. Most molds can grow over a wide pH range (2.0–8.5), but the greater number prefer an acidic medium. Foods protected from bacterial spoilage can be spoiled by the growth of molds, since some molds can develop, even at low temperatures, in foods with a high concentration of sugar or salt or in foods that are strongly acidic (all conditions that discourage the growth of bacteria). Such foods as preserved fruits and jellies, butter, pickles, sauerkraut, vinegar, and salted and smoked meats can become moldy, even though kept in a cool place.

Aflatoxins

Two common groups of storage fungi, the *Aspergilli* and *Penicillia,* produce aflatoxins. Aflatoxins are among the most potent mycotoxins (poisons produced by fungi) known. They are also highly carcinogenic (cancer causing). The molds that produce these toxins can grow on a wide range of foods. Their growth is not prevented by refrigeration, although it is retarded.

The danger of aflatoxin ingestion was first recognized in 1960 in England when an outbreak resulted in the death of an estimated 100,000 turkeys. Within a short time, outbreaks occurred in ducklings, chickens, swine, and young cattle. The outbreaks were attributed to toxic substances in the animal feeds. The common element in all the feeds was a peanut meal imported from Brazil.

Subsequent examination of animal feeds containing peanut meal revealed that peanut meal from at least 14 producing countries was contaminated with the toxic substance. Further studies showed that agricultural commodities other than peanuts were occasionally contaminated with the toxin. Aflatoxins have since been found in samples of cottonseed, soybeans, rice, wheat, barley, sorghum, peas, beans, rye, oats, cheese, hazelnuts, walnuts, Brazil nuts, apple juice, smoked bacon, peaches, figs, and other foods from various parts of the world.

The mold that produces this toxic substance is *Aspergillus flavus* and its toxic product is known as aflatoxin. There are many strains of *A. flavus,* and the mold is widespread. It is a common contaminant of many foods and food crops but, fortunately, many of the strains do not produce aflatoxin.

Toxin producing strains of *A. flavus* grow best at a temperature of 50°–112°F. (10°–45°C) and a relative humidity of 75 percent, or higher. Its optimum temperature for growth extends from 68°–95°F. (20°–35°C), depending on the strain and the growth medium. As the temperature decreases the rate of aflatoxin formation also decreases. Toxin production in most strains ceases at 50–55°F. (10–13°C) but the growth of toxin-producing strains continues until a temperature of 42°F. (6°C) is reached.

Van Walbeek et al. (1969) have suggested that aflatoxin-producing fungi may grow on refrigerated food without producing spores and would therefore be colorless and virtually impossible to detect. This could present a dangerous situation since some strains have produced toxin at 45.5°F. (7.5°C). Foods that support rapid aflatoxin production at room temperature include bread, cheese, apple juice, and various fruit drinks.

Van Walbeek et al. (1962) examined samples of foods associated with illness. Three of the four samples (meat pie, apple squares, and dry spaghetti) showed growth of aflatoxin-producing molds. The meat pie produced an illness, characterized by high fever and various complications, that required hospital treatment for approximately eight weeks. The apple squares were thought to be responsible for a digestive upset that was successfully treated at home. The dry spaghetti was implicated in the illness of two children who required hospital treatment.

Machinery Mold

Machinery mold is the name given to a slimy mold, *Geotrichium,* that commonly grows on processing equipment in canneries. Even though it is not associated with foodborne illness, it does represent a sanitation problem in the industry.

Splittstoesser et al. (1977) conducted surveys at nine different factories that processed tart cherries, peas, green beans, corn, or beets. Seven of the plants were canneries; the others engaged in freezing operations. A total of 230 food samples were examined for *Geotrichium.* Of these samples, 42 (18 percent) were positive, all from the canneries. Approximately 49 percent of the bean samples were contaminated with *Geotrichium.*

Molds are beneficial in some circumstances. Saprophytes that live in the soil help in the decomposition of organic matter into simpler compounds that can be reused by plants for food. Some species of *Penicillium* are responsible for the characteristic flavor of many cheeses. Mold-ripened cheeses include blue, Roquefort, Camembert, Brie, and Gammelost. Molds are also used in making Oriental foods, such as, soy sauce, miso, and sonti.

PROTOZOA

The protozoa are a phylum of microscopic, one-celled animals that occur universally. Some species are parasitic in man and cause serious disease. The phylum contains four classes distinguished by their method of locomotion. They are:

1. Rhizopoda. These protozoans move by thrusting their protoplasm outward against the cell membrane, forming projections called pseudopods. The amoeba is the commonly cited representative but the group does contain two pathogenic species that can be transmitted in food.
2. Mastigophora. This group contains the flagellated protozoa. They move by means of one or more hairlike projections called flagella. The group contains several pathogenic species, the most noted of which is the agent of African sleeping sickness.
3. Ciliata. Members of this group move by means of many short hairs called cilia. None are pathogenic to man.
4. Sporozoa. These forms are either immotile or move with a gliding motion. Malaria is caused by some species in this group.

Protozoans are mostly free-swimming forms that exist in large numbers, both in marine and fresh waters. They are mentioned here only because two species, *Entamoeba histolytica* and *Toxoplasma gondii* are pathogens that can be transmitted through food.

Entamoeba histolytica causes the disease known as amoebic dysentery. It is most common in warm regions and in areas where sanitation is poor. The organism causes ulceration of the intestinal tract, which results in diarrhea and abdominal pains.

Cysts are formed in the intestinal tract by the amoebae. These cysts leave the body with the feces. They are resistant to drying and can survive for long periods of time in the environment. The disease can be transmitted by fecal contamination of food or water.

Food can be contaminated through infected food handlers and through flies that have walked on sewage containing the cysts. Rats can also carry the organism, contaminating the environment with their feces. Foods grown on contaminated soil can be a source of the disease. All fruits and vegetables should therefore be washed thoroughly before being served raw.

Toxoplasma gondii is a parasite that infects a number of human organs, including the heart, liver, lymph nodes, spleen, and brain. The infection is called toxoplasmosis. It resembles mononucleosis in that it is characterized by swollen lymph glands, fever, and malaise. The disease is rarely fatal. *Toxoplasma gondii* can be transmitted through undercooked meat, particularly pork. Undercooked hamburger has been implicated on several occasions. The organism can be killed by thorough cooking and by freezing.

ROUNDWORMS

Most of the members of this phylum are either small or microscopic in size. They occur as both aquatic and terrestrial forms and can be either parasitic or free-living. The pathogenic forms are all parasites. Among the members of this group that infect humans are the ascaris worm, hookworm, pinworm, filaria (elephantiasis), and trichina.

The trichina worm (*Trichinella spiralis*) is the only member of this phylum that is commonly associated with food (Fig. 2-5). The illness produced is called trichinosis. It results from eating undercooked pork that is infected with trichina larvae.

Trichinella is a parasite of man and other mammals such as the rat and pig. Pigs frequently become infected through eating dead rats that are infected and through the ingestion of uncooked garbage containing infected meat scraps.

The incidence of trichinosis in the United States is currently on the decline, having fallen from 16 percent in the 1940s to about 5 percent

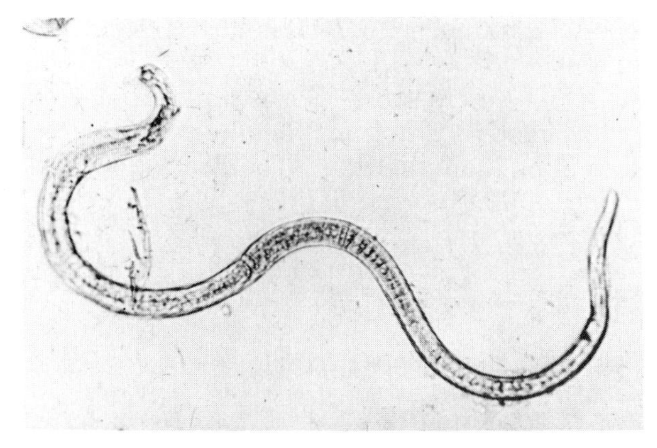

Figure 2-5 The trichina worm.

at present. As the habit of feeding garbage to pigs has decreased, mostly due to legislation, the incidence of trichinosis in man has fallen.

Man becomes infected through the ingestion of undercooked pork containing larval trichina worms (Fig. 2-6). The larval worms are microscopic and cannot be detected during routine inspection at the meat-packing plant. They are released from the meat into the small intestine. During the next four or five days, as they mature into adult worms, they commonly cause intestinal upset and diarrhea. The male trichina is about $1\frac{1}{2}$ mm in length; the female is about 3 mm.

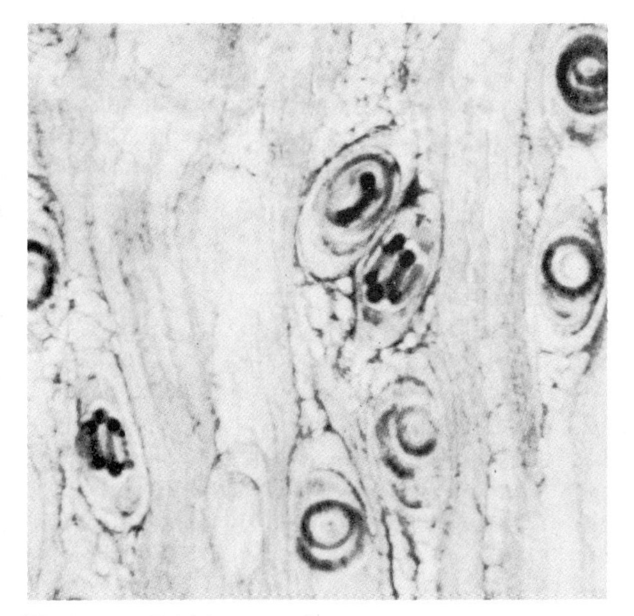

Figure 2-6 Trichina worm larvae, encapsulated in pig muscle. From *Elements of Biology* by Paul B. Weisz. Copyright © 1965 by McGraw-Hill Book Company. Used with the permission of McGraw-Hill Book Company.

Mating occurs in the small intestine. The female delivers live young in the intestinal lining. These larval forms reach the circulatory system and are carried to the skeletal muscles. The heart muscle is also invaded on occasion.

Damage is done when larvae enter the tissues and also during the subsequent two weeks or so when the larvae are mobile in the tissues. Eventually, the infected tissue walls off the larvae and they remain as a cyst. Temporary muscle pain and weakness, heart damage, eye damage or blindness, and even death (if the brain is invaded) can result from trichinosis. The life cycle of the worm in the pig and rat are the same as in man.

The principle methods for killing trichina worms in meat are heating and freezing. In heating, all parts of the pork must be heated to a temperature of at least 137°F. (58°C). When freezing pieces six inches or less in thickness, a temperature of 10°F. (−12°C) for 20 days will kill the larvae.

FLATWORMS

This phylum of worms contains two types that can be transmitted through food, tapeworms and flukes. The members of this group are flattened on the top and bottom sides, the tapeworms being ribbonlike and the flukes leaflike. All tapeworms and flukes are parasites.

Tapeworms

There are two important species of tapeworms in the United States—the beef tapeworm

(*Taenia saginata*) and the pork tapeworm (*Taenia solium*). Both forms are ingested as larvae and have similar life cycles in the human body. Both worms live as adults in humans and as larvae in either pigs or cattle, depending on the species of worm; in other words, two hosts are required for the development of these worms.

The pork tapeworm enters the human body in undercooked pork in an immature form called a bladderworm. Pork containing bladderworms is known as "measly pork." The bladderworms are visible as tiny white dots on the meat.

If ingested, the bladderworm is carried to the small intestine, where the cyst around it dissolves and the head of the worm attaches itself by means of suckers and hooks to the intestinal lining. From the head, series of segments or proglottids grow backward until a length of six to ten feet is reached in the adult stage. The tapeworm absorbs its nutrients from digested food in the small intestine and it releases waste products into it. The resulting symptoms can include loss of weight, ulcers, nausea, vomiting, intestinal colic, and diarrhea.

Each segment of the tapeworm contains male and female reproductive organs and mating can occur among different segments of the same worm. A tapeworm may produce more than 250 million eggs during a year and can live for ten years in the small intestine.

Eggs pass from the body with the feces. In order for the cycle to continue, a pig must ingest these eggs. They are carried to the intestine, where the young are released. These larval forms burrow into blood vessels and are carried to skeletal muscles in the pig's body. A cyst (the bladderworm) forms that must now be ingested by another organism such as man.

Bladderworms are usually detected by meat inspectors, so very little infected pork reaches the market. Thorough cooking will kill them, a temperature of 132°F. (56°C) being lethal. Freezing temperatures also destroy the organism.

The beef tapeworm has a similar life cycle and characteristics except that the cow replaces the pig as the larval host. The fish tapeworm, which attains a length of 25 feet, has been reported in the United States. Certain freshwater fish, which are eaten raw or undercooked, are the infected carriers.

Chinese Liver Fluke

Two species of flukes are of importance in foodborne illness: the Chinese liver fluke (*Clonorchis sinensis*) and the sheep liver fluke (*Fasciola hepatica*).

The Chinese liver fluke is a parasite in the Far East. In the adult stage, it lives in the bile duct of the liver sometimes blocking it, if large numbers are present, and causing inflammation and jaundice. Cirrhosis of the liver is a frequent result of a heavy infestation.

The eggs that result from mating pass from the bile duct, into the small intestine, and out of the body with the feces. They must reach water containing a certain type of snail. The eggs are ingested by the snail, hatching within it. The resultant larvae produce many more larvae by a type of asexual reproduction. The final larval forms leave the snail and seek out a certain type of fish, into which they burrow and encyst.

If the fish is caught and eaten raw or insufficiently cooked, the cysts are released in the small intestine and the larvae emerge and make their way up the bile ducts into the liver. Here they mature into adulthood. Thorough cooking of fish is the best defense against these worms.

Sheep Liver Fluke

The sheep liver fluke also resides in the bile ducts of the liver. It is primarily a disease of sheep, but outbreaks in man are not uncommon (Hardman et al.; 1970; Ashton et al., 1970).

The eggs are passed in the stool and must reach water, where they hatch into larval forms that penetrate the proper species of snail and reproduce asexually. The final larval forms emerge from the snail and encyst on water vegetation such as watercress.

If the vegetation is ingested unwashed, the cysts pass to the small intestine and release the larvae. These larval forms penetrate the wall of the small intestine and make their way to the liver. Here they reach adulthood, causing jaundice and cirrhosis of the liver.

Thorough washing of watercress and all water vegetation consumed by human beings is the best defense against these worms.

SUMMARY

Viruses are the smallest of all known organisms. They are not cellular and can develop only within the cells of a host organism. A virus consists, typically, of a single kind of nucleic acid, either RNA or DNA, enclosed within a protein coat. When it enters a cell, it causes the cell to make components of the virus rather than of the cell. These components are then assembled into new virus particles that escape from the cell, frequently causing its death. The harm done by a virus infection depends primarily on the type of cell or organ that is attacked.

Foods have served as a vehicle of transmission for viruses on many occasions. Viruses are parasites, however, reproducing only in the cells of a host organism; they do not grow and reproduce in food as bacteria do.

The resistance of viruses to heat is not nearly as great as that of spore-forming bacteria. Most viruses can be inactivated by heating at 149°F. (65°C) for one-half to one hour or by a brief heating at 165°F. (74°C). Some viruses, however, such as hepatitis and polio viruses, can withstand temperatures approaching 212°F. (100°C).

Yeasts are a group of microscopic, one-celled fungal organisms, usually spherical, ovoid, or elliptical in shape. They are nonmotile and colorless. Most yeasts grow best under aerobic conditions but one important group that causes fermentation can grow anaerobically.

The optimum temperature for yeast growth ranges from about 77°–86°F. (25–30°C) with the upper limit close to 120°F. (45°C). Some yeasts can grow at freezing temperatures, but most are killed or their growth retarded. Vegetative cells are usually killed by temperatures of 122°–136°F. (50–58°C) for 10 to 15 minutes; spores require 140°F. (60°C) for the same length of time. Yeasts grow best in an acidic environment of pH 4.0–4.5.

Yeasts commonly occur on grapes and other fruits and vegetables. They can be found in soil, dust, manure, and water. Flowers, insects, and honey are also frequent habitats. Foods with a high sugar content (fruits, jellies, honey, and syrup) are most likely to be spoiled by these organisms.

Molds include most of the fuzzy, cottony, or powdery growths that appear on such foods as bread, tomatoes, peaches, strawberries, and jellies and jams. They may be green, blue, black, white, or some other color.

A growing mold consists of a mass of branching, intertwined, hairlike filaments called hyphae. The entire mass of hyphae comprises the mycelium. Some of the hyphae penetrate into food, decomposing it for nutrients and causing spoilage. Molds reproduce by forming spores.

The optimal temperature for the growth of most molds is 77°–86°F (25–30°C). A number of molds can grow well at lower temperatures, however. Most molds are killed in 5 to 10 minutes by moist heat at a temperature of 140°F. (60°C). They can grow over a wide pH range.

The mold *Aspergillus flavus* produces an extremely potent toxin known as aflatoxin.

These toxins have been found in peanuts, cottonseed, Brazil nuts, walnuts, rice, wheat, and many other foods.

Two members of the protozoa cause foodborne illness. *Toxoplasma gondii* can infect a number of human organs, including the heart, liver, lymph nodes, spleen, and brain. Undercooked hamburger has been implicated as the transmitting vehicle on several occasions. *Entamoeba histolytica* causes the disease known as amoebic dysentery. The organism causes ulceration of the intestinal tract, resulting in diarrhea and abdominal pain.

The trichina worm is a roundworm that sometimes causes foodborne illness. Man becomes infected by ingesting undercooked pork that contains microscopic larval trichina worms. These worms mature in the small intestine and their larvae reach the muscles via the circulatory system. They can cause muscle damage, blindness, brain damage, and death.

Flatworms contain two types that are foodborne: tapeworms and flukes. Tapeworms are ingested in undercooked and infected pork, beef, or some species of fresh-water fish. They enter the body as a cyst and mature in the small intestine.

Two species of flukes are of importance in foodborne illness: the Chinese liver fluke and the sheep liver fluke. Both flukes live their adult stages in the bile ducts of the liver and cause damage to this organ. The Chinese liver fluke is ingested in certain species of undercooked fish; the sheep liver fluke gains entrance to the body on unwashed, infected water plants such as watercress.

Review Questions

1. List five kinds of biological agents that can cause foodborne illness.

2. How do viruses compare in size to other forms of life?

3. How does a virus reproduce?

4. How does a virus gain entry to a cell?

5. How did the existence of viruses become known even though they could not be seen?

6. What is the most frequent of the foodborne viral diseases?

7. List four ways in which food can become infected with viruses.

8. a. In what ways are yeasts desirable? b. In what ways are they undesirable?

9. Describe a yeast cell.

10. Where are yeasts found in nature?

11. What type of food is most likely to be spoiled by yeasts?

12. Describe a mold growth.

13. How are molds spread from one place to another?

14. Compare the optimum temperature range of yeasts and molds.

15. What types of food are most likely to support a mold growth?

16. What are aflatoxins?

17. What are the two foodborne illnesses that are transmitted by viruses?

18. Describe the life history of the trichina worm.

19. What damage is done to the human body by the trichina worm?

20. Describe the life history of the pork tapeworm.

21. What is the best defense against the trichina and the tapeworm?

References

Appleman, M. D., E. P. Hess, and S. C. Rittenberg. (1949). "An Investigation of a Mayonnaise Spoilage." *Food Technol.* 3:201–203.

Ashton, W. L. G., P. L. Boardman, C. J. D'Sa, P. H. Everall, and A. W. J. Houghton, (1970). "Human Fascioliasis in Shropshire." *Brit. Med. J.* 3:500–502.

Becker, M. E. (1966). "Water-Borne and Food-Borne Viruses." *J. Milk Food Technol.* 29:243–245.

Berg, G. (1964). "The Food Vehicle in Virus Transmission." *Health Lab. Sci.* 1:51–59.

Brock, T. D. (1970). *Biology of Microorganisms.* Prentice-Hall, Inc. Englewood Cliffs, New Jersey.

Buck, J. D., P. Bubucis, and T. J. Combs. (1977). "Occurrence of Human-Associated Yeasts in Bivalve Shellfish from Long Island Sound." *Appl. Environ. Microbiol.* 33:370–378.

Cliver, D. O. (1969). "Viral Infections." In *Food-Borne Infections and Intoxications,* ed. Hans Riemann. Academic Press, Inc. New York.

Fabian, F. W. and M. C. Wethington. (1950). "Spoilage in Salad and French Dressing due to Yeasts." *J. Food Sci.* 15:135–137.

Frank, H. K. (1968). "Diffusion of Aflatoxins in Foodstuffs." *J. Food Sci.* 33:98–100.

Frazier, W. C. (1967). *Food Microbiology.* McGraw-Hill Book Co. New York.

Hardman, E. W., R. L. H. Jones, and A. H. Davies. (1970). "Fasciolasis-A Large Outbreak." *Brit. Med. J.* 3:502–505.

Hesseltine, C. W., O. L. Shotwell, J. J. Ellis, and R. D. Stubblefield. (1966). "Aflatoxin Formation by Aspergillus flavus." *Biol. Rev.* 30:795–805.

Kachani, Z. F. (1965). "Propagation and Properties of Hepatitis Virus." *Nature.* 208:605–606.

Longree, K. (1972). *Quantity Food Sanitation.* John Wiley and Sons, Inc. New York.

Lynt, R. K., Jr. (1966). "Survival and Recovery of Enterovirus from Foods." *Appl. Microbiol.* 14:218–222.

McBride, M. E., W. C. Duncan, and J. M. Knox. (1977). "The Environment and the Microbial Ecology of Human Skin." *Appl. Environ. Microbiol.* 33:603–608.

Rose, A. and J. S. Harrison. (1969). *The Yeasts,* vol. 1. Academic Press, Inc. New York.

Splittstoesser, D. F., F. R. Kuss, W. Harrison, and D. B. Prest. (1971). "Incidence of Heat Resistant Molds in Eastern Orchards and Vineyards." *Appl. Microbiol.* 21:335–337.

Splittstoesser, D. F., M. Groll, D. I. Downing, and J. Kaminski. (1977). "Viable Counts Versus the Incidence of Machinery Mold (Geotrichium) on Processed Fruits and Vegetables." *J. Food Protection.* 40:402–405.

Stanier, R. Y., M. Doudoroff, and E. A. Adelberg. (1970). *The Microbial World.* Prentice-Hall, Inc. Englewood Cliffs, New Jersey.

Taylor, W. T. and R. J. Weber. (1961). *General Biology.* D. Van Nostrand Co. Princeton, New Jersey.

Torrey, G. S. and E. H. Marth. (1977). "Isolation and Toxicity of Molds from Foods Stored in Homes." *J. Food Protection.* 40:187–190.

Torrey, G. S. and E. H. Marth. (1977). "Temperatures in Home Refrigerators and Mold Growth at Refrigeration Temperatures." *J. Food Protection.* 40:393–397.

van Walbeek, W., P. M. Scott, and F. S. Thatcher. (1962). "Mycotoxins from Food-Borne Fungi." *Can. J. Microbiol.* 14:131–137.

van Walbeek, W., T. Clademenos, and F. S. Thatcher. (1969). "Influence of Refrigeration on Aflatoxin Production by Strains of Aspergillus flavus." *Can. J. Microbiol.* 15:629–632.

Wogan, G. N. (1969). "Alimentary Mycotoxicoses." In *Food-Borne Infections and Intoxications,* ed. Hans Riemann. Academic Press, Inc. New York.

3 Communicable Diseases

Sunday mornings were usually quiet and peaceful at the United States Air Force Academy. Sunday morning, April 28, 1968, started no differently than other Sundays at the Academy—until the first cadets began to appear at the hospital outpatient clinic, some singly, others in small groups. The problems were severe sore throats and high fevers.

As the day continued, so did the onslaught on the hospital and its staff, and by Sunday night over 600 cadets had inundated the facilities. An additional 360 cadets had been treated in their dormitories by a weary medical staff. By the next evening, the total had risen to over 1,000, with 74 of these being admitted to the hospital and 28 others to an emergency hospital set up in the gymnasium.

On Tuesday, April 30, a second wave of cases appeared. By the end of the epidemic, approximately 1,200 cadets had become ill, with 111 of these requiring hospitalization.

The clinical picture was that of severe streptococcal throat infection ("strep throat"). Eighty-two percent of the cadets interviewed experienced a sudden onset of chills and fever, sore throat, and headache. About one-third of the hospitalized cadets experienced incapacitating muscular weakness (prostration). Temperatures averaged 102°F. on admission.

Laboratory tests reported Group A beta-hemolytic (red blood cell destroying) streptococci M-Type 12 to be present in throat swab cultures obtained from patients. This strain has also been implicated in rheumatic fever and in kidney diseases.

Because of the explosiveness of the outbreak, a common source of the infection was immediately suspected. For several reasons, suspicion fell on the noon meal of Friday, April 26. A review of the foods eaten by the victims implicated tuna salad.

This salad was made from tuna fish, boiled eggs, mayonnaise, and relish. All of the ingredients, except the eggs, were taken from previously unopened containers. The eggs had been boiled and sliced on April 25, 24 hours before the implicated meal. The 96 dozen boiled eggs used in the salad were shelled and sliced by six men. These eggs were then placed in covered containers and left at approximately 58°F. (14°C) for 18 hours, a very dangerous practice. The other ingredients were then added, and the salad was refrigerated for the remaining six hours before serving.

Throat cultures taken on April 30 from 229 food handlers employed in the mess hall yielded three cultures of Group A streptococcus. One of the positive food handlers was a member of the group that directly assisted in the preparation of the boiled eggs.

In all probability, this food handler infected the eggs during the preparation of the tuna salad, either through his respiratory system or through hand contact with the eggs. The bacteria were then allowed an 18-hour period of growth before refrigeration occurred. This length of time easily would have allowed massive bacterial growth to take place.

Experiments have shown that nasal carriers frequently discharge hundreds of millions of pathogenic streptococci into handkerchiefs when blowing their noses. Many of the bacteria pass through the handkerchief and contaminate the hands. Hamburger and Robertson reported that coughing and sneezing could expel streptococci-laden droplets as far as 9½ feet from the subject (Fig. 3-1). Infected persons disseminate the streptococci in highly contaminated nasal secretions that reach the environment chiefly via the hands when the carrier blows his nose. Streptococci discharged in this manner contaminate handkerchiefs, clothing, bedclothing, and dust, and are released into the air when these reservoirs are agitated.

Strep throat is a well-known communicable disease. There are many ways that communicable diseases can be passed from one person

Figure 3-1 An uncovered cough or sneeze can spread droplets containing microorganisms as far as 9½ feet.

to another. In the case just documented it was passed through food, with the result that 1,200 Air Force Academy cadets became ill and 111 of these required hospitalization.

Foodservice workers have a special interest as well as a real involvement in studying communicable diseases. About 40 percent of the communicable diseases that physicians must officially report to health departments are associated with the business of providing food and/or drink to the public.

In order to understand the place of communicable diseases in the disease classification system, it is necessary to become acquainted with the various categories of disease.

Disease

The word *disease* is a broad term. Human diseases include most conditions that interfere with the normal state of the body. Originally, disease meant chiefly those illnesses for which the cause was unknown. It was not until 1683, when microorganisms were first observed by Leeuwenhoek, that their existence was recognized. Another 180 years passed before Pasteur implicated them in disease, and it was not until 1878 that Koch conclusively proved the germ theory of disease.

Disease is one of man's greatest enemies. It has killed and crippled more persons than all the wars ever fought and has cost billions of dollars in lost manpower and medical bills. There are so many different causes and types of diseases, however, that the term disease sometimes describes very little. It is important, therefore, to learn something about the different kinds of diseases so that communicable diseases can be seen in their proper perspective.

NONINFECTIOUS DISEASES

Some diseases are not caused by microorganisms and therefore cannot be spread from person to person. Foodservice workers do not need to be concerned with these diseases on the job since they cannot be spread personally in any known circumstance. These diseases can generally be placed into one of five types, although some of the types overlap.

Congenital Diseases

These diseases are present at birth. Genetic diseases such as hemophilia and PKU fall into this category. Other conditions may be due to unfortunate accidents in development during time in the womb. Some heart and spinal cord defects, cleft palate, "harelip," and other developmental defects are congenital, as are some forms of mental retardation.

Environmental Diseases

Our present society is marked, not only by a wealth of new products, but also by a wealth of new and highly dangerous chemicals used in making many of these products. Some chemicals used in making pesticides, herbicides, and plastics can cause severe injury to nerve cells and to parts of the body, such as, skin, lungs, brain, and liver.

Radiation poisoning is another environmentally caused disease. The Three-Mile Island nuclear reactor accident presented a threat to the health of thousands of persons. During the same year (1979), a factory in Arizona that used radioactive materials contaminated an entire neighborhood, including a school lunch plant.

Cancer is also considered an environmental disease by an increasing number of people in the medical profession. About 80 percent of current cancers are thought to be caused by such environmental factors as cigarette smoking, chemicals in food, drink, and medicines, and chemicals resulting from air and water pollution. Emphysema, silicosis, and the "black lung" disease of coal miners are other environmentally caused diseases.

Deficiency Diseases

These diseases result when the body fails to receive the required amount of certain necessary nutrients. Among these diseases are rickets (vitamin D), scurvy (vitamin C), pellagra (niacin), beriberi (thiamine), anemia (iron), and kwashiorkor (protein).

Allergic Diseases

Allergies result when the body becomes hypersensitive to some foreign substance, usually a protein. Hay fever and asthma are two common allergic diseases. Among the substances most frequently reacted to are pollen, animal hairs, some foods, bee stings, and penicillin. Emotional disturbances can also cause allergic reactions in some people.

Organic or Functional Diseases

These diseases are due to some abnormality of the body tissues or organs. Included in this category are heart diseases, muscular dystrophy, multiple sclerosis, glaucoma, hardening of the arteries, cerebral palsy, epilepsy, and diabetes. Some of these conditions are congenital and some result from physical damage, but in all cases some tissue or organ is permanently impaired.

INFECTIOUS DISEASES

Diseases caused by microorganisms (microbes) are called infectious diseases. Some microorganisms are injurious because they destroy or damage body tissues. Most infectious diseases, however, are caused by toxins secreted by the microorganisms.

There are two types of infectious disease—communicable and noncommunicable. Communicable diseases are spread, in some way, from person to person; noncommunicable diseases, although caused by microorganisms, require an animal agent, such as an insect or rodent, for transmission. Foodservice workers should be acquainted with the mechanisms by which both man and animals spread disease.

Communicable Diseases

Infectious diseases passed from one person to another by direct or indirect contact are referred to as communicable diseases. Some examples are diphtheria, mumps, measles, whooping cough, typhoid fever, influenza, infectious hepatitis, and some food poisoning illnesses.

The germs of a communicable disease are often spread from a sick person to a well person. This can happen at any stage of the disease—incubation, acute, or convalescent. However, the greatest danger of transmission occurs during the acute stage or when the person is actually sick. Because of this danger, it is recommended that persons in the foodservice industry, who have any communicable disease, remain at home until the illness is totally cured.

Even after a person has clinically recovered from a disease, he may still harbor some of its germs. Although no longer sick, he can spread these germs to others. Such a person is called a carrier of the disease. The carrier state is only temporary in most people, lasting for a few days to a few weeks. The carriers of most gastrointestinal illnesses are temporary carriers. Some diseases, however, such as typhoid fever, can result in permanent carriers.

Almost anything that leaves the body of a sick person or a carrier may have germs in it. They may be in the air that he breathes out or the discharges from his mouth, nose, throat, lungs, or intestine. Clothes, dishes, and other articles that he has used or touched may have germs on them (Fig. 3-2). Germs from his body may get into water, milk, or food. The dangers of an ill or recovering person working in a kitchen are illustrated by the following case that occurred in southeast Texas.

Figure 3-2 (a) If cups are carried in this manner, communicable diseases can be spread through contact with the drinking surface of the bottom cup. (b) Proper method of carrying cup and saucer.

Early in the school year, a sixth-grade boy developed infectious hepatitis. Within the next two months, 26 other cases appeared: one cluster of four cases in siblings and classmates of the student with the initial case, 21 cases in high school students who attended the same centralized school as the original case, and one teacher.

All of the high school students were found to have attended the same lunch period. The single school-operated cafeteria served 700–800 lunches per day in approximately 350 compartmented trays. The cafeteria was operated by a staff of adult employees but part-time student helpers assisted with dishwashing and cleaning in order to earn lunches.

The fact that the 21 high school students attended the same lunch period assumed even greater significance when it was learned that the student implicated as the source for the first cases worked regularly in the cafeteria during this lunch hour.

In spite of his illness, characterized by fever, nausea, vomiting, and jaundice, the boy with the initial case continued in school and worked regularly in the cafeteria. His duties included wiping tables, sweeping, and occasionally working in the dishwashing area. When the demand for trays and flatware was heavier than the usual processing could supply, he sometimes dried them by hand. On one such day, he developed nausea and vomiting and had to be sent from the kitchen. It seems probable that he contaminated trays or utensils in some manner after they had been washed. These trays were then distributed through the cafeteria serving line.

Another case, in eastern Massachusetts, involved the transmission of scarlet fever through food. The first two cases to appear were children. On the following day five more cases were reported, these in a neighboring town. All of the victims had attended a church reunion and lunch two days prior to the appearance of the first cases. Approximately 200 persons had attended the reunion and 182 of these were reported to have eaten lunch at the church. Of these 182, 102 were reported ill. Of the 102 patients, 24 developed scarlet fever, 56 had sore throats, 7 diarrhea, 4 vomiting, 3 nausea, and 8 miscellaneous complaints.

The food served after the service consisted of coffee and cream, a variety of homemade cakes contributed by the ladies of the church, and ground ham sandwiches. The investigation revealed that all persons who were ill, with one exception, had eaten ham sandwiches or ground ham.

Of the 140 persons who ate the ham, four ate it immediately after it was cooked and did not become ill. One hundred and sixteen persons ate ham sandwiches at the lunch, 82 of whom became ill. Fourteen others ate leftover sandwiches brought home from the meeting, 13 of whom became ill. Six other individuals ate some of the ground ham left over from making the sandwiches. All of these people became ill. The progressive buildup of the bacteria as time elapsed is obvious.

The ground ham was prepared by grinding and mixing two cooked hams. They were prepared by two ladies from the church. One of the ladies, who naturally handled the ham she cooked, deboned, and sliced, felt perfectly well until the day of the church lunch. On that day, she became ill and went home immediately after the church service. After she arrived home, she vomited and developed diarrhea. The following day she awoke covered with the red rash of scarlet fever.

The evidence indicates that the lady who became ill was in the early stage of scarlet fever at the time she prepared the ham. Since only the ham that she prepared was responsible for the outbreak, it can be assumed that she infected the meat, either directly by her respiratory discharges or indirectly by her hands that had become contaminated by respiratory discharges. The warm ham and its accompanying juice incubated the streptococcus for about 12 hours and allowed the formation of toxin, which subsequently was distributed throughout the ground ham and caused the illness.

Spread of Communicable Diseases

Communicable diseases can be spread through either direct or indirect contact with an infected person. In direct contact, germs are picked up from an infection on the body surface or body opening of an infected person. This requires actually touching or being touched by an infected person, and includes hand-to-body and hand-to-mouth contact, kissing, and inhalation of air containing droplets of infective materials (within a range of three feet or less). A cough or sneeze directly into the face is also considered direct contact. Spread of disease in this manner can usually be avoided by guarding against contact with infected people. Among diseases spread by direct contact are boils and abscesses, impetigo, pinkeye, venereal disease, and athlete's foot.

Since most communicable diseases are spread through indirect contact, the foodservice worker must be knowledgeable in this area. In this type of germ transference, pathogenic microorganisms travel from an infected person to a well person by way of contaminated food or drink, dishes, drinking glasses, eating utensils, or dust. Very large numbers of bacteria-laden dust particles are liberated into the air from the skin and clothing as a result of normal bodily activities. Many of these particles may remain airborne for more than one-half hour. Dust particles from clothing can transmit communicable diseases. The spouts of drinking fountains, coins and currency, doorknobs, and many other objects that are handled every day can also be sources of infection through indirect contact.

Procedures followed by food handlers are vitally important to the patrons of a foodservice establishment. Incorrect habits or lack of knowledge can have dire consequences, both to the public and to the foodservice establishment. Some unsafe practices that contribute to the spread of communicable diseases are:

1. Unsanitary hands. The hands of a foodservice worker contact many things during a work day. A large number of such contacts are involved with job procedures. Many others, unfortunately, include personal habits of long

standing that may be undesirable in a food operation. Proper and thorough handwashing with soap, especially after going to the restroom, is essential (Fig. 3-3). Carelessness

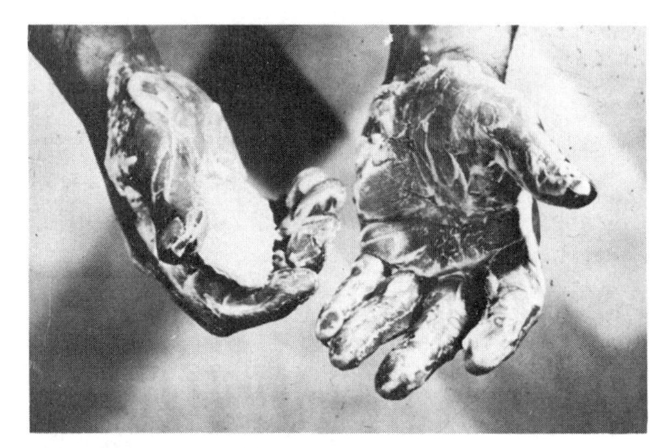

Figure 3-3 Thorough handwashing at necessary times is vital for the protection of food. Handwashing reduces the number of microorganisms to a safe level.

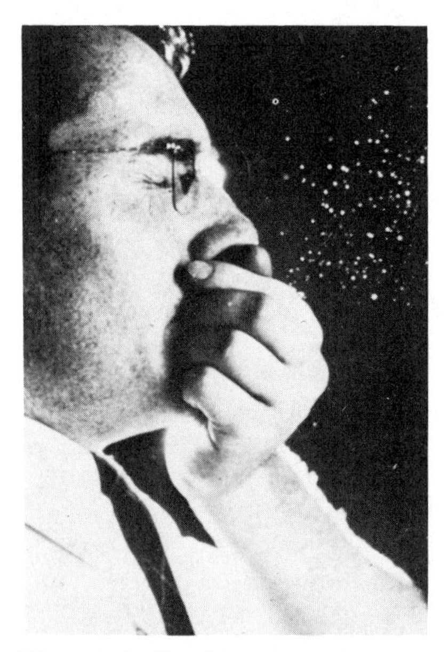

Figure 3-4 Hand to nose contact can contaminate the hands and render them dangerous to handle food, utensils, or equipment.

in this matter can result in the spread of salmonellosis, *Clostridium perfringens*, and streptococcus bacteria, dysentery, and many viral diseases, including infectious hepatitis. Any mouth- or nose-to-hand contact should be followed by washing the hands with soap (Fig. 3-4). The nostrils are the reservoir of Staphylococcus aureus in the human, and the hands of 40 percent of the human population carry this bacterium due to discharge from the nostrils. *Staphylococcus aureus* easily passes through the pores of a handkerchief and contaminates the hands.

2. Improper handling of eating utensils. Handling forks by the tines or spoons by the bowls also represents a potential hazard to anyone who uses them. Hands should not contact the eating part of any utensil that will later contact a patron's food (Fig. 3-5).

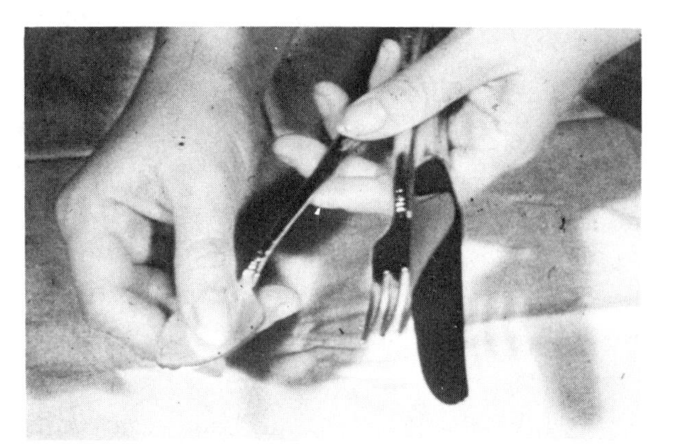

Figure 3-5 This type of mishandling can result in the spread of communicable diseases.

3. Mishandling glasses. A very common and dangerous practice is the habit of picking up two or three used glasses at one time, sometimes allowing the fingers to contact fluid remaining in the glasses, and then picking up clean glasses in the same manner. Hands should not contact the eating or drinking surfaces of glasses, cups, plates, or bowls (Fig. 3-6).

Among the dangerous practices just discussed is working with foods or utensils with hands that are contaminated with some patho-

Figure 3-6 Improper method of handling (a) glasses and (b) cups can result in the spread of communicable diseases.

genic microorganism. Contaminated hands of a person handling food or utensils, who is in some stage of a communicable disease, are a major cause of food poisoning. Proper care of the hands with regard to communicable diseases cannot be stressed too strongly.

During a recent five-year period, 21 foodborne or waterborne outbreaks of shigellosis (bacillary dysentery) were reported to the National Communicable Disease Center. In each of these outbreaks, the source of contamination was a human carrier who, through carelessness and faulty hygiene, contaminated the food that became the vehicle of infection.

During September and October of 1968, at least 98 persons were afflicted with food poisoning in a prosperous industrial community in an eastern state. The resulting investigation uncovered four separate outbreaks of shigellosis. All of the outbreaks were traced to a single catering establishment.

Shigella sonnei was the bacterial organism identified as causing the outbreaks. Six food handlers were responsible for preparing the salads that were incriminated. At least two of these food handlers had been sick with diarrheal illnesses when the salads were prepared. Their hands had considerable contact with the salads, which they mixed and blended, and improper handwashing after restroom visits undoubtedly left the hands contaminated with shigellae. The hands then contaminated the salads.

Some insects are also responsible for the spread of communicable diseases. The housefly and cockroach are carriers of many diseases. These insects travel from a source of infection, such as sewage, to a person or to foods or articles the victim contacts. Germs are carried on their feet or bodies and also in their intestinal tracts. Flies have been known to carry typhoid fever, shigellosis, anthrax, salmonellosis, cholera, tuberculosis, ophthalmia, staphylococcal food poisoning, and poliomyelitis in this manner. Dingman reported an outbreak of five cases of polio that resulted from flies transmitting the virus from a four-year-old patient on a dairy farm to milk that was sold and consumed raw.

Noncommunicable Diseases

These diseases are also caused by microorganisms, but their transmission to man is through animals rather than through other humans. Some of these infections are passed through food but there is little that the foodservice worker can do to prevent this type of disease. The two most common noncommunicable diseases associated with foods in the United States are trichinosis and tapeworm infestation.

The tapeworm is a member of a phylum that includes all flat and nonsegmented worms. The adult tapeworm has a flat, ribbonlike body, is grayish white in color, and lives in the small intestine. The head possesses suckers. Extending back from a slender neck are a number of segments, the size and number depending on the species. The largest and oldest segments, which contain fertilized eggs, are at the terminal end of the tapeworm. Segments containing fertilized eggs break off the main body of the tapeworm and pass from the intestine with the feces. If these eggs are ingested by the proper animal they are carried to the small intestine where they hatch into microscopic larvae that penetrate the intestinal blood vessels of the host animal and are carried to the muscles. They burrow into the muscles and form cysts. If the animal is butchered and the meat ingested undercooked, the cysts survive, pass to the intestine, grow into adult tapeworms, and fasten to the intestinal lining. The most common sources of tapeworms in the United States are cattle, hogs, and some fish, primarily from Lake Superior.

The trichina worm enters the body in undercooked pork. The cysts pass to the small intestine, hatch, and grow into small mature worms. The worms mate in the small intestine. The resulting larvae penetrate the intestinal blood vessels and are carried to the muscles. They penetrate the muscles and form cysts. If the eye muscles are penetrated, blindness can result. Occasionally, they burrow into the brain and cause death. Most cases arise as a result of tasting raw, homemade sausage being prepared for festivals by certain ethnic groups.

Other noncommunicable diseases are malaria, yellow fever, typhus, and Rocky Mountain spotted fever. These diseases are transmitted by insects and ticks.

SUMMARY

Foodservice workers have a special interest as well as a real involvement in studying communicable diseases. About 40 percent of the communicable diseases that physicians must officially report to health departments are associated with the business of providing food and/or drink to the public.

Human diseases include most conditions that interfere with the normal state of the body. Disease is one of man's greatest enemies. It has killed or crippled more persons than all the wars ever fought and costs billions of dollars.

Diseases are classified as infectious or noninfectious. Infectious diseases are caused by microorganisms such as bacteria and viruses. Noninfectious diseases are not caused by microorganisms and therefore cannot be spread from person to person. There are five generally accepted types of noninfectious diseases.

1. Congenital diseases. These diseases are present at birth. They include hemophilia, PKU, some heart and spinal cord defects, and many forms of mental retardation.
2. Environmental diseases. These diseases are caused by factors in our environment and in-

clude radiation poisoning, most cancers, and black lung disease of coal miners.

3. Deficiency diseases. These diseases result when the body fails to receive the required amount of certain necessary nutrients. Rickets, scurvy, and anemia are examples.
4. Allergic diseases. Allergies result when the body becomes hypersensitive to some foreign substance. Hay fever and asthma are the most common allergic diseases.
5. Organic diseases. These diseases are due to some abnormality of the body tissues or organs. Included in this category are heart diseases, muscular dystrophy, multiple sclerosis, and epilepsy.

There are two major types of infectious diseases: communicable and noncommunicable.

1. Communicable diseases. Communicable diseases are those diseases that are passed from person to person, from a sick person or a carrier to a well person. The transfer can occur at any stage of a disease and the contact can be direct or indirect. In direct contact, germs are picked up from an infection on the body surface or body opening.

   Most communicable diseases are spread through indirect contact. In this type of germ transference, pathogenic microorganisms travel from an infected person to a well person via contaminated food or drink, dishes, eating utensils, or dust.

   Procedures followed by food handlers are of vital importance to the patrons of foodservice establishments. Improper procedures or lack of knowledge can have dire consequences. Three unsafe practices are: (1) working with unsanitary hands, (2) improper handling of eating utensils, and (3) mishandling glasses. Contaminated hands are a major factor in the spread of communicable diseases. Some communicable diseases can be spread by insects, particularly the housefly and the cockroach.
2. Noncommunicable diseases. These diseases are also caused by microorganisms but their transmission to man is through animals rather than through other humans. The two most common noncommunicable diseases associated with food in the United States are trichinosis and tapeworm infestation.

Review Questions

1. What is meant by the term *disease?*

2. a. What are noninfectious diseases? b. Describe five types of noninfectious diseases and give several examples of each.

3. What are infectious diseases?

4. a. Distinguish between a communicable and noncommunicable disease. b. List five communicable diseases. c. A person employed in a foodservice establishment who has a communicable disease should remain at home until recovery is complete. Why?

5. What is a carrier?

6. Describe the two types of contact whereby communicable diseases can be spread and give several examples of each.

7. List four unsafe practices that can contribute to the spread of communicable diseases.

8. In what ways can insects spread communicable diseases?

References

Becker, M. E. (1966). "Water-Borne and Food-Borne Viruses." *J. Milk Food Tech.* 29:243–245.

Berg, G. (1964). "The Food Vehicle in Virus Transmission." *Health Lab. Sci.* 1:51–59.

Cliver, D. O. (1966). "Implications of Foodborne Hepatitis." *Pub. Health Rep.* 81:159–165.

Cliver, D. O. (1967). "Food-Associated Viruses." *Health Lab. Sci.* 4:213–221.

Dingman, J. C. (1916). "Report of Possibly Milk-borne Epidemic of Infantile Paralysis." *N.Y. State J. Med.* 16:589–590.

Donadio, J. A. (1969). "Foodborne Shigellosis." *J. Infect. Dis.* 119:666–668.

Duguid, J. P. and A. T. Wallace. (1948). "Air Infection with Dust Liberated From Clothing." *Lancet* 2:845–849.

Dull, H. B., T. C. Doege, and J. W. Mosley. (1963). "An Outbreak of Infectious Hepatitis Associated with a School Cafeteria." *Southern Med. J.* 56:475–480.

Getting, V. A., S. M. Wheeler, and G. E. Foley. (1943). "A Food-Borne Streptococcus Outbreak." *Amer. J. Pub. Health* 33:1217–1223.

Hamburger, M., Jr. and O. H. Robertson. (1948). "Expulsion of Group A Hemolytic Streptococci in Droplets and Droplet Nuclei by Sneezing, Coughing, and Talking." *Amer. J. Med.* 4:690–701.

Hardy, A. V. and J. Watt. (1948). "Studies of the Acute Diarrheal Diseases. XVIII. Epidemiology." *Pub. Health Rep.* 63:363–378.

Hill, H. R., R. A. Zimmerman, G. V. K. Reid, E. Wilson, and Major R. M. Kilton. (1969). "Food-Borne Epidemic of Streptococcal Pharyngitis at the United States Air Force Academy." *New Eng. J. Med.* 280:917–921.

Kloos, W. E. and M. Musselwhite. (1975). "Distribution and Persistence of Staphylococcus and Micrococcus Species and Other Aerobic Bacteria on Human Skin." *Appl. Microbiol.* 30:381–395.

Mason, J. O. and W. R. McLean. (1962). "Infectious Hepatitis Traced to the Consumption of Raw Oysters." *Amer. J. Hyg.* 75:90–98.

Moorehead, S. and H. H. Weiser. (1946). "The Survival of Staphylococci Food Poisoning Strain in the Gut and Excreta of the House Fly." *J. Milk Technol.* 9:253–259.

Nicholas, G. E. (1873). "The Fly in its Sanitary Aspects." *Lancet* 2:724.

Ostrolenk, M. and H. Welch. (1942). "The Common Housefly (Musca domestica) as a Source of Pollution in Food Establishments." *Food Research* 1:192–200.

Otto, J. H., C. J. Julian, and J. E. Tether. (1971). *Modern Health.* Holt, Rinehart, and Winston, Inc. New York.

Schifferes, J. J. (1970). *Healthier Living*. John Wiley and Sons, Inc. New York.

Stewart, W. H., L. J. McCabe, Jr., E. C. Hemphill, and T. DeCapito. (1955). "IV. Diarrheal Disease Control Studies." *Amer. J. Trop. Med.* 4:718–724.

Williams, R. E. O. (1963). "Healthy Carriage of Staphylococcus aureus: Its Prevalence and Importance." *Biol. Rev.* 25:56–71.

Winchester, A. M. (1969). *Biology and its Relation to Mankind*. Van Nostrand Reinhold Co. New York.

I. The Illness
  A. Foodborne Illness
  B. Incidence
  C. Symptoms
  D. History
II. The Bacteria
III. Foods
  A. Cooked Foods
  B. Uncooked Meat and Seafood
  C. Milk and Milk Products
IV. Reservoirs
V. Toxins
  A. Types
  B. Characteristics
  C. Antigenicity
VI. Case studies
  A. Case 1: Shrimp Salad
  B. Case 2: Doughnut
  C. Case 3: Chicken Salad
VII. Summary

4 The Staphylococci

On January 26, 1967, a 39-year-old housewife and mother purchased some barbecued chicken legs from the warm display case in a supermarket in Ottawa, Canada. Two hours later the chicken was served for lunch. The mother ate only one mouthful because the chicken did not taste good. The 3-year-old child spat out the first bite because of the taste.

About two and one-half hours after eating the one bite of chicken, the mother became nauseous and dizzy. A half-hour later she began to vomit. The vomiting sessions continued until midnight, increasing in intensity as time went by. At that time she entered a hospital. By then she was nearly prostrate and quite ill. Two days later she was discharged from the hospital.

An examination of the chicken revealed the astounding number of 408 million *Staphylococcus aureus* per gram of chicken (1 gram = 1/28 oz.). It could not be determined where the contamination occurred, but the results were apparent. One mouthful of the tainted chicken resulted in hospitalization.

THE ILLNESS

Foodborne Illness

There are two generally recognized types of foodborne illness—foodborne intoxication and foodborne infection. These terms should be used properly, although it must be recognized that the term *food poisoning* is still in widespread use to mean foodborne illness. Therefore, it is used occasionally in this text.

If the foodborne illness is caused by a toxin or poison that is in the food previous to ingestion, the illness is properly termed a foodborne intoxication. *Staphylococcus aureus* causes foodborne intoxication, since it produces toxin in food during the course of its growth. When the unfortunate victim ingests the food, he also ingests the toxin and the first symptoms of staphylococcal food intoxication appear, usually within a few hours or less. Other common causes of foodborne intoxication are the toxins of *Clostridium botulinum* (botulism) and *Clostridium perfringens.*

Foodborne infection is caused by the ingestion of food containing large numbers of bacteria, which release toxins in the body (gastrointestinal tract) rather than in the food.

In a foodborne intoxication, then, the illness is caused by the ingestion of toxin already in the food. The bacteria, themselves, are rather incidental, although some additional toxin may be produced in the intestinal tract. In a foodborne infection, the illness is caused by the ingestion of the pathogenic bacteria that are in the food and that produce toxins in the intestinal tract rather than in the food.

Incidence

In 1975, there were 497 reported outbreaks of foodborne disease. At least 10 to 20 times that many probably went unreported. These outbreaks affected 28,260 people. In 191 (38 percent) of the 497 outbreaks, the cause was determined. Most of the outbreaks (64 percent) were due to bacteria. The bacteria most frequently implicated were staphylococci. Forty-five staphylococcal outbreaks involving 4,067 people were reported. The staphylococcus most frequently implicated was *Staphylococcus aureus.*

Eighteen of these 45 reported outbreaks involved ham. Incriminated foods also included tuna casserole, mashed potatoes (military base), turkey (school), lasagne (camp), Mexican food (military base), beef, fish, rice ball, roast beef sandwich, turkey salad (school), shrimp salad, jambalaya, chicken/rice casserole, potato salad (school), lobster bisque, sausage, and salad dressing. Table 4-1 lists some of the larger outbreaks in 1975.

Table 4-2 shows where most of the reported staphylococcal outbreaks occurred in 1975. Homes, as usual, led the way.

Table 4-1
Major staphylococcal foodborne outbreaks in 1975. CDC Annual Summary.

Food	Number Sick	Locality	State
Ham	336	Church	Minnesota
Turkey salad	324	School	Illinois
Barbecue pork	275	School	South Carolina
Ham	200	Picnic	North Carolina
Shrimp salad	200	Cafeteria	Louisiana
Chicken salad	126	Church	Florida
Ham	100	School	Tennessee
Ham	83	Fire hall	Pennsylvania
Chicken salad	81	Restaurant	Georgia
Potato salad	74	School	Missouri

Table 4-2
Localities of some reported staphylococcal foodborne outbreaks in 1975. CDC Annual Summary.

Locality	Number of Reported Outbreaks
Home	12
Restaurant	8
School	6
Church	4
Military Base	2
Cafeteria	2
Picnic	2

Symptoms

The symptoms of staphylococcal food intoxication appear within about three hours after ingestion. The onset time can vary, however, from one-half hour to more than seven hours. One of the most characteristic features of this illness is the short time between the ingestion of the toxin-containing food and the appearance of the symptoms marking the illness.

The time that it takes for the victim to become ill depends on the amount of enterotoxin (enteron—intestine; toxin—poison) that has been consumed and the susceptibility of the person to the toxin. The amount of toxin consumed depends on the quantity of food eaten and the amount of toxin present in the food. Susceptibility to the toxin varies from person to person.

The symptoms of staphylococcal food intoxication begin with increased salivation, soon followed by nausea, vomiting, retching, abdominal cramps, and diarrhea. Headache and sweating sometimes occur. A severe attack can result in dehydration, prostration, and shock. Subnormal temperatures are more common than fever. Recovery usually occurs in two days or less. Staphylococcal food intoxication is rarely fatal, but it is very unpleasant.

History

Staphylococci were first observed in pus by Robert Koch in 1878 and by Louis Pasteur in 1880. At that time, they were not associated with foodborne illness. A. Ogston, in 1881, is credited with designating the name staphylococcus (from the Greek **staphyle**—grape). Ogston had observed the typical grapelike clusters on many occasions and thereby arrived at the name. In 1884, Rosenbach was able to differentiate between two separate kinds of staphylococci. He named them *Staphylococcus aureus* and *Staphylococcus albus*.

Although the staphylococci were not discovered until the time of Koch and Pasteur, they have undoubtedly caused foodborne illness for hundreds of years. A report of a food-poisoning outbreak in Paris appeared in the *Edinburgh Medical and Science Journal* as early as 1830. A father, his 27-year-old daughter, and a 9-year-old child suffered stomach cramps, diarrhea, and vomiting about three hours after consuming a ham pie for dinner. These symptoms and the short period of time before illness are typical of staphylococcal food intoxication. Several other patrons of the bakeshop where the pie was purchased also suffered the same illness at that time.

The comment was also made in this report that some foods occasionally acquired poisonous qualities for no apparent reason. Cheese, ham, bacon, and sausage were mentioned as problem foods.

In 1862, a food-poisoning episode was reported in the English journal, *Lancet*. An entire family of ten became ill shortly after eating rabbit pie for dinner. The sickness was marked by diarrhea, vomiting, and violent abdominal cramps, all indicative of possible staphylococcal food intoxication.

In 1873, an outbreak involving pork brawn was reported in the *British Medical Journal*. Sixteen persons became ill, some violently so, with vomiting, diarrhea, abdominal cramps, and a sense of burning and constriction in the throat. The symptoms occurred within two to three hours after eating the brawn.

By the late 1870s, cream puffs were recognized as a food capable of causing foodborne illness. Several cases of food poisoning due to cream puffs had been reported. In 1879, Henry Leffmann, a toxicologist (a specialist in the field of poisons) on the staff of Jefferson Medical College summed up the state of knowledge at that time, stating, "I have made chemical examinations of portions of cream puffs that have caused trouble but have not found anything that could be assigned as the cause of the action."

At this time, the staphylococci had barely been related to human conditions. Only a year had passed since Koch's discovery of staphylococci in pus. They were not even remotely suspected of causing food poisoning and their toxins were as yet unknown. It was no wonder, then, that investigators such as Professor Leffmann were unsuccessful in their search for the cause of food-poisoning episodes involving seemingly wholesome foods.

A major breakthrough occurred with the report of Professor Victor Vaughan, which appeared in 1884 following a six-month period marked by the occurrence of nearly 300 cases of cheddar cheese poisoning in the state of Michigan. All of the implicated cheese had come from one factory.

Vaughan obtained samples of incriminated cheese. He covered small pieces with alcohol, agitated the container, and then filtered the mixture. He then evaporated the filtrate (the liquid that passed through the filter) and ingested some of the remaining substance. Within a short time he became nauseous. The experiment was repeated several times with the same result.

Vaughan then examined the residue from the evaporated filtrates with a microscope and found spherical bodies that he speculated might be coccus type bacteria. He concluded from these experiments, ". . . that the poisonous material, whatever it may be, is contained in the alcoholic extract. This would indicate a chemical poison and not a bacteric one. The production of this poisonous material is due to the rapid growth of some bacterium."

The Michigan cheddar cheese outbreak is probably the first recorded outbreak in which a definite type of bacteria was implicated and a possible mechanism, a poison produced by the bacteria, was suggested.

In 1894, staphylococcus was identified as the culprit in a food-poisoning outbreak in which all the members of a family except one child became sick. The child did not eat any of the incriminated beef. One death was recorded from this outbreak.

By 1901, research into the properties of the staphylococcus toxin was under way. It was then called tyrotoxicon, and research at that time indicated that it could be decomposed at a temperature of 194°F. (90°C).

In 1914, M. A. Barber made the most thorough investigation done up to that time of a food-poisoning outbreak. The outbreak was actually one of a series of outbreaks that had occurred on a farm in Luzon, Philippine Islands. For several years, episodes of acute gastroenteritis (stomach and intestinal upset) had repeatedly occurred on this farm. These episodes were marked by vomiting and diarrhea, sometimes accompanied by muscle cramps and faintness. During the course of his investigation, Barber, himself, was afflicted three times. Examination of the kitchen, food, and methods of food preparation yielded nothing suspicious. It was finally noted that each attack of illness followed the drinking of fresh cream.

Samples were drawn from the farm's two milk cows. The samples were refrigerated for twenty-four hours. Barber then drank some of the cream from one sample and left the remainder of the sample in the kitchen. No illness followed the drinking of the refrigerated cream. Five hours later he drank some of the unrefrigerated cream. Within two hours he became nauseous and was afflicted with diarrhea. Cream from the other cow was sampled in the same manner but no illness resulted.

Agar cultures made from milk drawn from the incriminated cow yielded a white staphylococcus. These staphylococci were transferred into a flask of milk from another source and incubated at 98°F. (37°C) for about eight and one-half hours. It was then refrigerated until the next day. When Barber drank some of the milk, he was sick in less than two hours.

Biochemical tests revealed that the staphylococci fermented mannite and maltose and, to some degree, lactose. When innoculated under the skin of guinea pigs and monkeys, abscesses sometimes formed. Barber summarized, "Acute attacks of gastroenteritis were produced in milk by a toxin elaborated by a white staphylococcus which occurred in almost pure culture in the udder of a cow. The fresh milk was harmless, and the toxin was produced in effective quantities only after the milk had stood some hours at room temperature."

This classic piece of investigation, like that of Gregor Mendel's work with the genetics of peas, went little read and unappreciated for years. The trail was not really picked up again until the work of the noted food-poisoning investigator, Professor G. M. Dack, appeared in 1930. Dack had been given two Christmas cakes for examination, both of which were obtained from the same bakery. They were three-layer sponge cakes with cream filling. One cake had made eight people sick, the other three. Through the use of human volunteers the cake, itself, was determined responsible for the problem rather than the icing, filling, or decoration of pistachio nuts and maraschino cherries.

Culture plates revealed 19 different types of bacteria present, three of which were present in large numbers. Three human volunteers drank different amounts of a filtrate prepared from one of the three cultures. Within three hours they were suffering nausea, chills, and stomach cramps. None of the other cultures caused symptoms in the volunteers. The culprit on this occasion was revealed by the microscope to be a yellow staphylococcus. It exhibited many of the same characteristics as Barber's white staphylococcus.

From this time on, research on the staphylococci and their toxins shifted into high gear in university and public health laboratories. As food-poisoning reports increased in number, particularly from cream-filled and custard-filled bakery goods, the staphylococci, especially *Staphylococcus aureus*, became recognized as our most common cause of foodborne illness.

The Bacteria

The name *Staphylococcus* is derived from the Greek words *Staphyle* (bunch of grapes) and *Kokkus* (seed or berry). The genus name

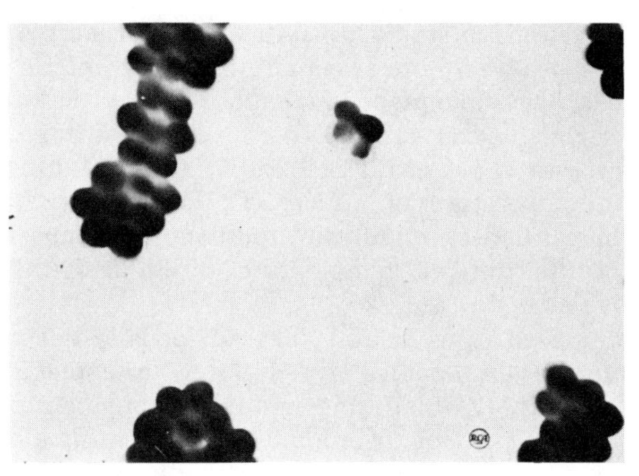

Figure 4-1 *Staphylococcus aureus*

Staphylococcus thus describes the irregular clusters of spherical cells, which this type of bacterium can form in a liquid medium. However, staphylococci can also occur singly, in pairs, and in tetrads (groups of four). The genus *Staphylococcus* is a member of the family Micrococcaceae.

Staphylococci are gram-positive, nonmotile (not capable of movement) cocci, about 0.8–1.0 micron in diameter. They do not require oxygen but they grow better aerobically than anaerobically. Many food-poisoning strains produce a yellow pigment. They do not produce spores. The food-poisoning members are coagulase-positive; that is, they are capable of producing an enzyme that can clot rabbit or human blood. *Staphylococcus aureus* produces acid from glucose and ferments mannitol. It does not form indole.

The optimum growth temperature of *Staphylococcus aureus* is 99°F. (37°C) but its growth range extends from 44°–114°F. (7–46°C). A temperature of 150°F. (66°C) maintained for at least 12 minutes will effectively kill this organism. The staphylococci are capable of surviving many food processing treatments. They withstand drying and freezing well; these processes result in only slight reductions in numbers. High levels of salt and sugar in foods prevent the growth of many bacteria but the staphylococci grow well in their presence. *Staphylococcus aureus* is very salt-tolerant, growing in sodium chloride solutions that approach saturation. They can therefore grow in curing solutions and on curing and cured meats.

Staphylococcus aureus is capable of growing over a wide pH range. It grows best in foods with a near neutral pH of 6–8, but it can grow throughout a range of 2–10. In general, the better the food medium, the wider the range of pH and temperature over which growth can take place.

Several characteristics have been suggested as indicators of the ability of a suspected staphylococcus to cause food poisoning. The most commonly used are the positive coagulase reaction (coagulase-positive) and the fermentation of mannitol.

FOODS

Cooked Foods

The type of food most likely to be involved in a staphylococcus outbreak is one that is high in protein and that has been handled extensively after cooking. The more a food is handled, the more likely it is to become contaminated with bacteria found on hands and equipment.

Foods endangered are those that have been cooked and then cut, sliced, or mixed and subsequently used in sauces or salads (for example, egg, tuna, potato, or chicken); foods that have been cooked and then allowed to cool slowly, thereby allowing bacterial growth; and foods that have been rewarmed without getting hot enough to kill any staphylococci present.

Hodge, in his investigation of 95 outbreaks of foodborne illness, reported that 94 of the outbreaks were due to cooked foods containing large amounts of protein. In 63 outbreaks (67 percent), the food involved was a mixture, such as, tuna salad, turkey salad, creamed chicken, potato salad, meat loaf, chicken pie, egg salad, or cream-filled pastries. These foods are usually handled considerably after cooking. Leftovers were involved in 94 percent of the outbreaks and unrefrigerated food in 89 percent.

Ham is probably the single food that has contributed most heavily to staphylococcal foodborne illness statistics. Other foods include custards, meat sauces, gravies, meats and meat products, meat-filled sandwiches such as barbecued pork and roast beef, roast turkey and dressing, and imitation cream filling. This is not, by any means, a complete listing of foods that have been involved in staphylococcal food intoxication cases. It should be apparent, therefore, that with such a wide variety of foods suitable and available as growth media, extreme cleanliness and attention to proper food-handling procedures must be followed unceasingly.

Turkey meat is a food that has ranked high on the list of foods incriminated in staphylococcal foodborne illness. During a recent ten-year period, this food was responsible for 11 percent of 425 cases of staphylococcal food intoxication. Turkey, like ham, is an excellent growth medium for several of the food-poisoning bacteria.

Rasmussen and Strong surveyed prepared salads obtained from retail stores in Madison, Wisconsin. The salads purchased were chicken, ham, tuna, shrimp, egg, kidney bean, and baked bean. Twenty-six percent of the 244 bacterial colonies (isolated from the salads) were coagulase-positive staphylococci.

Growth of staphylococci in many foods can begin at temperatures as low as 44°F., (7°C) although this minimum temperature varies with the food and the strain of *Staphylococcus aureus*. One study has indicated for example, that a 100-fold increase in numbers of staphylococci can take place in cream pie filling that is refrigerated at 41°F. (5°C). On the other hand, growth in custard is not readily apparent until a temperature of 46°F. (8°C) is reached. In ham salad no significant increase in numbers occurs until a temperature of 50°F. (10°C) is reached.

Toxin production does not usually occur until a temperature of 64°F. (18°C) or above is reached. By that time enough staphylococci may be present to produce toxin in sufficient quantity to cause illness. Growth occurs at temperatures up to 114°F. (45°C) at which point growth ceases and the bacteria begin to die. Growth and toxin formation can take place in steam tables in cafeterias and restaurants and in food-vending machines that keep food heated for extended periods of time, if the temperature is not regulated properly.

Uncooked Meat and Seafood

In a survey of market meats, Jay found that 32 percent contained coagulase-positive staphylococci. The meats from which staphylococci were recovered, in decreasing order of frequency, were: chicken, pork liver, fish, spiced ham, ground beef, steak, hamburger, beef liver, pork chops, veal steak, and lamb chops.

Commercial chicken was tested by Messer et al. The group rinsed chickens obtained from stores and examined the rinse solution. They found *Staphylococcus aureus* in the rinse solution of 80 percent of the chickens examined. The same group found staphylococci in 35 percent of hamburger tested.

Chicken and turkey pies have been examined on many occasions. In one such study, Canale-Parola and Ordal examined the pies manufactured by five companies. All samples were purchased from retail stores. Only five samples out of the 40 purchased were negative for staphylococci. Pies from all five manufacturers contained the bacteria.

Duitschaever et al., in 1977, investigated the incidence of *Staphylococcus aureus* in fresh ground beef and in hamburger patties. They examined 108 samples of fresh ground beef obtained from retail store display cases and 99 samples of commercial frozen hamburger patties, also obtained from display cases. *Staphylococcus aureus* was present in 46 percent of the hamburger samples. Of the 99 samples of frozen beef patties, 93 (94 percent) contained *Staphylococcus aureus*. Surkiewicz et al. (1975) found *Staphylococcus aureus* in 85 percent of 74 sets of raw beef patties collected in 42 federally inspected establishments.

Christiansen and King examined sandwiches and salads obtained from commercial retail outlets. They found that 39 percent of the salads and 60 percent of the sandwiches contained coagulase-positive staphylococci. For the salads, the incidence of these organisms was highest in chicken salad (60 percent), followed by pimiento cheese (46 percent), potato salad (45 percent), barbecued pork (26 percent), ham salad (25 percent), and cole slaw (20 percent). For the sandwiches, coagulase-positive staphylococci were isolated from ham salad (71 percent), chicken salad (60 percent), and sliced and chopped ham (53 percent).

Surkiewicz et al. (1968) investigated the occurrence of staphylococci in retail frozen fish. This group found that 20 percent of their samples of frozen, raw, breaded fish and 10 percent of their samples of frozen, fried, breaded fish contained coagulase-positive staphylococci.

Foster et al. examined 597 samples of fresh and frozen seafood products obtained from retail stores. Among the fresh items were clams, oysters, shrimp, and salmon. Frozen items included catfish, flounder, and sole. All samples contained small numbers of *Staphylococcus aureus*.

Milk and Milk Products

Milk and milk products have also been thoroughly investigated. Raw milk and cheese have caused many outbreaks of foodborne illness in the past. Butter was the carrier in an outbreak involving several restaurants in Illinois in 1978.

Barber's report in 1914 showed that enterotoxigenic staphylococci could be transmitted to human beings through raw milk from an apparently healthy cow. Minnet reported in 1936 that staphylococci found in the udder of cows were capable of causing food poisoning in man.

In 1940, Williams obtained samples from 387 cows comprising 10 herds. From these cows, Williams isolated 119 strains of coagulase-positive *Staphylococcus aureus*. A 1952 study showed that 25 of 37 cultures of staphylococci isolated from cow udders produced enterotoxin.

In 1960, Foltz et al. reported on the incidence of potentially pathogenic staphylococci in dairy products at the consumer level. They obtained 207 samples of pasteurized dairy products from retail outlets. These products had been produced in 42 different plants. Seven classes of dairy products were examined: pasteurized milk, low-fat milk, chocolate milk, buttermilk, half and half, coffee cream, and whipping cream. Four of these classes, pasteurized milk, buttermilk, half and half, and whipping cream, contained samples possessing coagulase-positive staphylococci.

Mickelson et al. investigated 125 samples of cheese purchased from retail stores. Twenty varieties were included. Eighty-eight samples (70 percent) contained *Staphylococcus aureus;* coagulase-positive staphylococci were isolated from samples of cheddar, bondost, blue, and brick cheese. Hendricks et al. describes an outbreak in Iowa, in 1959, at a state institution. Two hundred persons became ill from eating cheddar cheese infected with *Staphylococcus aureus*.

RESERVOIRS

The largest reservoir of *Staphylococcus aureus* is man. A large percentage of healthy persons have been shown to be carriers of pathogenic staphylococci. Getting et al. demonstrated that staphylococcal food-poisoning outbreaks can sometimes be traced to specific food handlers who were apparently free of any infection. The group reported that 18 percent of 122 food handlers, who were apparently free of any infection, were carriers of the same strains of staphylococci that were isolated from foods suspected to have caused foodborne illness in places where they worked.

The nose is the main site for multiplication of these organisms. It becomes colonized in infancy and remains the major human reservoir into adulthood. Williams has reported a 40 to 44 percent adult carrier rate in the nose and similar percentages for the hands. In 1975, Kloos reported that staphylococci composed more than 50 percent of the bacteria isolated from the head, nose, and axillae (armpits) and from 10 to 70 percent of the bacteria isolated from the hands and arms.

There are three kinds of nasal carriers: chronic carriers are constant long-term carriers; temporary or intermittent carriers periodically harbor the bacteria for a few weeks; and carriers who rarely harbor these organisms. Staphylococci are often found in the sinuses and are frequently associated with sinus infections. They are abundant in post-nasal drip associated with colds. Foodservice workers should, therefore, remain at home if they are ill.

Many outbreaks of foodborne illness have also been caused by foodservice personnel with various skin infections, such as, pimples, boils, and sores. These infections contain prodigious numbers of pathogenic staphylococci and should be completely covered.

Staphylococci are also carried by animals. The most important animal carrier is probably the cow. Crabtree and Litterer reported a series of outbreaks at a boys preparatory school in Tennessee. The outbreaks extended from July through October and affected 233 students and 9 faculty members. The outbreaks were traced to a cow that had chronic mastitis. Many cases with a similar cause have been reported by other investigators. Gwatkin examined 260 cows for the presence of bacteria. Nineteen cases of mastitis were found, all of which were apparently caused by staphylococci. One-hundred-ninety strains of potentially pathogenic staphylococci were isolated from these cows. Twenty-five of the strains were isolated from apparently healthy cows.

Moorehead and Weiser fed houseflies a strain of *Staphylococcus aureus* that had been isolated from a cake that caused an outbreak in Milwaukee, Wisconsin. Staphylococci were periodically recovered from fly vomit and fecal material for 5 days and from the digestive tracts of 18 of 32 flies for 28 days.

TOXINS

The staphylococcal enterotoxins are a common cause of foodborne intoxication in many countries. The ingestion of these toxins causes illness with a wide variety of symptoms, the most common of which are vomiting and diarrhea. Salivation, nausea, retching, and abdominal cramping also occur frequently. The first symptoms usually occur two to six hours after eating the enterotoxin-containing food.

Death rarely occurs in otherwise healthy humans as a result of staphylococcal food poisoning, although some fatalities have resulted; some in young children, some in elderly people suffering from other conditions, and some in instances where food poisoning was not diagnosed and improper treatment resulted. If misdirected antibiotic treatment is prescribed, the normal intestinal flora can be destroyed, resulting in the uninhibited growth of the enterotoxin-producing staphylococci and the liberation of additional toxin into the intestine. Not all strains of staphylococci produce enterotoxin but it has not yet been determined what percentage do.

Types

By 1960, it was realized that there were actually two different staphylococcal enterotoxins capable of causing inflammation of the lining of the stomach and intestine. They became known as enterotoxins A and B. In 1965, enterotoxin C was identified; and in 1967, a fourth enterotoxin, designated D, was isolated. The most recently discovered toxin, type E, was identified in 1971. Of these enterotoxins, type A has been the most frequent cause of foodborne staphylococcal intoxication. Some of the staphylococci produce more than one type of enterotoxin.

Characteristics

Enterotoxins are proteins that are produced in foods by the staphylococci when proper conditions exist. They are simple proteins, that is, they contain only amino acids. The purified enterotoxins are fluffy snow-white materials that are hydroscopic (takes water from the air) and readily soluble in water and salt solutions. They have molecular weights between 25,000 and 35,000.

The toxins do possess some degree of heat stability. Temperatures that will kill staphylococci will not necessarily inactivate their enterotoxin. The toxin survives boiling for 20 to 60 minutes and to destroy it requires a temperature of 240°F. (116°C) for 11 minutes. It gradually loses its potency during such heating.

Conditions that favor the growth of staphylococci also favor the production of enterotoxin. Toxin can be produced at temperatures between 60° and 114°F. (16–46°C) but production is best between 70° and 97°F. (21–36°C). At these temperatures enough toxin to cause illness can be produced in four to six hours. The lower the temperature during growth, the longer it will take to produce a sufficient amount of enterotoxin to cause illness. The importance of prompt and adequate refrigeration is obvious.

Antigenicity

As early as 1914, Barber, in his investigation of the milk-poisoning outbreaks in Luzon, Philippine Islands, remarked that some of the inhabitants of the farm seemed to be immune to the illness. It was not until 1938, however, that Davison et al. demonstrated that the staphylococcal enterotoxin had the ability to act as an antigen, that is, to cause antibody formation. Davison injected filtrates from staphylococcal cultures under the skin of monkeys and kittens, causing them to produce antibodies in their blood. In 1944, Dolman was able to produce immunity to enterotoxin in cats and humans.

In 1952, Surgalla et al. made a major advancement in this research. His group was able to observe the reactions between staphylococcal enterotoxin and antibodies that had been produced in response to the toxin. The reactions took place in an agar culture. It was apparent from these reactions that the toxin contained several antigens.

In 1955, cats were immunized by Thatcher and Matheson with filtrates obtained from different *Staphylococcus aureus* strains. These different strains caused the production of different antibodies, suggesting that there were actually several enterotoxins rather than just one. We have seen, with the resultant isolation of the staphylococcal enterotoxins, that five types of toxin are now known to exist.

CASE STUDIES

Hopefully, the following cases will serve as a practical guide to some proper and improper procedures associated with the preparation of foods.

Case 1: Shrimp Salad

This outbreak occurred aboard a United States Navy ship during a cruise for navy dependents. Twenty-eight individuals were affected. The incriminated food, which had the following history of preparation, was shrimp salad.

The shrimp were removed from the freezer at approximately 8:30 A.M. on the day previous to the outbreak. They were cooked, peeled, and placed under refrigeration by 10:45 A.M. The morning of the outbreak, at about 9:00 A.M. they were removed from the refrigerator, hand-chopped, and returned to the refrigerator about one hour later. At 10:45 A.M., the shrimp were again removed from the refrigerator, hand-mixed with onion, celery, pimiento, lettuce, and mayonnaise. The shrimp salad remained unrefrigerated at room temperature until it was served buffet style between the hours of 12:00 and 1:00 P.M.

An inspection of food preparation facilities failed to reveal any conditions that could have contributed to the contamination of the shrimp. An examination of the foodservice personnel, however, uncovered one man who had a draining sore on his right thumb. This man had prepared the shrimp salad. Laboratory analysis of drainage from the thumb revealed the same strain of *Staphylococcus aureus* present in the shrimp salad.

The following points should be noted:

1. The steward who prepared the salad should not have worked in the food area until the sore had healed. Most skin infections (boils, pimples, and sores) contain *Staphylococcus aureus*. This organism can easily be transmitted to food.
2. All foodservice personnel should wear disposable gloves and use the proper utensils when actually handling food.
3. After the final mixing procedures, the shrimp should have been refrigerated until served.

Case 2: Doughnut

On Sunday morning, December 16, 1962, two employees in the flight-control section of a large commercial airline each consumed a cream-filled doughnut from an unrefrigerated vending machine in the office. Less than three hours later, both men became ill with nausea, vomiting, and diarrhea. A doughnut taken from the vending machine yielded 215 million *Staphylococcus aureus* per gram of filling.

The internal temperature of this vending machine was hot enough to melt the chocolate coating on candy bars in adjacent compartments. The vendor stated that since the filling was imitation cream made from artificial ingredients, it was considered incapable of supporting the growth of organisms causing foodborne illness.

Lack of knowledge is the important point here. The imitation cream filling was not thoroughly researched and so the manufacturer and vendor were ignorant of the possible consequences of placing the imitation cream-filled products in unrefrigerated vending machines. The lesson to be learned is that lack of knowledge or education can result in illness and business reputations can be jeopardized. All persons associated with food processing, manufacturing, preparation, and servicing must, in the interest of public health, be aware of the possibilities of contamination of their products.

Case 3: Chicken Salad

This outbreak involved 1,364 children who had eaten a lunch served at 16 elementary schools in Texas. The lunches were prepared in a central kitchen and transported to the schools by truck. Chicken salad was the implicated carrier of the staphylococci.

The chickens used in the salad were boiled for three hours on the afternoon before the meal was served. After cooking, the chickens were deboned, fan-cooled to room temperature, ground into small pieces, placed into 12-inch-deep aluminum pans, and stored overnight in a walk-in refrigerator at 42°–45°F. (5.5–7.0°C).

The following morning, the remaining ingredients were added and blended. The food was then placed in thermal containers and transported to the schools, where it was kept at room temperature until the noon meal. Within two to three hours the illness developed.

The chickens probably became contaminated with staphylococci during the deboning process. Growth and toxin production would have been rapid during the cooling process due to the size of the aluminum pans. Growth and toxin production would have continued while the chicken salad remained at room temperature at the schools.

The following points should be noted:

1. Improper handling of the chickens during deboning was the probable cause of contamination. Gloves were not worn during this process.
2. The chickens should not have been cooled in 12-inch-deep pans. In a pan of this depth the chicken could not cool rapidly, and sufficient time existed for heavy bacterial growth and toxin production before the temperature finally reached the minimum growth level.
3. The chicken salad should have been refrigerated in the schools. If this was not possible, the item should not have been on the menu. A food item that cannot be prepared and maintained safely until serving should not be offered.

We have here two common causes of foodborne illness: contamination of a food, probably by a food handler, and inadequate refrigeration until serving. Both causes occur all too frequently.

In summary, the conditions that are necessary for a staphylococcal foodborne outbreak are:

1. A food item containing enterotoxin-producing staphylococci. The food will probably have become contaminated through the transmission of staphylococci from a human source. *Prevention:* contamination of foods can be reduced by proper sanitation. No employee should handle food, utensils, or equipment if he is sick, if he has an infection such as a carbuncle, or if he has an unhealed cut or sore.
2. The food must be a good growth medium for the staphylococci that are present. High protein foods are particularly susceptible.
3. The food must remain within the growth range of the staphylococci (44°–114°F.) [7–46°C] for several hours or longer. *Prevention:* prompt and adequate refrigeration is the best way to prevent growth of staphylococci.

SUMMARY

There are two recognized types of foodborne illness—foodborne intoxication and foodborne infection. Staphylococci cause foodborne intoxication since they produce toxin in food during the course of their growth.

Staphylococcus aureus is implicated more frequently in foodborne illness than any other organism. Ham is the most frequently implicated food while homes, followed by restaurants and schools, are the places where most outbreaks occur.

The symptoms of staphylococcal food intoxication usually appear in about three to five hours. The attack frequently begins with increased salivation, followed by nausea, vomiting, retching, abdominal cramps, and diarrhea. Headache and sweating sometimes occur. The disease is rarely fatal.

Staphylococci were first observed in pus by Robert Koch in 1878 but were not associated with foodborne illness at that time. In 1914,

M. A. Barber made a thorough investigation of a series of food-poisoning episodes that had occurred on a farm in Luzon, Philippine Islands. He attributed the outbreaks to a toxin produced by a white staphylococcus residing in the udder of a cow producing milk for the farm residents. In 1930, G. M. Dack investigated a food-poisoning outbreak caused by a Christmas cake. He found a yellow staphylococcus present that produced a toxin that caused gastroenteritis when ingested.

Staphylococci are gram-positive, nonmotile, parasitic cocci, about 0.8–1.0 micron in diameter, that usually occur in grapelike clusters. They do not require oxygen but they do grow better aerobically than anaerobically. Many food-poisoning strains produce a yellow pigment. Staphylococci do not form spores. The food-poisoning staphylococci are coagulase-positive.

The optimum growth temperature of *Staphylococcus aureus* is 99°F. (37°C) but its growth range extends from 44°–114°F. (7–46°C). A temperature of 150°F. (65°C), maintained for at least 12 minutes, will result in an effective kill. *Staphylococcus aureus* is salt-tolerant and will grow in sodium chloride solutions that approach saturation.

The type of food most likely to be involved in a staphylococcus outbreak is one high in protein, that has been handled extensively after cooking. Endangered foods are those that have been cooked and then cut, sliced, or mixed and subsequently used in sauces or salads (such as, egg, tuna, potato, or chicken); foods that have been cooked and then allowed to cool slowly, thereby allowing bacterial growth; and foods that have been rewarmed without reaching a temperature high enough to kill existing staphylococci. Ham is probably the food that has contributed most heavily to staphylococcal foodborne illness statistics.

The largest reservoir of *Staphylococcus aureus* is man. A large percentage of healthy persons are carriers of pathogenic staphylococci. The nose is the main site for multiplication of these organisms, having become colonized in infancy. Staphylococci are also carried by animals, the most important being the cow. Flies can also serve as carriers.

Five enterotoxins are presently known: types A, B, C, D, and E. Of these enterotoxins, type A has been the most frequent cause of staphylococcal intoxication. The toxins are water-soluble proteins produced in foods by the staphylococci when proper conditions exist.

Review Questions

1. How does *Staphylococcus aureus* cause foodborne illness?

2. a. What is characteristic about the time it takes for staphylococcal food poisoning to appear? b. What factors affect this time element?

3. What factors determine the amount of toxin that is ingested?

4. Describe the typical symptoms of staphylococcal food poisoning.

5. What was significant about the investigation of Victor Vaughan into the cheddar cheese food-poisoning outbreaks?

6. Describe several characteristics of *Staphylococcus aureus.*

7. What is the temperature range of S. *aureus*?

8. What pH range does S. *aureus* prefer for growth?

9. In what type of food is S. *aureus* most likely to grow?

10. What is the relationship between the handling of food and the possibilities of contamination?

11. What foods are likely to become contaminated with S. *aureus*?

12. What is the danger from staphylococcal infections in dairy herds?

13. Where, in humans, is the main reservoir of staphylococci?

14. Describe the three kinds of human nasal carriers.

15. What is the most important animal carrier of staphylococci?

16. Describe the staphylococcal enterotoxins.

17. Describe the heat stability of the staphylococcal enterotoxins.

18. What is the temperature range for the production of staphylococcal enterotoxins?

References

Angelotti, R. (1969). "Staphylococcal Intoxications." In *Foodborne Infections and Intoxications*, ed. Hans Riemann. Academic Press, Inc. New York.

Angelotti, R. (1970). "Food and Drug Administration View of Staphylococcal Contamination of Food." *J. Milk Food Technol.* 33:358–360.

Angelotti, R., M. J. Foster, and K. H. Lewis. (1961). "Time-Temperature Effects on Salmonellae and Staphylococci in Foods. I. Behavior in Refrigerated Foods. II. Behavior at Warm Holding Temperatures." *Amer. J. Pub. Health* 51:76–88.

Barber, M. A. (1914). "Milk Poisoning due to a Type of Staphylococcus Albus Occurring in the Udder of a Healthy Cow." *Philippine J. Sci.* 9:515–519.

Bell, W. B. and M. O. Veliz. (1952). "Production of Enterotoxin by Staphylococci Recovered from the Bovine Mammary Gland." *Vet. Med.* 47:321–322.

Bergdoll, M. A. (1970). "Enterotoxins." In *Microbial Toxins*, vol. 3. Eds. T. C. Montie, S. Kadis, and S. J. Ajl. Academic Press, Inc. New York and London.

Bergdoll, M. S., C. R. Borja, and R. M. Arena. (1965). "Identification of a New Enterotoxin as Enterotoxin C." *J. Bacteriol.* 90:1481–1485.

Bryan, F. L. (1968). "What the Sanitarian Should Know About Staphylococci and Salmonellae in Non-Dairy Products. I. Staphylococci." *J. Milk Food Technol.* 31:110–116.

Bryan, F. L. and T. W. McKinley. (1974). "Prevention of Foodborne Illness by Time-Temperature Control of Thawing, Cooking, Chilling, and Reheating Turkeys in School Lunch Kitchens." *J. Milk Food Technol.* 37:110–116.

Buckenham, J. (1862). "Extraordinary Cases of Poisoning" *Lancet* 2:297.

Canale-Parola, E. and Z. J. Ordal. (1957). "A Survey of the Bacteriological Quality of Frozen Poultry Pies." *Food Technol.* 11:578–582.

Casman, E. P. (1960). "Further Serological Studies of Staphylococcal Enterotoxin." *J. Bacteriol.* 79:849–856.

Casman, E. P. (1969). "Staphylococcal Food Poisoning." *FDA Papers* 3:15–16, 21.

Casman, E. P., R. W. Bennet, A. E. Dorsey, and J. A. Issa. (1967). "Identification of a Fourth Staphylococcal Enterotoxin, Enterotoxin D." *J. Bacteriol.* 95:1875–1882.

Christiansen, L. N. and N. S. King. (1971). "The Microbial Content of Some Salads and Sandwiches at Retail Outlets." *J. Milk Food Technol.* 34:289–294.

Chu, F. S., K. Thadhani, E. J. Schantz, and M. S. Bergdoll. (1966). "Purification and Characterization of Staphylococcal Enterotoxin A." *Biochemistry* 5:3281–3289.

Crabtree, J. A. and W. Litterer. (1934). "Outbreak of Milk Poisoning due to a Toxin-Producing Staphylococcus Found in the Udders of Two Cows." *Amer. J. Pub. Health.* 24:1116–1122.

Dack, G. M. (1956). *Food Poisoning,* 3rd ed. University of Chicago Press. Chicago.

Dack, G. M., W. E. Carey, O. Woolpert, and H. Wiggers. "An Outbreak of Food Poisoning Proved to be due to a Yellow Hemolytic Staphylococcus." *J. Prev. Med.* 4:167–175.

Davison, E., G. M. Dack, and W. E. Carey. (1938). "Attempts to Assay the Enterotoxic Substance Produced by Staphylococci by Parenteral Injection of Monkeys and Kittens." *J. Infect. Dis.* 62:219–223.

Denny, C. B., P. L. Tan, and C. W. Bohrer. (1966). "Heat Inactivation of Staphylococcal Enterotoxin A." *J. Food Sci.* 31:762–767.

Dolman, C. E. (1944). "Antigenic Properties of Staphylococcus Enterotoxin." *Can. J. Pub. Health.* 35:337–351.

Duitschaever, C. L., D. H. Bullock, and D. R. Arnott. (1977). "Bacteriological Evaluation of Retail Ground Beef, Frozen Beef Patties, and Cooked Hamburger." *J. Food Protect.* 40:378–381.

Felsenfeld, O. and V. M. Young. (1949). "A Study of Human Salmonellosis in North and South America." *American J. Trop. Med.* 29:483–491.

Foltz, V. D., R. Mickelsen, W. H. Martin, and C. A. Hunter. (1960). "The Incidence of Potentially Pathogenic Staphylococci in Dairy Products at the Consumer Level. I. Fluid Milk and Fluid Milk By-Products." *J. Milk Food Technol.* 23:280–284.

Foster, J. F., J. L. Fowler, and J. Dacey. (1977). "A Microbial Survey of Various Fresh and Frozen Seafood Products." *J. Food Protect.* 40:300–303.

Getting, V. A., A. D. Rubenstein, and G. E. Foley. (1944). "Staphylococcus and Streptococcus Carriers. Sources of Food-borne Outbreaks in War Industry." *Amer. J. Pub. Health.* 34:833–840.

Gwatkin, R. (1937). "Further Observations on Staphylococcic Infections of the Bovine Udder." *Can J. Pub. Health.* 28:185–191.

Hendricks, S. L., R. A. Belknap, and W. J. Hausler, Jr. (1959). "Staphylococcal Food Intoxication due to Cheddar Cheese. I. Epidemiology." *J. Milk Food Technol.* 22:313–317.

Hodge, B. E. (1960). "Control of Staphylococcal Food Poisoning." *Pub. Health Rep.* 75:355–361.

Jay, J. M. (1962) "Further Studies on Staphylococci in Meats. III. Occurrence and Characteristics of Coagulase-Positive Strains from a Variety of Nonfrozen Market Cuts." *Appl. Microbiol.* 10:247–251.

Kloos, W. E. and M. Musselwhite. (1975). "Distribution and Persistence of Staphylococcus and Micrococcus Species and Other Aerobic Bacteria on Human Skin." *Appl. Microbiol.* 30:381–395.

La Chapelle, Lt. N. C., Capt. E. H. Joy, and Lt. C. W. Halverson. (1966). "A Gastroenteritis Outbreak of Staphylococcus Aureus, Type 29." *Amer. J. Pub. Health.* 56:94–96.

McKinley, T. W. and E. J. Clarke, Jr. (1964). "Imitation Cream Filling as a Vehicle of Staphylococcal Food Poisoning." *J. Milk Food Technol.* 27:302–304.

Mackey, E. (1873). "Symptoms of Irritant Poisoning from Pork Brawn." *Brit. Med. J.* 1:533.

Messer, J. W., J. T. Peeler, R. B. Read Jr., J. E. Campbell, H. E. Hall, and H. Haverland. (1970). "Microbiological Quality Survey of some Selected Market Foods in two Socioeconomic Areas." *Bacteriol. Proc.* 1970:12.

Mickelson, R., V. D. Foltz, W. H. Martin, and C. A. Hunter. "The Incidence of Potentially Pathogenic Staphylococci in Dairy Products at the Consumer Level. II. Cheese." *J. Milk Food Technol.* 24:342–345.

Miller, W. A. and M. L. Small. (1955). "Efficiency of Cooling Practices in Preventing Growth of Micrococci." *J. Amer. Diet. Assoc.* 31:469–473.

Minnet, F. C. (1936). "Staphylococci from Animals with Particular Reference to Toxin Production." *J. Path. Bacteriol.* 42:247–263.

Minor, T. E. and E. H. Marth. (1971). "Staphylococcus Aureus and Staphylococcal Food Intoxications. A review. I. The Staphylococci: Characteristics, Isolation, and Behavior in Artificial Media." *J. Milk Food Technol.* 34:557–564.

Minor, T. E. and E. H. Marth. (1972). "Staphylococcus Aureus and Staphylococcal Food Intoxications. A Review. II. Enterotoxins and Epidemiology." *J. Milk Food Technol.* 35:21–29.

Minor, T. E. and E. H. Marth. (1972). "Staphylococcus Aureus and Staphylococcal Food Intoxications. A Review. III. Staphylococci in Dairy Foods." *J. Milk Food Technol.* 35:77–82.

Minor, T. E. and E. H. Marth. (1972). "Staphylococcus Aureus and Staphylococcal Food Intoxications. A Review. IV. Staphylococci in Meat, Bakery Products, and other Foods." *J. Milk Food Technol.* 35:228–241.

Moorehead, S. and H. H. Weiser. (1946). "The Survival of Staphylococci Food Poisoning Strain in the Gut and Excreta of the House Fly." *J. Milk Food Technol.* 9:253–259.

Mossel, D. A. A. (1962). "Attempt in Classification of Catalase-Positive Staphylococci and Micrococci." *J. Bacteriol.* 84:1140–1147.

Parfentjev, I. A. and A. R. Catelli. (1964). "Tolerance of Staphyloccus Aureus to Sodium Chloride." *J. Bacteriol.* 88:1–3.

Pivnick, H., T. R. B. Barr, I. E. Erdman, and J. I. Pataki. (1968). "Staphylococcus Food Poisoning from Barbecued Chicken." *Can. J. Pub. Health.* 59:30–31.

Rasmussen, C. A. and D. H. Strong. (1967). "Bacteria in Chilled Delicatessen Foods." *Pub. Health Rep.* 82: 353–359.

Stone, R. V., Sr. (1943). "Staphylococcic Food-Poisoning and Dairy Products." *J. Milk Food Technol.* 6:7–16.

Sternberg, G. (1901). *A Textbook of Bacteriology.* William Wood and Co. Baltimore.

Surgalla, J. J., M. S. Bergdoll, and G. M. Dack. (1952). "Use of Antigen-Antibody Reactions in Agar to Follow the Progress of Fractionation of Antigenic Mixtures: Application to Purification of Staphylococcal Enterotoxin." *J. Immunol.* 69:357–365.

Surkiewicz, B. F., R. J. Groomes, and L. R. Shelton, Jr. (1968). "Bacteriological Survey of the Frozen Prepared Foods Industry. IV. Frozen Breaded Fish." *Appl. Microbiol.* 16:147–150.

Surkiewicz, B. F., R. W. Johnston, A. B. Moran, and G. W. Krumm. (1969). "A Bacteriological Survey of Chicken Eviscerating Plants." *Food Tech.* 23:1066–1069.

Surkiewicz, B. F., M. E. Harris, R. P. Elliott, J. F. Macaluso, and M. M. Strand. (1975). "Bacteriological Survey of Raw Beef Patties Produced at Establishments Under Federal Inspection." *Appl. Microbiol.* 29:331–334.

Thatcher, F. S. and B. H. Matheson. (1955). "Studies with Staphylococcal Toxins. II. The Specificity of Enterotoxin." *Can. J. Microbiol.* 1:382–400.

U.S. Department of Health, Education and Welfare. (1976). *Foodborne and Waterborne Disease Outbreaks Annual Summary.* 1975. Center for Disease Control. Atlanta, Georgia.

Vaughan, V. C. (1884). "Poisonous or 'Sick' Cheese." *Pub. Health Pap. Rep.*, Amer. Pub. Health Assoc. 10:241–245.

Williams, R. E. O., (1946). "Skin and Nose Carriage of Bacteriophage Types of Staphylococcus Aureus." *J. Pathol. Bacteriol.* 58:259–268.

Williams, R. E. O. (1961). "Healthy Carriage of Staphylococcus Aureus: Its Prevalence and Importance." *Biol. Rev.* 27:56–71.

Williams, W. L. (1941). "Staphylococcus Aureus Contamination of a Grade 'A' Raw Milk Supply." *J. Milk Technol.* 4:311–313.

5 The Salmonella

Between June 24 and 27, 1964, 80 of the 173 residents at an institution for the mentally retarded in Tennessee were stricken with an acute gastrointestinal illness marked by vomiting, fever, and diarrhea. One death resulted. An examination of patient's stool samples revealed the presence of *Salmonella schwarzengrund*.

The bacteria were traced to a commercially prepared food supplement that was added to the meals of some residents as part of their diet. The food supplement was a dry meal prepared by adding hot water. It was contained in 25-pound cartons that were stored in a walk-in refrigerator at approximately 40°F. (4°C). On the day before the outbreak, however, the refrigerator was not functioning properly and the 12-to 14-hour period of elevated temperature was sufficient for heavy growth of the salmonellae. Examination of the supplement cartons after the outbreak revealed four species of salmonella, including *Salmonella schwarzengrund*.

Inquiry by Food and Drug Administration (FDA) investigators into the distribution list of the supplement manufacturer revealed that this supplement was used by 22 institutions for the mentally retarded in nine states. Three of these institutions had also reported outbreaks of gastrointestinal illness.

An investigation of the plant producing the supplement was undertaken. The plant, itself, and all of the employees were negative when tested for salmonellae, but two of the ingredients used in preparing the supplement yielded salmonellae. The two ingredients, brewer's dried yeast and cottonseed protein, were purchased from other manufacturers. The brewer's yeast contained two of the Salmonella isolated from the supplement. The cottonseed protein contained the other two. Both manufacturing plants changed their production methods and the problem ended.

During the past 20 years, the number of reported isolations of salmonellae from humans has steadily increased. Accompanying the rise in occurrence of this disease is an increased difficulty in recognizing some epidemics or related outbreaks when they do occur. This difficulty has been caused by a change in the pattern of the salmonella epidemic. In many current outbreaks, the source of the contamination can be hundreds of miles from the outbreak.

Another example of this type of outbreak occurred in late 1973. One hundred fifteen cases of salmonellosis were reported by the United States and Canada from 23 states and 7 provinces. At least 18 persons, most of whom were children, were hospitalized. Christmas wrapped chocolate candy was the implicated food item. The candy supply in the United States was traced from the retail stores to two distributors, one in New Jersey and the other in Pennsylvania. From these distributors, the trail led to a manufacturer in Quebec Province, Canada.

As mentioned previously, there are two types of foodborne illness—foodborne intoxication and foodborne infection. Foodborne intoxication describes the condition resulting from the consumption of food containing toxin or a poison. Foodborne infection describes the condition resulting from the consumption of food containing microorganisms (microscopic living things, usually bacteria). The microorganisms multiply in the body and release toxins that inflame and irritate the lining of the digestive tract. They sometimes invade and injure other organs of the body such as the liver and spleen. The Salmonella group causes foodborne infection.

THE ILLNESS

Incidence

In 1975, there were 497 reported outbreaks of foodborne illness. These outbreaks affected 18,260 persons. In 191 (38 percent) of the 497 outbreaks, the cause was determined. Thirty-

Table 5-1
Major salmonella foodborne outbreaks in 1975. CDC Annual Summary.

Food	Number Sick	Locality	State
Potato salad	232	Picnic	Minnesota
Ham	205	Church	Wisconsin
Unknown	176	Club	California
Chicken salad	168	Wedding reception	Louisiana
Spaghetti meat sauce	82	Camp	New Jersey
Roast beef, turkey	80	Church	Tennessee
Turkey	60	Home	Rhode Island
Lettuce	50	Nursing home	Arkansas
Milk	47	Home	Louisiana
Turkey	35	Home	Georgia

eight of these outbreaks, involving 1,573 persons, were caused by salmonellae. In 1975, these bacteria ranked second to the staphylococci in both number of outbreaks (38) and in number of cases (1,573).

Foods with high protein content are most frequently implicated. Among the foods implicated in 1975 were lettuce, baked goods, snow cones, barbecued pork, banana pudding, turkey, ice cream, milk, beef sandwiches, chicken salad, potato salad, beef, pork, spaghetti, roast duck, chicken, tomato sauce, meat sauce, roast beef, Mexican food, ham, and ground beef. Table 5-1 lists some of the larger outbreaks in 1975.

Table 5-2 shows where most of the reported salmonella outbreaks occurred in 1975. Homes, as usual, led the way followed by restaurants, church gatherings, picnics, camps, and nursing homes.

A five-year survey reported by Aserkoff showed that through the years 1963–1967, approximately 20,000 isolations of salmonellae from humans were reported annually. A constant seasonal pattern was noted, with the highest number of isolations reported from July through October. The majority of isolations (64 percent) were in persons under 20, with no sex preference shown. The six types of salmonella recovered most frequently were *S. typhimurium, S. heidelberg, S. newport, S. infantis, S. derby,* and *S. enteritidis.*

Yearly losses in work time and physician's payments due to salmonellosis have been estimated at between $10 and $100 million dollars. There are an estimated 400,000 to 2 million

Table 5-2
Localities of some reported salmonella outbreaks in 1975. CDC Annual Summary.

Locality	Number of Reported Outbreaks
Home	12
Restaurant	7
Church	5
Picnic	2
Camp	2
Nursing home	2

cases of salmonellosis per year in the United States.

Symptoms

Salmonellae are frequently the cause of symptomless infection and minor illnesses in man. They occasionally cause major disability and death. These problems may be caused by any of a large number of salmonella serotypes, most of which bear the name of the location where the first culture of the organism was obtained. This foodborne infection is due to the ingestion of large numbers of salmonellae that multiply within the small intestine, irritating and inflaming the lining of the intestinal tract. The symptoms, in general, resemble those of staphylococcal food intoxication and a salmonella outbreak can be just as explosive.

Usually the infecting salmonellae have grown to very high numbers in the contaminated food that has been ingested. The severity of the illness depends on the number of salmonellae present, the resistance of the infected person, and the virulence of the salmonella serotype that has gained entrance to the body.

In general, the very young, the elderly, the undernourished, and the ill are the most susceptible to a severe attack. One study has indicated that there may also be a genetic susceptibility to diarrheal diseases. A person with a high resistance can ingest salmonellae without becoming ill but can instead become a carrier for several weeks.

Salmonellae can produce four varying clinical conditions: acute gastroenteritis or the average food-poisoning syndrome; typhoidal or typhoidlike patterns (enteric fever); focal infections; and the carrier state. Of these, gastroenteritis is the most common condition, comprising about 70 percent of salmonellosis cases. It most frequently follows the consumption of contaminated food. *Salmonella typhimurium* and *Salmonella enteritidis* are the most frequently isolated salmonellae in the United States.

Nearly all of the salmonella serotypes are capable of causing gastroenteritis. The illness usually begins with headache and chills, followed by nausea, vomiting, abdominal pain, and diarrhea. A mild fever, normally not over 100°F. usually follows the diarrhea. Many of the symptoms resemble those of staphylococcal food intoxications (Table 5-3).

Symptoms appear from 6 to 72 hours after ingestion of infected food. Most cases appear in 18 to 48 hours. Severity can vary from mild diarrhea to prostration. Dehydration due to vomiting and diarrhea sometimes occurs, occasionally resulting in death. The disease usually

Table 5-3
Comparison of major symptoms of staphylococcus food intoxication and salmonella food infection.

Symptoms	Staphylococcal Food Intoxication	Salmonella Food Infection
Vomiting	Common	Common
Abdominal cramps	Common	Common
Diarrhea	Common	Common
Fever	Absent	Common
Prostration	Common	Rare

runs its course in one to seven days. The person may be a carrier for several weeks after convalescence and during this time can cause secondary cases of salmonellosis. For this reason, the carrier represents a potential hazard in any foodservice profession.

Acute gastroenteritis sometimes occurs with greater frequency and severity than usual in patients who have undergone stomach surgery, particularly if it results in less than normal production of gastric acid in the stomach. With this killing agent reduced or stopped, salmonellae can pass through the stomach unharmed. The resultant gastroenteritis can be severe.

Typhoid fever occurs when the pathogen involved is *Salmonella typhi*. This disease is no longer common in countries that practice good sanitation, but it may occasionally occur due to leaking sewage pipes or septic tanks, if the contaminated sewage comes into contact with food or water supplies. The disease can also be transmitted by a carrier, usually through the handling of food. One of the more dangerous public health aspects of salmonella food poisoning is the occasional production of a lifetime salmonella carrier.

Typhoid fever is characterized by a high, sometimes continuous fever, severe headache, and enlargement of the spleen. Prostration and dehydration are common. Rose spots sometimes appear on the body. Constipation is more common than diarrhea.

A major typhoid epidemic with an interesting history occurred in Aberdeen, Scotland, in 1964. The outbreak involved a total of 507 persons and was traced to sliced corned beef purchased from a supermarket. The corned beef had arrived at the supermarket in a six-pound tin. As it was sliced it contaminated the slicing machine which, in turn, contaminated other cold meats that were sliced.

The six-pound tin had been canned in South America. As sometimes happens, the can was defectively sealed. The canning factory cooled their tins in river water, which received the untreated sewage from a city of 600,000 people. The city had an incidence rate of 10,000 cases of typhoid fever per year. The infection undoubtedly took place during the cooling process and growth continued during shipment, finally culminating in the Aberdeen outbreak.

The typhoidlike syndrome or enteric fever caused by salmonellae is usually milder than typhoid fever. It can be caused by many types of salmonella but *Salmonella paratyphi A*, *Salmonella paratyphi B* (*Salmonella schottmuelleri*), and *Salmonella paratyphi C* (*Salmonella hirschfeldii*) are among the most common causes. Fever and malaise are often the only symptoms, but they can last from one to three weeks. *Salmonella paratyphi B* is the most common cause of enteric fever in the United States.

Focal infections are caused by an organism that gains entrance to some particular site in the body. It is usually carried to this site through the blood stream. Salmonellae acting in this manner can cause pneumonia, meningitis, osteomyelitis, pleurisy, aneurysm, endocarditis, and urinary tract infection.

Bone infection can result as a complication of salmonellosis. About .85 percent of typhoid fever infections and .2 percent of paratyphoid B infections result in bone disease caused by these salmonellae. Bone and joint infections resulting from salmonella food poisoning have also been reported.

Salmonella infections have also occurred in the aorta, the body's main artery. In a series of 23,000 autopsies performed at Boston City Hospital, bacterial aneurysms comprised 2.6 percent of 338 aortic aneurysms. Salmonellae were among the most common causative agents.

Salmonellae are one of the most common causes of meningitis in children. Three-fourths of the reported cases of salmonella meningitis have occurred among children under two. An epidemic of 21 cases occurred among infants in a maternity home in Havana, Cuba; the resulting mortality rate was 100 percent. *Salmonella havana*, a food-poisoning species, was the implicated agent. The types of salmonellae that are most successful in gaining entrance to the circulatory system, and thence to the central nervous system, are *Salmonella enteritidis, Salmonella paratyphi B, Salmonella cholerasuis, Salmonella typhimurium,* and *Salmonella panama.*

The carrier state results from the ingestion of salmonellae, usually in food or water. A gastrointestinal illness may or may not follow, but the future carrier retains some salmonellae in the digestive system proper or an accessory part such as the gall bladder. Because of his own resistance, he will not become ill, but through his excretion he may sometimes infect others.

The carrier state lasts for varying periods of time. It has been estimated that the carrier rate in the general population is about 0.2 percent. Permanent typhoid carriers frequently harbor salmonellae in their gall bladders; they multiply and pass down the bile duct into the intestine and are excreted.

Temporary carriers are more common. In 500 outbreaks in the United States, Caribbean Islands, Venezuela, and Columbia, symptomless human carriers, employed as food handlers, caused approximately 56 percent of all traceable outbreaks. Felsenfeld and Young investigated 56 outbreaks of food poisoning due to salmonellae. Twenty-six (46 percent) were caused by human carriers.

The carrier state occurs much more frequently in food handlers than in the general population. It is undoubtedly an occupational hazard for those who continually handle uncooked meats and carcasses.

Salmonella infections in animals are similar to those in man. Both gastrointestinal and typhoidal conditions occur. However, two salmonella diseases found in animals have no counterpart in man. These are ovarian infections in chickens, caused by *Salmonella pullurom*, and abortion in mares, caused by *Salmonella abortus equi.*

History

Salmonella is the generic name applied to a group of bacteria named for Dr. D. E. Salmon who, in 1885, was the first person to describe them. They have been known agents of foodborne illness for many years. Many cases attributed to them in early times, however, were later realized to be caused by such bacteria as *Staphylococcus aureus* or *Clostridium perfringens.*

It had become apparent by the late 1880s that salmonellae could cause illness in man, but they were regarded mainly as animal pathogens and their study was left to the veterinarians. Most of the approximately 1,400 serotypes can cause gastrointestinal illness in man, although a few species, such as *Salmonella gallinarium,* which causes fowl typhoid in chickens and turkeys, are host specific.

One of the first reported outbreaks of salmonellosis occurred in 1888 in the village of Frankenhausen, Germany. Fifty-seven persons became sick with acute gastroenteritis. One person, who consumed well over one pound of raw beef, died about a day and a half later. *Salmonella enteritidis* was isolated from the internal organs of the victim and from the meat of the infected cow responsible for the outbreak.

In 1895, a salmonellosis outbreak took another life, this time in a Belgium village. Several sausages from a number that were alleged to have caused illness were presented to the inspector of meat for examination. The inspector, on the basis of suitable color, odor, and general appearance, pronounced the sausages fit for consumption. To back up his opinion, the inspector ate three slices of one of the sausages. The director of the slaughterhouse that manufactured the sausages ate slices from several sausages and three of the plant workers ate varying amounts. These events occurred on a Saturday.

Ten to twelve hours later, the inspector had developed severe diarrhea and on Monday he was forced to call his physician due to violent abdominal pains, continuous diarrhea, increasing weakness, and a fever of 102°F. By Wednesday he was delirious and prostrated. His condition became progressively worse and on Friday morning he died.

The others who ate the sausages became ill, in varying degrees, on the day after eating the sausages. The director of the slaughterhouse suffered from diarrhea, headache, and blurred vision. The others had milder symptoms. All recovered within a week.

By the turn of the century, the housefly had been incriminated as a vector of food-poisoning bacteria. Salmonella epidemics, such as typhoid fever, were traced to flies during the Boer (1899–1902) and Spanish-American Wars (1898). In some of the epidemics, more than 20,000 cases occurred; flies walked and fed on infected fecal material and then transferred the bacteria to food in the kitchens.

On April 9, 1923, a new carrier of salmonellae appeared on the scene. Fifty-nine persons became ill after consuming cream-filled crumb cake and eclairs purchased from a bakery in New York City. Suspicion centered on the cream-filling as the common food item.

The bakery passed its initial inspection with flying colors. It was clean, orderly, and free of insects. Three cats were kept to control rodents. The employees were neat and clean. All employees were tested for salmonellae through stool samples. Even the cats were tested. All were negative.

Leftover samples of cake and eclairs from the April 5 baking did contain *Salmonella typhimurium*, however. This species or serotype matched the stool samples obtained from some of the victims of the outbreak.

Finally, after moving some of the bulk supplies from a board serving as a shelf, rat excreta containing *Salmonella typhimurium* was found. The shelf was almost over the place where the pail of cake and eclair filling was placed to cool. It was thought that infected excreta was bounced or rolled from the shelf into the pail by the cats jumping on the board to hunt the rats using it as a trail.

In 1923, Savage and White (and later, Khalil) isolated four species of salmonellae from rats: *Salmonella enteritidis*, *Salmonella typhimurium*, *Salmonella newport*, and *Salmonella thompson*.

In 1934, guinea pigs joined the list of animal carriers of salmonellae. Salmonellae were found in three animals from one cage in a breeding colony in New York. Salmonellae were recovered intermittently from guinea pigs on this breeding farm until 1949, when a major epidemic killed much of the stock. *Salmonella newport* was the predominant species involved.

In 1943, the same breeding farm reported the presence of salmonellae in their mouse-breeding colonies. The mouse colonies also suffered extensive losses from *Salmonella newport* in the 1949 epidemic. The source of the infection was finally located when Griffin (1952) published the results of his study of the prepared feed purchased for use at the breeding farm. He found *Salmonella newport* in two of the bags of dog feed that were supplied for the animals.

Barto (1938) and Stafseth (1944) reported on the presence of salmonellae in dogs. Barto isolated *Salmonella typhimurium*, a common human infectant, from an adult dog and Stafseth isolated *Salmonella cholerasuis* from pups. In 1948, Wolff et al. isolated salmonellae from 18 of 100 dogs examined.

Griffin's report on the presence of salmonellae in dog food was confirmed by later studies, and by 1976 Morse estimated that the prevalence of canine salmonellosis could be as high as 27 percent. A number of investigators had already speculated that some of these canine strains could be transmitted to humans.

The final link in this chain was forged in 1973. On December 30, an adult dachsund became ill with salmonellosis. The owner of the dog operated a baby-sitting service in her home. One of her charges was a 16-month-old girl. The child was exposed to the sick dog and the contaminated environment for several days. She became ill on January 6 with a temperature of 105°F., vomiting, and diarrhea. It was necessary to admit her to the hospital for treatment and recovery. *Salmonella enteritidis* was recovered from the stools of both the dog and the child.

The finding of salmonellae in dry dog food led to the investigation of other animal feeds. In 1955, Galton, in his investigation of dehydrated bone meals, found that 26.5 percent of the samples examined contained *Salmonella typhimurium* and *Salmonella seftenberg*, both of which cause salmonellosis in human beings. Bone meal is a common constituent of many animal feeds, including cattle, poultry, and dog feeds. In 1955, Erwin reported the presence of salmonellae in poultry feed.

In 1947, wild birds joined the lengthy list of carriers of salmonellae. During that year, Cass isolated *Salmonella pullorum* from a pheasant in Minnesota and Raush found the same serotype in a coot in Ohio. Since then, salmonellae have been recovered from the common tern, house sparrow, brown-headed cowbird, herring gull, white-throated sparrow, grackle, starling, and red-winged blackbird. *Salmonella typhimurium* and *Salmonella enteritidis* were among the species recovered.

THE BACTERIA

The genus *Salmonella* belongs to the family Enterobacteriaceae. This family includes *Shigella*, *Proteus*, and *Escherichia*, in addition to other genera. Salmonellae identification is readily made by laboratories equipped to perform the necessary biochemical tests.

Salmonellae are gram-positive, rod-shaped bacteria that do not form spores. Their natural habitat is the intestinal tract of man and animals. They can exist under aerobic and anaerobic conditions. Many are motile (capable of movement). They are susceptible to heat, extreme cold, radiation with gamma rays and x rays, ultraviolet light, and most disinfectants. Salmonellae survive longest in the lower ranges of relative humidity. They grow best in nonacid foods with a pH range of 5.5–8.0, but they can survive for varying periods of time in higher and lower pH foods. Salmonellae have survived in tomato juice with a pH of 4.4 for up to 30 days.

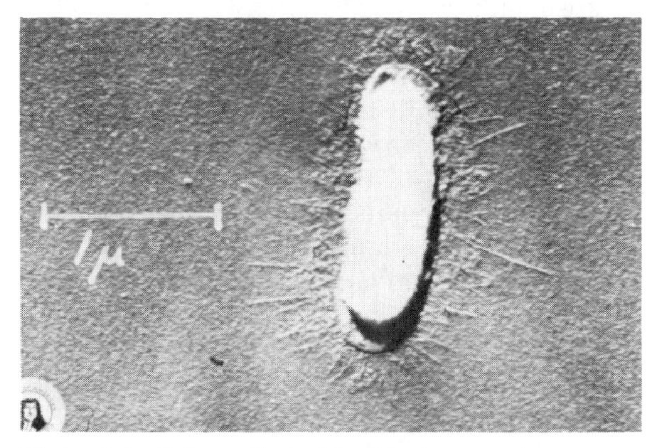

Figure 5-1 *Salmonella typhimurium*

Like the staphylococci, the salmonella possess antigens and therefore cause the formation of antibodies. Individual members of the group are identified by serological recognition of their main antigens. Thus, the members of this genus are also referred to as serotypes rather than species. Approximately 1400 serotypes are known. Unlike the staphylococci, which actually "poison" the food with their toxins, the salmonella do not release harmful toxins into the food. Rather, the food serves as a vehicle into the body for the bacteria (food borne infection), not a vehicle for a toxin (foodborne intoxication). Their toxin is released inside the body, in the intestinal tract. It is an endotoxin (polysaccharide-protein-lipid substance) rather than an exotoxin (protein).

The salmonella ferment mannitol and usually dulcitol. They do not ferment lactose or sucrose, nor do they produce indole. Their fermenting processes usually produce hydrogen sulfide (H_2S). The growth range of the salmonella extends from 44°–114°F. (7–46°C). Their optimum growth temperature of 98°F. (37°C) can result in a generation time as low as 15 minutes. They are generally killed by exposure to a temperature of 122°F. (50°C) for one hour or 140°F. (60°C) for 20 minutes. All are killed by the usual pasteurization procedures.

The growth range varies somewhat with the food. Angelotti et al. (1961, III) noted a decrease in salmonellae numbers at 112°F. (44°C) in ham salad, but no decrease at that temperature in chicken a la king or custard. It required a temperature of 116°F. (47°C) in chicken a la king and 120°F. (49°C) in the custard to produce a decrease in population.

The same variability occurs at the low end of the temperature range. Angelotti et al. detected growth in chicken a la king at 44°F. (6°C) but no growth was observed in custard or ham salad up to 50°F. (10°C).

The results of these experiments indicate that perishable foods should be held at a temperature lower than 44°F. (6°C) and above 114°F. (46°C) in order to prevent growth of these organisms. Perishable food should be allowed to remain within this temperature range for only a brief period of time. Any other course of action is potentially dangerous.

FOODS

Market Foods

It has been mentioned that foods with a high protein content are the most frequently infected. Protein foods make an excellent medium for the growth of salmonellae. Meat and poultry and their products are usually incriminated, but almost all foods are susceptible to these bacteria.

In one survey, a total of 9,851 samples were taken during a 4½-year period in large, federally inspected processing plants in California. Of this total, 4,254 turkeys yielded 177 positive isolations, an incidence of 5.25 percent. A total of 1,996 chicken hens yielded 13 isolations (0.65 percent) and 3,601 chicken fryers yielded 241 isolations (2.45 percent). *Salmonella typhimurium* was the most prevalent salmonella in turkeys; *Salmonella infantis* was the most prevalent in chicken fryers.

Surkiewicz examined rinse samples of eviscerated carcasses, samples of chill tank waters, and swab samples of carcasses from nine federally inspected chicken eviscerating plants. Salmonellae were isolated from 20.5 percent of the eviscerated birds. Felsenfeld et al. (1950) reported that the occurrence of salmonellae in uninspected fowl was about seven times higher than in federally inspected meats.

Although the number of infected fowl arriving at processing plants is usually low, salmonellae can be spread easily to other carcasses during processing. Carcasses are frequently contaminated by defeathering machines and the contact surfaces of other machines. Rubber or plastic gloves can become contaminated, thus spreading infection to other market-bound fowl. Surveys of dressed poultry collected from poultry processing plants or from retail markets have shown varying, but frequently high levels of contamination (4 percent–50 percent).

Within one 4½-year period, 31 outbreaks involving 6,742 cases of salmonellosis occurred at banquets, where poultry dishes and egg-containing desserts were allegedly responsible for the illness of 92 percent of the victims.

Pork products also have a high incidence of salmonellae. Galton et al. (1954) examined 217 samples of fresh pork sausage and 127 samples of smoked sausage. A total of 51 (23 percent) fresh sausage and 16 (12.5 percent) smoked sausage samples were positive. The proportion positive varied from a high of 57.5 percent for the products of local abattoirs to a low of 7.5 percent for the samples from national distributors.

Felsenfeld et al. found salmonellae in 14 percent of samples from 573 specimens of U.S. inspected pork. They found uninspected pork harbored salmonellae about twice as often as U.S. inspected pork.

Poultry and pork and their products are the main carriers of salmonellae among meats. The incidence of salmonellae in lamb and beef is low, with the exception of hamburger. In their investigation of market meats, Felsenfeld et al. found salmonellae in 18 percent of the hamburger samples tested. Other investigators have reported similar results.

Fish taken from open waters are free from salmonellae, but fish caught in sewage-polluted waters have harbored these organisms. Smoked whitefish has been responsible for several outbreaks. In an outbreak in 1934, 34 persons developed gastrointestinal symptoms and one death resulted. A 1940 outbreak due to *Salmonella typhimurium* resulted in 47 cases of food poisoning and two deaths. *Salmonella newport* was incriminated in a 1955 Memorial Day holiday outbreak of 37 cases. In 1968, *Salmonella newport* caused 300 cases of food poisoning from smoked whitefish. The source of contamination was traced to Hay River, Northwest Territory, Canada.

Salmonellae contamination in manufactured dairy products became a serious concern in 1966 when 11 serotypes were isolated from nonfat dry milk from 9 states. Ray investigated 8 plants producing dried milk products. All 8 contained salmonellae in their plant environment. An outbreak in 1965, involving instant nonfat dry milk contaminated with *Salmonella new brunswick*, had resulted in 29 cases in 17 states. Half of the victims were infants.

Outbreaks of salmonellosis attributable to cheese were reported as early as 1923. In Great Britain, persons contracted paratyphoid fever from eating cream cheese. In all instances for which background information is available, food-poisoning cheeses were made from raw milk or from milk that was recontaminated after pasteurization.

Survival of salmonellae in colby cheese has been reported for up to 302 days at 43°–48°F. (6–9°C). Goepfert et al. have reported survival of the same species in cheddar cheese for at least 12 weeks at 7.5°–13°C (46–55°F.). Hargrave found survival of *Salmonella typhimurium* for two to nine months in cheddar and colby cheese. Since then, salmonellosis, nearly always caused by *Salmonella typhimurium* or *Salmonella typhi,* has been associated with cheddar, colby, Camembert, Romano, Jack Quarg, cream, cottage, and other cheeses. Park et al. reported that *Salmonella typhimurium* can survive in coldpack cheese food for up to 27 weeks.

Among other products that have been found to contain salmonellae are chocolate candy, candy coatings, muffin mix, food seasonings, barbecued turkey roll, imitation ice cream, cocoa, vegetable gums, gelatin, enzymatic drain cleaners, carmine dye, and chicken and duck eggs.

Cooked Foods

In 1963, because of the increasing numbers of salmonella outbreaks and an increase in the variety of foods yielding salmonella isolations, the Salmonella Surveillance Program was established jointly by the Center for Disease Control (CDC) and the Association of State and Territorial Epidemiologists and Laboratory Directors. In 1966, the Food and Drug Administration (FDA) launched an extensive salmonella surveillance program that included intensive testing of processed foods and certain pharmaceutical products of animal origin.

This increased emphasis placed on salmonella reporting and investigating resulted in a marked increase in the number of food items incriminated as carriers. Three types of food have long been recognized as carriers; poultry, meat (beef and pork), and eggs, although outbreaks due to eggs have virtually disappeared since 1974.

During the period 1963 to 1975, there were 651 reported outbreaks of salmonellae involving 38,811 cases. The causative food was identified in 463 (71 percent) of the outbreaks; of these, poultry (primarily turkey rolls and precooked barbecued chicken) accounted for 99 (21 percent), meat (primarily precooked, packaged roasts of beef, prepared beef sandwiches, and sausage) for 69 (15 percent), and eggs for 53 (11 percent). Together, these three food types caused approximately half of the outbreaks of known cause.

Cooked foods suffer contamination from two sources. The food may be contaminated at the time of entry into the retail store, foodservice establishment, or some intermediate manufacturer who mixes ingredients for resale; or the food may become contaminated on the premises through a human factor.

A large variety of foods have caused outbreaks of salmonella infections. Many of these have already been mentioned. Meat products, such as meat pies, hash, sausage, cured meats (ham, bacon, and tongue), sandwiches, and chili, often are allowed to stand at room temperature, permitting the growth of salmonellae.

Milk and milk products, including fresh milk, fermented milks, ice cream, and cheese have caused salmonella infections. Since eggs can carry salmonellae, undercooked foods made with unpasteurized eggs, for example, cream- or custard-filled pastries, cream cakes, baked Alaska, and eggnog, may cause salmonella infections.

Precooked foods, which have become more readily available, have caused increased concern in recent years. Several outbreaks that occurred in 1975 and 1976 illustrate the problem. In early July, 1975, the Health Department of Edison Township reported an outbreak of salmonellosis to the New Jersey State Department of Health. Eleven individuals had become ill after eating roast beef sandwiches at a luncheonette. The precooked roast beef used in the sandwiches originated out-of-state. *Salmonella saint-paul* was isolated from stool samples of the victims.

The investigation of this outbreak led to the discovery of an outbreak that had occurred about a week earlier. Precooked roast beef from the same company was again incriminated. The outbreak resulted from the consumption of this beef at a graduation party and *Salmonella saint-paul* was again the agent. Seven cases of food poisoning resulted during this outbreak.

In mid-July, another food-poisoning outbreak occurred in the area. The precooked roast beef that was incriminated was traced to the same out-of-state company involved in the previous outbreaks. This food-poisoning episode occurred at a summer camp and once again *Salmonella saint-paul* was isolated from stool samples.

Three different distributors for the out-of-state company had supplied the roast beef in these outbreaks. Although USDA investigators were unable to determine the origin of the salmonellae in the meat, it was determined that they were surviving because of inadequate cooking in the company's kitchen. The investigation resulted in recalls of lots of affected roast beef from Connecticut, Florida, Michigan, New Jersey, New York, Ohio, Pennsylvania, Tennessee, and Virginia.

Outbreaks of salmonellosis in New Jersey, Pennsylvania, and Connecticut during August, 1976, was also traced to precooked, packaged, ready-to-eat beef. Six of the 21 patients had to be hospitalized. Most had eaten at delicatessens and sandwich shops. The outbreak was first recognized when inquiries were received from the New Jersey and Pennsylvania departments of health concerning an increase in the number of isolates of *Salmonella bovis-morbificans*. Initial questioning of cases revealed prominent consumption of roast beef.

The New Jersey and Pennsylvania cases had eaten roast beef at several different delicatessens, which served precooked, packaged roast beef from Company A. Roast beef consumed by four of seven cases in Connecticut was obtained from a single grocery chain delicatessen. It had received its precooked, packaged roast beef from its commissary in Boston, Massachusetts (Company B). Sources of raw meat for both companies were imported beef. *Salmonella bovis-morbificans* is a rare serotype, which suggests that a common overseas source was responsible for the contamination of the beef.

RESERVOIRS

The primary reservoir of salmonellae is the vertebrate intestine. The only vertebrates from which salmonellae have not been isolated under natural conditions are aquatic vertebrates living in unpolluted waters.

The list of carriers includes not only poultry, cattle, and pigs, but also the cat, dog, buffalo, sheep, donkey, seal, whale, mouse, rat, guinea pig, hamster, dove, pigeon, gull, parrot, sparrow, tortoise, lizard, snake, whitefish, housefly, flea, cockroach, tick, louse, oyster, and clam. Salmonellae have also been recovered after cases of human illness from Easter chicks, ducks, rabbits, and from pet turtles.

Salmonellae can be found in various types of livestock on farms. The incidence of salmonellae in the intestine of farm pigs is rather low, however, and the most common state exhibited is the carrier state (rather than the active infectious state). Extensive research with pigs has shown that as the animals proceed from the farm to the stockyards, and from the holding pens to the abattoirs, there is a progressive increase in the number of salmonella isolations. Galton et al. (1954) has shown a sevenfold difference in the proportion of infected hogs on the farm and in the abattoir.

Transfer of infection within the pens can occur directly by contact with infected feces. It can take place through the inhaling of infected droplets from the splash of feces or from the cattle licking one another. It can occur indirectly by fecal contamination of food and water. Transfer of infection between pens occurs by moving pigs from infected pens to replace those sent for slaughter and by the introduction of new pigs for fattening. Thus, all of the conditions needed for a cycle of continuous infection exist within the holding pens and abattoirs.

Salmonellae can be recovered from the feces and mesenteric lymph nodes of pigs. Of 2100 slaughtered pigs in different slaughterhouses in Holland, 25 percent were found to harbor salmonellae in the portal or mesenteric lymph nodes, in the feces, or in both. Salmonellae were isolated from 20 percent and from 50 percent, respectively, of two groups of 50 pigs from Arizona. Stress factors, including handling, crowding, cold, lack of food or water, and transporting may trigger a mechanism that changes an infected, but nonexcreting pig into an excreter.

Nearly all stages of the slaughtering process provide sources of possible contamination (handling, slaughtering, dressing). The problem is very serious in certain operations, such as, sticking, hide removal, scalding and dehairing, head skinning, brisket opening, bung dropping, and evisceration. Washing of the carcass has been incriminated as a major mechanism of spreading contamination.

The same general pattern exists in cattle. A small percentage of farm cattle are active excreters. As the cattle progress through the various holding pens on the way to the slaughterhouse the incidence of salmonella infection increases.

Two serotypes, *Salmonella typhimurium* and *Salmonella dublin* are frequently found in cattle. It has been estimated that 1 percent of the adult cattle on farms excrete salmonellae. These cattle form the main reservoir from which infection is transferred to calves. Calves born on farms where there are excreting adults are exposed to infection from birth and some of these calves become carriers.

Salmonellosis in fowl has long been recognized as an important economic burden on the poultry industry. Transmission of *Salmonella pullorum* infection from the adult hen to the chick via the infected egg has been known for more than 90 years. Many other types of salmonellae are now known to be transmitted.

In addition to being susceptible to salmonella infection, domestic poultry and wild birds carry salmonellae for long periods of time without apparent illness. Most wild birds can apparently be carriers and distribute the disease through their droppings. These droppings can contaminate animal feeds, crops, water supplies, pastures, and barnyards. On at least one occasion, dessicated coconut, which had been open-air dried, was implicated in a salmonellosis outbreak. The contamination was thought to have occurred through bird droppings.

Small numbers of salmonellae can establish infection in chicks. Infected poultry have been reported to excrete salmonellae for periods ranging from 15 days to 18 months. Sadler and Corstvet found 5.25 percent of turkeys, 1.85 percent of chicken fryers, and 0.54 percent of chicken hens to be contaminated with salmonellae.

Salmonellae can be transmitted to egg shells during laying from droppings or from fecally contaminated nests. The pores in the shell are large enough to admit salmonellae. These organisms can also penetrate the membranes. Cracked or "checked" eggs (eggs with hairline cracks in the shell) are even easier to invade. Surveys have shown that although only a fraction of 1 percent of individual shell eggs are contaminated in this way, the later pooling and mixing of yolks and whites distributes the infection through the mixture. Liquid, dried, and frozen whole egg mixtures are frequently contaminated with salmonellae.

Dogs rank second to pigs in the frequency of salmonellosis and the number of serotypes harbored. Wolff et al. recovered salmonellae from 18 of 100 dogs examined in Michigan. Sixteen different salmonella types were isolated, the most frequent being *Salmonella oranienburg*. Galton et al. (1952) collected 8,157 rectal swabs from dogs and found 27.6 percent of them to be positive for salmonellae. Mackel examined 1,626 dogs in Florida and reported an incidence of 244 (15 percent) positive isolations. Isolations as high as 70 percent have been reported from canine hospitals.

Many salmonella infections in dogs are the result of eating dog food containing salmonellae. In one survey, Galton et al. (1955) isolated salmonellae from 26 of 98 (26.5 percent) samples of dog meal. In another investigation Pace recovered salmonellae from seven of eight packages of dried dog food. Eleven serotypes were recovered from the seven packages.

Rats are also carriers of salmonellae, particularly *Salmonella enteritidis* and *Salmonella typhimurium*. A food-poisoning incident involving rats was discussed in this chapter. As early as 1923, Savage and White reported finding that 6 of 96 rats examined from a slaughterhouse contained *Salmonella enteritidis*. Felsenfeld (1949) reported an incidence of 53 percent among the Puerto Rican rats examined.

Ludlam conducted an extensive examination of rats killed in the Nottingham area of England. He reported an incidence of 4.4 percent among the general rat population and a varying rate of 6.4 percent to 40 percent among the rats from a butcher's by-products factory in

Nottingham. *Salmonella enteritidis* was the most frequently isolated of the salmonellae.

Animal feeds and their ingredients have been extensively investigated since Griffen's report in 1952 implicated dog feed in a series of salmonellosis outbreaks among mice and guinea pigs in a New York breeding colony. A study was conducted to determine the rate of salmonella contamination in each lot of animal protein meal delivered to a large feed mill in the southeastern part of the United States over a 10-month period. Of 311 samples examined, 211 (68 percent) contained one or more salmonella serotypes. In 206 lots of meat meal, 178 (86 percent) contained salmonellae. A total of 21 of 37 (57 percent) feather meal samples and 12 of 68 (18 percent) fish meal samples also contained salmonellae.

Surveys have repeatedly found salmonellae in bone meal used as a base for fertilizer and cattle feed. Of complete foods manufactured from these raw materials, protein concentrate meals to be diluted with grain products on farms proved to be the most heavily contaminated (10.5 percent). Pelleted foods showed the least contamination (0.5 percent) due to the heat used in pelleting.

Erwin collected samples from 206 bags of poultry feed. A total of 60 percent of the mash samples, 25 percent of pelleted samples, and 15 percent of the granule samples yielded salmonellae or other enteric organisms. Clise examined 11 offal reduction plants in Maryland. He found that eight of them were producing supplements containing salmonellae.

The effluents from meat-packing plants have also been examined for salmonellae. Vanderpost investigated the waste treatment facilities and final effluents of 11 meat-packing plants in the province of Alberta, Canada. He found salmonellae in the final effluents of 75 percent of the plants.

In the insect world, flies and cockroaches are probably the main carriers of salmonellae. It has been mentioned that flies carrying salmonellae have caused major epidemics among troops during wartime. Infected flies can be carriers during the course of their entire life and, through their droppings, can contaminate not only food but uninfected flies as well. Experiments have shown that *Salmonella oranienburg* can survive for 10 to 20 days in different species of cockroaches.

The following cases will illustrate some unsafe practices in the preparation and handling of foods.

CASE STUDIES

Case 1: Turkey

This outbreak occurred at an annual supper for the parent-teacher association of a North Carolina public school. The menu included turkey, one of the most frequently implicated foods in salmonellosis outbreaks. Approximately 405 plates were served at the banquet. On the following day, many people who had attended the banquet appeared at local hospitals and physicians' offices. Several hundred cases of food poisoning, 32 requiring hospitalization, resulted from the banquet.

The turkeys used at the dinner weighed 25 to 30 pounds and were supplied by a local packing plant from a poultry wholesaler in a midwestern state. They had been processed and thawed in a local dining establishment, which had been denied a Grade A rating by local sanitarians and where food-handling practices were highly questionable. This establishment also had limited refrigeration facilities. It was thought likely that the turkey was recontaminated following cooking, when it was placed in the same pan in which the frozen birds had been thawed. *Salmonella heidelburg* was isolated from leftover turkey and dressing and from stool samples of the hospitalized patients.

In this case, a cardinal rule of food preparation was broken; never place cooked food in utensils that have held raw food. Adequate cooking will destroy bacteria that may be present but recontamination can occur if the cooked food is returned to its original utensil.

Turkeys frequently contain salmonellae at the time they leave the processing plant. Some of the turkeys used for the banquet were probably contaminated when they arrived at the restaurant, thereby contaminating the pans in which they were thawed. When the cooked turkeys were placed back in these pans they were recontaminated. As a result, several hundred people became ill, some seriously enough to require hospitalization.

Case 2: Imitation Ice Cream

Salmonellosis resulting from the use of unpasteurized egg mixtures has virtually disappeared from the United States at the commercial level. The danger of using unpasteurized egg mixtures is illustrated by this case, as is the cycle of salmonella contamination.

In mid-April, 1967, reports of four outbreaks of suspected food poisoning following Jewish social functions were received by the New York Department of Health. A review of recent food-poisoning reports located two more such episodes. Menus for each event were obtained and compared. Imitation ice cream called chiffonade was common to all six menus. The caterers confirmed that in each case the dessert had been purchased from the same food manufacturer.

An embargo was then placed on the dessert item and the manufacturer was closed down. Further review of past food-poisoning cases uncovered eight additional outbreaks that had resulted from this imitation ice cream. All 14 outbreaks occurred within a two-week period and resulted in 1,790 cases of food poisoning. Stool samples taken from victims resulted in isolation of *Salmonella typhimurium* and *Salmonella braenderup*.

The chiffonade implicated in the outbreaks was manufactured in New York City by a company specializing in kosher frozen desserts. Cultures of ingredients, other than egg yolks, used in making the chiffonade yielded no salmonellae. However, four unopened cans of unpasteurized, frozen, sugared egg yolks grew *S. typhimurium*, *S. braenderup*, and *S. siegburg*. These egg yolks were added to the chiffonade mixture during a slow cooling process, and the chiffonade was not subsequently heated before being consumed.

The contaminated frozen egg yolks came from an egg breaker in New York City. This company was using several unsafe manufacturing procedures, including the use of checked eggs as a source of yolks, allowing the egg and sugar mixture to stand as long as seven hours in the mixing vat, and shipping frozen egg yolks in unrefrigerated trucks.

These procedures allowed the bacteria to spread throughout the egg mixture. They also provided incubation periods that permitted further growth of salmonellae present in the raw egg yolks.

Environmental samples were taken from the egg-breaking plant. Twenty-seven of 40 samples yielded salmonellae. Eleven serotypes were found, including *S. typhimurium*, *S. braenderup*, and *S. siegburg*.

Shell eggs were obtained from the three suppliers of the egg-breaking company. Salmonellae were isolated from cultures of these eggs. Eggs yielding *S. braenderup* were traced to an egg broker in Connecticut, but attempts to identify the specific farm or farms that furnished the contaminated eggs were unsuccessful.

Although it was impossible to trace the infection to its original source, salmonella contamination of poultry feed is well documented. Starting at this point, the probable course of events was: (1) chickens were fed contaminated feed, (2) infected feces contaminated the shells of the eggs, (3) checked eggs, more likely to be penetrated by salmonellae, were used in the chiffonade, (4) the egg-breaking and mixing processes distributed the contaminated egg yolks throughout large volumes of the sugared egg yolks used to make the chiffonade, (5) the long period of time the contaminated egg yolk mixture spent in the vats allowed extensive salmonellae growth, and (6) the chiffonade mixture was not cooked after the unpasteurized yolks were added.

The outbreak could have been prevented by following proper sanitary practices at any one of these points. The current practice of using only pasteurized eggs on a commercial scale has solved this problem adequately in the United States.

As a final note, the attack rate (the percentage of persons eating the contaminated food who became ill) in this case was 52 percent. The chiffonade that led to the 14 documented outbreaks came from at least six production lots, which contained a minimum of 18,000 servings and was served at numerous banquets and parties in six eastern states. If 52 percent of all those served the contaminated chiffonade desserts had become ill, an estimated 9,360 people would have suffered from the unsafe manufacturing practices of one company.

SUMMARY

There are two types of recognized foodborne illness—foodborne intoxication and foodborne infection. Salmonellae cause foodborne infection since they release their toxins within the body of the victim rather than in the food that they infect.

Foods of high protein content are most frequently implicated in salmonellosis outbreaks. Meat and poultry and their products are the most likely to be infected.

Salmonellosis is usually caused by the ingestion of large numbers of salmonellae, which then multiply within the small intestine, irritating and inflaming the lining of the intestinal tract. Salmonellae can produce four varying clinical conditions; acute gastroenteritis, typhoid or typhoidlike conditions, focal infections, and the carrier state. Of these, gastroenteritis is the most common condition, comprising about 70 percent of salmonellosis cases.

Nearly all of the salmonella serotypes are capable of causing gastroenteritis. The illness usually begins with headache and chills, followed by nausea, vomiting, abdominal pains, and diarrhea. A mild fever, normally not over 100°F. usually follows the diarrhea. Symptoms appear within 6 to 72 hours after ingestion of infected food. Most cases appear in 18 to 48 hours. The disease usually runs its course in one to seven days, but the person may be a carrier for several weeks.

Salmonella is the generic name applied to a group of bacteria named for D. E. Salmon, who, in 1885, was the first to describe them. They are gram-negative, rod-shaped bacteria that do not form spores. Their natural habitat is the intestinal tract of man and animals. Many are motile. They are susceptible to heat, extreme cold, radiation with gamma rays and x-rays, ultraviolet light, and most disinfectants.

Salmonellae grow best in nonacid foods with a pH range of 5.5 to 8.0, but they can survive for varying periods of time in higher and lower pH foods.

Their optimum growth temperature of 98°F. (37°C) can result in a generation time as low as 15 minutes. The growth range of these organisms varies somewhat with the food. In general, it extends from 44°–114°F (7–46°C). All are killed by the usual pasteurization processes.

Poultry and pork and their products are the main carriers of salmonellae among meats. The incidence of salmonella in lamb and beef is low, with the exception of hamburger. Precooked foods have caused increasing concern in recent years.

The primary reservoir of salmonellae is the vertebrate intestine. The major carriers are poultry, cattle, pigs, and dogs. They frequently become infected through contaminated animal feed. Among our common household pests, rats, mice, and cockroaches can carry salmonellae and transmit them to man by contaminating his food and premises with their droppings.

Review Questions

1. What type of food is most frequently implicated in salmonella outbreaks?

2. What seasonal pattern of salmonella outbreaks has been noted?

3. a. What clinical conditions can salmonellae produce? b. Which is the most common?

4. a. Describe the typical symptoms of salmonella food poisoning. b. On what does the severity of the illness depend?

5. a. What are the symptoms of typhoid fever? b. What are the symptoms of enteric fever?

6. List several focal infections that can be caused by salmonellae.

7. Compare the occurrence of the carrier state in food handlers with that of the general population.

8. List four animal carriers of salmonellae.

9. What animal feeds have been implicated as transmitters of salmonellae?

10. a. List several characteristics of salmonella. b. What is their natural habitat in man?

11. What agents are lethal to salmonellae?

12. Within what pH range do salmonellae grow best?

13. How do salmonellae cause foodborne illness?

14. Within what temperature range do salmonellae grow best?

15. a. Name several foods that are highly susceptible to the growth of salmonellae. b. Why are these foods highly susceptible?

16. Why is the incidence of salmonella infection much lower in farm livestock than in processed meat?

17. Why is it necessary to use pasteurized egg mixtures when preparing foods?

References

Angelotti, R., M. J. Foster, and K. H. Lewis. (1961). "Time-Temperature Effects on Salmonellae and Staphylococci in Foods. I. Behavior in Refrigerated Foods. II. Behavior at Warm Holding Temperatures." *Amer. J. Public Health.* 51:1 76–83, 83–88.

Angelotti, R., M. J. Foster, and K. H. Lewis. (1961). "Time-Temperature Effects on Salmonellae and Staphylococci in Foods. III. Thermal Death Time Studies." *Appl. Microbiol.* 9:4 308–315.

Appleman, M. D. and M. D. Appleman, Jr. (1969). "Cross-Contamination in Food Handling and Acute Salmonella Gastroenteritis in the Hospital." *Hospital Management* 108:34–39.

Armstrong, R. W., T. Fider, G. T. Curlin, A. B. Cohen, G. K. Morris, W. T. Martin, and J. Feldman. (1970). "Epidemic Salmonella Gastroenteritis due to Contaminated Imitation Ice Cream." *Amer. J. Epidemiol.* 91:300–307.

Aserkoff, B., S. A. Schroeder, and P. S. Brachman. (1970). "Salmonellosis in the United States—A Five Year Review." *Amer. J. Epidemiol.* 92:13–24.

Baldwin, R. E., M. L. Fields, W. C. Poon, and B. Korschgon. (1971). "Destruction of Salmonellae by Microwave Heating of Fish with Implications for Fish Products." *J. Milk Food Technol.* 34:467–470.

Bate, J. G. and U. James. (1958). "Salmonella Typhimurium Infection Dust-Borne in a Children's Ward." *Lancet* ii:713–715.

Bryan, F. L. (1968). "What the Sanitarian Should Know About Staphylococci and Salmonellae in Non-Dairy Products. II. Salmonellae." *J. Milk Food Technol.* 31:131–137.

Cass, J. S. and J. E. Williams. (1947). "Salmonella Pullorum Recovered From a Wild Pheasant in Minnesota." *J. Amer. Med. Assoc.* 111:282.

Childers, A. B. and E. E. Keahey. (1970). "Sources of Salmonella Contamination of Meat Following Approved Livestock Slaughtering Procedures." *J. Milk Food Technol.* 33:10–12.

Clise, J. D. (1965). "Salmonellae From Animal By-products." *Public Health Rep.* 80:899–902.

Cohen, M. L. and P. A. Blake. (1977). "Trends in Food-borne Salmonellosis Outbreaks: 1963–1975." *J. Food Protection* 40:798–800.

Collins, R. N., M. D. Treger, J. B. Goldsby, J. R. Boring III, D. B. Coohon, and R. N. Barr. (1968). "Interstate Outbreak of Salmonella Newbrunswick Infection Traced to Powdered Milk." *J. Amer. Med. Assoc.* 203:838–844.

Dack, G. M. (1956). *Food Poisoning* 3rd ed. University of Chicago Press. Chicago.

Datta, N. and R. B. Pridio. (1960). "An Outbreak of Infection with Salmonella Typhimurium in a General Hospital." *J. Hyg. Camb.* 58:229–241.

Erwin, L. E. (1955). "Examination of Prepared Poultry Feeds for the Presence of Salmonella and Other Enteric Organisms. *Poultry Sci.* 34:215–216.

Faddoul, G. P., G. W. Fellows, and J. Baird. (1966). "A Survey on the Incidence of Salmonellae in Wild Birds." *Avian Diseases* 10:89–94.

Felsenfeld, O. and V. M. Young. (1949). "A Study of Human Salmonellosis in North and South America." *Amer. J. Trop. Med.* 29:483–491.

Felsenfeld, O., V. M. Young, and T. Yokimura. (1950). "A Survey of Salmonella Organisms in Market Meat, Eggs, and Milk." *J. Amer. Vet. Med. Assoc.* 116:17–21.

Foster, E. M. (1968). "Microbiology in Today's Foods." *J. Amer. Dietitic Assoc.* 52:485–489.

Foster, E. M. (1969). "The Problem of Salmonellae in Foods." *Food Tech.* 23:74–78.

Foster, E. M. (1971). "The Control of Salmonellae in Processed Foods: A Classification System and Sampling Plan." *J. Assoc. Official Anal. Chem.* 54:259–266.

Galton, M. M. and M. S. Quan. (1944). "Salmonella Isolated in Florida During 1943 with the Combined Enrichment Method of Kauffman." *Amer. J. Public Health.* 34:1071–1075.

Galton, M. M., M. J. E. Scatterday, and A. V. Hardy (1952). "Salmonellosis in Dogs. I. Bacteriological, Epidemiological, and Clinical Considerations." *J. Infect. Dis.* 91:1–5.

Galton, M. M., W. D. Lowery, and A. V. Hardy. (1954). "Salmonella in Fresh and Smoked Pork Sausage." *J. Infect. Dis.* 95:232–235.

Galton, M. M., W. V. Smith, H. B. Mcelrath, and A. V. Hardy. (1954). "Salmonella in Swine, Cattle, and the Environment of Abattoirs." *J. Infect. Dis.* 95:236–245.

Galton, M. M., M. Harless, and A. V. Hardy. (1955). "Salmonella Isolations from Dehydrated Dog Meals." *J. Amer. Vet. Med. Assoc.* 126:57–58.

Gangarosa, E. J., A. A. Bisno, E. R. Eichner, M. D. Treger, M. Goldfield, W. E. DeWitt, T. Fodor, S. M. Fish, W. H. Dougherty, J. B. Murphy, J. Feldman, and H. Vogel. (1968). "Epidemic of Febrile Gastroenteritis due to Salmonella Java Traced to Smoked Whitefish." *Amer. J. Public Health.* 58:114–121.

Gayler, G. E., R. A. Macready, J. P. Reardon, and B. F. McKernan. (1955). "An Outbreak of Salmonellosis Traced to Watermelon." *Public Health Reports* 70:311–313.

Gibson, E. A. (1961). "Salmonellosis in Calves." *Vet. Res.* 73:1, 284.

Goepfert, J. M., N. F. Olson, and E. H. Marth. (1968). "Behavior of Salmonella Typhimurium During Manufacture and Curing of Cheddar Cheese." *Appl. Microbiol.* 16:862–866.

Griffin, C. A. (1952). "A Study of Prepared Feeds in Relation to Salmonella Infection in Laboratory Animals." *J. Amer. Vet. Med. Assoc.* 121:197–200.

Hansen, E., R. Rogers, S. Emge, and N. J. Jacobs. (1964). "Incidence of Salmonella in the Hog Colon as Affected by Handling Practices Prior to Slaughter." *J. Amer Vet. Med. Assoc.* 145:139–140.

Hargrove, R. E., F. E. McDonough, and W. A. Mattingly. (1969). "Factors Affecting Survival of Salmonella in Cheddar and Colby Cheese." *J. Milk Food Technol.* 32:480–484.

Heard, T. W. (1969). "Housing and Salmonella Infections." *Vet. Rec.* 85:482–484.

Kampelmacher, E. H. (1963). "Public Health and Poultry Products." *British Vet. J.* 119:110–124.

Kaufmann, A. F., C. R. Hayman, F. C. Heath, and M. Grant. (1968). "Salmonellosis Epidemic Related to a Caterer-Delicatessen-Restaurant." *Amer. J. Public Health* 58:764–771.

Kunz, L. J., and O. T. G. Ouchterlony. (1955). "Salmonellosis Originating in a Hospital." *New Eng. J. Med.* 253:761–763.

Lang, D. J., L. J. Kunz, A. R. Martin, S. A. Schroeder, and L. A. Thomson. (1967). "Carmine as a Source of Nosocomial Salmonellosis." *New Eng. J. Med.* 276:829–832.

Longree, K. (1967). *Quantity Food Sanitation.* Interscience. John Wiley and Sons, Inc. New York.

Ludlam, G. B. (1954). "Salmonella in Rats, with Special Reference to Findings in a Butcher's By-Products Factory." *Monthly Bull.* Ministry Health Lab. Serv. 13:196–202.

McCall, C. E., R. N. Collins, D. B. Jones, A. F. Kaufmann, and P. S. Brachman. (1966). "An Interstate Outbreak of Salmonellosis Traced to a Contaminated Food Supplement." *Amer. J. Epidemiol.* 84:32–39.

MacCready, R. A., J. P. Reardon, and I. Saphra. (1957). "Salmonellosis in Massachusetts." *New England J. Med.* 256:1121–1128.

Mackel, D. C., L. F. Langley, and C. J. Prehal. (1965). "Occurrence in Swine of Salmonellae and Serotypes of Escherichia Coli Pathogenic to Man." *J. Bacteriol.* 89:1434–1435.

Magwood, S. E., J. Fung, and J. L. Byrne. (1965). "Studies on Salmonella Contamination of Environment and Product of Rendering Plants." *Avian Diseases* 9:302–308.

Marth, E. H. (1969). "Salmonellae and Salmonellosis Associated with Milk and Milk Products." *J. Dairy Sci.* 52:283–315.

Morse, E. V., M. A. Duncan, D. A. Estep, W. A. Riggs, and B. O. Blackburn. (1976). "Canine Salmonellosis: A Review and Report of Dog to Child Transmission of Salmonella Enteritidis." *Amer. J. Pub. Health.* 66:82–84.

Newell, K. W., and L. P. Williams. (1971). "The Control of Salmonellae Affecting Swine and Man." *J. Amer. Vet. Med. Assoc.* 158:89–98.

Olson, T. A., and M. E. Rueger. (1950). "Experimental Transmission of Salmonella Oranienburg Through Cockroaches." *Pub. Health Rep.* 65:531–540.

Ostrolenk, M. and H. Welch. (1942). "The House Fly as a Vector of Food Poisoning Organisms in Food Producing Establishments." *Amer. J. Pub. Health* 32:487–494.

Pace, P. J., K. J. Silver, and H. J. Wisniewski. (1977). "Salmonella in Commercially Produced Dried Dog Food: Possible Relationship to a Human Infection Caused by Salmonella Enteritidis Serotype Havana." *J. Food Protection* 40:317–321.

Park, H. S., E. H. Marth, and N. F. Olson. (1970). "Survival of Salmonella Typhimurium in Cold-Pack Cheese Food During Refrigerated Storage." *J. Milk Food Protection* 33:383–388.

Rausche, R. (1947). "Pullorum Disease in the Coot." *J. Wildl. Mangt.* 11:189.

Ray, B., J. J. Jezeski, and F. F. Busta. (1971). "Isolation of Salmonellae from Naturally Contaminated Dried Milk Products." *J. Milk Food Technol.* 34:389–393.

Sadler, W. W. and R. E. Corstvet. (1965). Second Survey of Market Poultry for Salmonella Infection." *Appl. Microbiol.* 13:348–351.

Salthe, O. and C. Krumwiede. (1924). "Studies on the Paratyphoid-Enteritidis Group." *Amer. J. Hyg.* 4:23–32.

Saphra, I. and J. W. Winter. (1957). "Clinical Manifestations of Salmonellosis in Man." *New Eng. J. Med.* 256:1128–1134.

Savage, W. G. and P. B. White. (1923). "Rats and Salmonella Group Bacilli." *J. Hygiene Camb.* 21:258–261.

Silliker, J. (1967). "Good Manufacturing Practices Required to Beat Salmonella." *Natl. Provisioner* 157:99–103.

Snoeyenbos, G. H., E. W. Morin, and D. K. Wetherbee. (1967). "Naturally Occurring Salmonella in Blackbirds and Gulls." *Avian Dis.* 11:642–646.

Stewart, H. C. and W. Litterer. (1927). "An Outbreak of Gastro-Enteritis." *J. Amer. Med. Assoc.* 89:1584–1587.

Surkiewicz, B. F., R. W. Johnston, A. B. Moran, and G. W. Krumm. (1969). "A Bacteriological Survey of Chicken Eviscerating Plants." *Food Tech.* 23:1066–1069.

Taylor, J. (1965). "Modern Life and Salmonellosis." *Proc. Roy. Soc. Med.* 58:167–170.

Thatcher, F. S. and J. Montford. (1962). "Egg-Products as a Source of Salmonellae in Processed Foods." *Can. J. Pub. Health* 53:61–69.

Tucker, C. B., G. M. Cameron, and M. P. Henderson. (1946). "Salmonella Typhimurium Food Infection from Colby Cheese." *J. Amer. Med. Assoc.* 131:1119–1120.

U.S. Department of Health, Education and Welfare. (1976). *Foodborne and Waterborne Disease Outbreaks Annual Summary 1975.* Center for Disease Control. Atlanta, Georgia.

U.S. Department of Health, Education and Welfare. Communicable Disease Center. (1974). *Morbidity, Mortality Weekly Reports* 23:85–86.

U.S. Department Health, Education and Welfare. Communicable Disease Center. (1976). *Morbidity, Mortality Weekly Reports* 5:34–35.

U.S. Department Health, Education and Welfare. Communicable Disease Center. (1976). *Morbidity, Mortality Weekly Reports* 42:333–334.

U.S. Department Health, Education and Welfare. (1965). *Salmonella Surveillance Reports* 38:5–6.

U.S. Department Health, Education and Welfare. Center for Disease Control. (1966). *Salmonella Surveillance Reports* 46:4.

Vanderpost, J. M. and J. B. Bell. (1977). "Bacteriological Investigation of Alberta Meat-Packing Plant Wastes with Emphasis on Salmonella Isolation." *Appl. Environ. Microbiol.* 33:538–545.

Walker, J. H. C. (1957). "Organic Fertilisers as a Source of Salmonella Infection." *Lancet* ii:283–284.

Walker, W. (1965). "The Aberdeen Typhoid Outbreak of 1964." *Scottish Med. J.* 10:466–479.

Welch, H., M. Ostrolenk, and M. T. Bartram. (1941). "Role of Rats in the Spread of Food Poisoning Bacteria of the Salmonella Group." *Amer. J. Pub. Health* 31:332–340.

Williams, L. P., J. B. Vaughn, A. Scott, and V. Blanton. (1969). "A Ten-Month Study of Salmonella Contamination in Animal Protein Meals." *J. Amer. Vet. Med. Assoc.* 155:167–174.

Wilson, E., R. S. Paffenbarger, Jr., M. J. Foter, and K. H. Lewis. (1959). "Salmonellae in the Environment: Results of Tests on Market Meats." *Bacteriol. Proced.* 12:9.

Wolff, A. H., N. D. Henderson, and G. McCallum. (1948). "Salmonella from Dogs and the Possible Relationship to Salmonellosis in Man." *Amer. J. Pub. Health* 38:403–408.

6

Clostridium perfringens

What could be a more pleasant way to spend an evening than at a company banquet with good friends and good food? That is probably what 1,800 company employees thought on the night of September 14, 1968. By the next morning, more than 900 of these people undoubtedly had second thoughts after attacks of nausea, abdominal cramps, headache, and diarrhea.

Food histories of the victims implicated roast beef as the probable culprit. On the day of the banquet, 90 ribs of roast had been prepared. After cooking, the roasts were boned, trimmed, and sliced by machine. They were prepared successively, with no break for cleaning the work surfaces. The meat was next placed in warming cabinets until serving time. Tests of cooked, unsliced roast beef yielded no *Clostridium perfringens*, but tests of roast beef scraps from trimmed and sliced roasts were positive.

It is probable that one or several of the beef ribs were contaminated when they arrived at the hotel. The infected ribs contaminated the work counter, and this subsequently infected other ribs, since no cleaning was done between the trimming and slicing of any of the roasts after they were cooked. Temperatures in the warming cabinets provided an ideal climate for rapid reproduction of the bacteria.

Clostridium perfringens is a relative newcomer to the United States. It was incriminated in English reports but was not reported in the United States until 1945, when McClung described four outbreaks; it was not until 1959 that official epidemiological reports were received of food-poisoning outbreaks in which *Clostridium perfringens* was identified as the agent. Since that time, the number of persons identified as victims of this organism has risen to the point at which it is now recognized as one of the major food-poisoning bacteria.

THE ILLNESS

Incidence

In 1976, there were 438 reported outbreaks of foodborne illness. These outbreaks affected 12,463 persons. In 132 (30 percent) of these outbreaks the cause was determined. Ninety-two of these outbreaks, affecting 3,270 people, were of bacterial origin. *Clostridium per-*

Table 6-1
Major *Clostridium perfringens* foodborne outbreaks in 1975 and 1976. CDC Annual Summary.

Food	Number Sick	Locality	State
Turkey	217	Restaurant	New York
Bread and gravy	150	Cafeteria	Wisconsin
Roast beef, gravy	125	School	New Jersey
Roast beef, turkey	63	Home	California
Roast beef	61	Restaurant	Hawaii
Gefilte fish	55	Home	Illinois
Roast beef	43	Restaurant	Connecticut
Chili	43	Picnic	Utah
Meat loaf	30	Nursing home	Connecticut
Roast beef	28	Restaurant	Wisconsin

fringens was responsible for 6 (6.5 percent) of the 92 outbreaks and 509 (15.6 percent) of the 3,270 cases of bacterial origin. In 1973, *Clostridium perfringens* was responsible for 1,424 cases of food poisoning of identified origin, second highest among the bacteria causing foodborne illness.

From 1970 through 1975, *Clostridium perfringens* was responsible for 154 outbreaks and 13,687 cases of foodborne illness. This represented 17 percent of all outbreaks and 24 percent of all cases of foodborne illness confirmed by the Center for Disease Control. In some previous years, *C. perfringens* has been responsible for more reported cases of foodborne disease than any other agent. In England and Wales, *C. perfringens* has been responsible for as high as 33 percent of the reported yearly outbreaks.

Foods involved in outbreaks of *Clostridium perfringens* foodborne illness are usually meat or poultry, stews, sauces, gravies, dressings, or casseroles that have been cooked and held at room temperature for excessive periods of time or refrigerated in large masses. Growth is initiated by resistant spores, which germinate after exposure to cooking temperatures, or by vegetative cells that are introduced into the food after cooking has occurred. Outbreaks frequently follow banquets or meals prepared at hospitals, schools, or some foodservice establishment involved in mass feeding. Table 6-1 lists some of the larger outbreaks in 1975 and 1976.

All of the six verified outbreaks in 1976 and 13 of the verified 16 outbreaks in 1975 were due to improper food handling procedures in a foodservice establishment. In the remaining three cases, the food was mishandled in private homes. Table 6-2 lists the major contributing factors to some *Clostridium perfringens* outbreaks.

Among the foods incriminated in recent years are roast beef, turkey and gravy, beef burrito, meat loaf, gefilte fish, bread and gravy, chicken and gravy, barbecued pork, chili, corned beef, chop suey, and tenderloin tips.

Table 6-2
Major contributing factors to *Clostridium perfringens* outbreaks in 1975 and 1976. CDC Annual Summary.

Contributing Factor	Number
Improper holding temperatures	16
Inadequate cooking	5
Poor personal hygiene	3
Contaminated equipment	2

The majority of outbreaks and cases are associated with mass feeding establishments, with restaurants accounting for the largest number of cases. Beef has been implicated most often in the outbreaks, followed by poultry and their products.

Symptoms

Clostridium perfringens foodborne illness is caused by the ingestion of large numbers of the organism. The incubation period is 8 to 22 hours, with an average of 12 hours. The body seems to react to an exotoxin produced in the small intestine by the bacteria as they sporulate. The illness is usually mild and short-lived, and does not produce permanent carriers.

The principal symptoms of food poisoning due to *Clostridium perfringens* are cramps, abdominal pain, diarrhea, and gas. Nausea and vomiting are not common. Fever, shivering, headache, and other signs of infection seldom occur. Recovery is usually complete within 24 hours. Table 6-3 shows a comparison of the symptoms of three of the major food-poisoning illnesses.

In addition to food poisoning, *Clostridium perfringens* produces several other diseases in man, including necrotic enteritis, post-abortal uterine infection, and infections of soft tissues.

Table 6-3
Comparison of the major symptoms of *Staphylococcus aureus*, *Salmonella* species, and *Clostridium perfringens* foodborne illness.

Characteristics	*Staphylococcus aureus*	*Salmonella* species	*Clostridium perfringens*
Diarrhea	common	common	common
Abdominal pain	present	present	present
Vomiting	common	rare	rare
Fever	rare	common	rare
Incubation period	2–6 hours	12–24 hours	8–22 hours
Length of illness	6–24 hours	1–14 days	12–24 hours

Necrotic enteritis is marked by sudden onset, noticeable loss of intestinal mucosa, inflammation of the intestinal lining, and intestinal bleeding. The disease was first recognized in Germany in 1946, when several hundred cases occurred over a two-year period with a mortality rate of about 40 percent. Since then the disease has been reported in several other countries, including the United States, England, France, and New Guinea.

The cause of the illness is *Clostridium perfringens* type C. Type C strains occur in the pig, and many outbreaks have been caused by pork. The disease, which is most severe in the young and elderly, is characterized by acute abdominal pain, bloody diarrhea, and vomiting.

Type A strains are sometimes responsible for severe and often fatal infections of the uterine wall. These infections usually follow mechanically induced abortions performed under unclean conditions or with nonsterile instruments. The toxin that is released by the bacteria destroys red blood cells and platelets, causing jaundice and kidney malfunction. This results in high levels of urea in the blood and can lead to coma and eventually death if medical treatment is not successful.

Clostridium perfringens type A is the cause of, or an accessory in, several types of wound infection. Such infections are rarely fatal, however. The most frequently encountered and least serious type of infection is simple contamination of a cut, wound, or bruise. Sometimes the wound becomes foul-smelling or exudes a watery fluid resulting from the digestion of dead tissue by the bacteria.

Anaerobic cellulitis can result from the invasion of *Clostridium perfringens* into deeper wounds. In these cases, the bacteria digest the dead tissue at a deeper level, producing gas in the process. The overlying tissues often become discolored from the gas but normally no lasting damage occurs.

Gas gangrene results from the invasion of healthy tissues by certain type A strains. Curtailment of the blood supply to the damaged area results in the anaerobic conditions necessary for bacterial growth and gas production. The growth then spreads throughout the healthy muscle tissue. Antibiotic treatment for gas gangrene is highly successful.

Clostridium perfringens, mostly type A strains, has been involved in infections in many parts of the human body. It has been identified in infections resulting in appendicitis, peritonitis, meningitis, and brain abscess, and in heart, kidney, urinary bladder, and liver infections.

Many animal diseases are caused by *Clostridium perfringens*. Type A strains, occurring in the intestine and producing a toxin that enters the blood (enterotoxemia), have caused disease and death in cattle, lambs, goats, and reindeer. Subcutaneous infections in chickens, which spread to the underlying tissues, are also caused by these strains. Type A strains are worldwide in occurrence in most types of soil.

Type B strains also cause enterotoxemia in animals. The illness has been reported in lambs, calves, and foals. Type B strains are not worldwide in distribution and although cases have been reported in England, Europe, South Africa, and the Middle East, no North American cases have been reported. Type B *Clostridium perfringens* does not normally occur in soil samples and seems to be restricted to a parasitic existence in animal intestines.

Type C strains cause necrotic enteritis in man and animals. Pigs, chickens, cattle, sheep, and goats are affected. The disease primarily occurs in the newborn animal. Healthy animals act as carriers.

Type D enterotoxemia occurs mainly in sheep and sometimes goats and cattle. It is worldwide in distribution. Type E strains cause enterotoxemia in calves.

History

The first report of a food-poisoning outbreak that was probably due to *Clostridium perfringens* appeared in 1895 in England, when Klein published the results of his investigation into the first of several outbreaks that occurred at St. Bartholomew's Hospital in London. He found an anaerobic spore-forming bacillus in the stools of patients. He named the organism *Bacillus enteritidis sporogenes*.

In 1899, Andrewes reported on this series of outbreaks stating: "Within the past three years the wards of St. Bartholomew's hospital have been thrice visited by a mild form of epidemic diarrhea. The outbreaks have all been of similar character, occurring suddenly in a single night and passing off as suddenly. In one instance as many as 146 persons were made ill." Andrewes also remarked that the organism was a widely distributed, gram-positive saprophyte and could be found in cultured soil, animal feces, sewage, and sometimes in man. Its spores were heat resistant.

The first outbreak in the series occurred in 1895 with 59 cases. The second occurred in March of 1898 with 146 cases and the third in August of 1898 with 86 cases. The symptoms were abdominal pain and diarrhea with no nausea or vomiting, all typical of *Clostridium perfringens* food poisoning. Milk was the suspected food.

In 1943, Knox and McDonald reported on food-poisoning episodes among school children in England. Once again the symptoms were mild, consisting of abdominal pain and diarrhea of rather short duration. Gravy made on the previous day at a central kitchen was found on several occasions to be heavily contaminated with anaerobic spore-forming bacteria.

The first United States report appeared in 1945 when McClung discussed a series of four outbreaks of foodborne illness due to *Clostridium perfringens*. Three of the outbreaks, extending from December 1943 to February 1945, resulted from food eaten in a large public cafeteria. The fourth outbreak resulted from food eaten in a private dining club. The symptoms included nausea, intestinal cramps, and diarrhea. In the majority of cases the symptoms began 8 to 12 hours after the meal. By the following day, most of the people involved, although weak, were able to pursue their normal activities.

In all instances, the cases were traced to dishes prepared from chicken that was cooked one day and served on the following day. The bacteria apparently survived in the chicken broth and subsequently multiplied during a storage period, since microscopic examination of the broth revealed large numbers of *Clostridium perfringens.*

A rigid cleanup system in the kitchen, plus the boiling of all chicken broth after refrigeration and before mixing with any other food, ended this problem in the cafeteria. This case illustrates the extreme importance of these procedures in any foodservice establishment.

In 1953, Betty Hobbs and associates of the Food Hygiene Laboratory in England issued the most comprehensive report to that time on *Clostridium perfringens,* its history, and characteristics. From their study of past cases, this group concluded that the food involved was ". . . almost invariably a cold or warmed-up meat dish made from meat, boiled, braised, steamed or stewed for 1–3 hr. on the day before it is required, and allowed to cool slowly overnight." Most of the strains identified were heat resistant. Eighteen *C. perfringens* outbreaks (from September 1949 to February 1952) were listed in the report, showing the meat dishes responsible and some results of serological typing of strains from food and feces. In all cases (13) in which food samples were received, heat resistant strains were found to be the causative agent. The production of heat resistant spores came to typify food-poisoning strains of *C. perfringens* in England and Europe.

Although *Clostridium perfringens* was reported as a cause of food poisoning in the United States in 1945, and Hobbs et al. had published their report in 1953, no marked increase in the recognition of this organism occurred in the United States for some years. The organism was suspected in a few cases but rarely reported officially as the identified food-poisoning agent.

In 1959, Hart et al. described a *Clostridium perfringens* outbreak (page 3) that affected nearly half of the 450 passengers who had been served a turkey dinner aboard a special railway train. Browne et al. also described an outbreak among persons who had attended a reunion dinner. Roast beef was the incriminated food item. After the beef had been cooked it was sliced on a cutting board and left at room temperature for three hours. Just before the meat was sliced, two frozen turkeys were prepared on the same cutting board, contaminating it with the liquid from the plastic bags enveloping the turkeys. The cutting board, in turn, passed the *Clostridium perfringens* on to the roast beef. The beef was then placed in an oven at 450°F. (232°C) for 20 minutes. The hostess arrived, earlier than expected, to pick up the meat, thus preventing most of the internal portions of the meat from reaching a temperature that would kill either spores or vegetative cells.

Two other outbreaks were reported in the United States during 1959. From 1959 to 1964, there were relatively few reported outbreaks due to *Clostridium perfringens.* After 1964, the number of yearly reported outbreaks began to increase substantially, due primarily to increased awareness of the existence of the organism. At the present time, it is one of the leaders in the number of cases of food poisoning reported yearly.

As the number of recorded cases of food-borne illness due to *Clostridium perfringens* began to rise in the United States, it was noticed that many of these outbreaks were due to heat-sensitive strains rather than to the classical heat-resistant strains described by Hobbs et al. In 1963, Hall et al. called attention to the fact that many United States food-poisoning outbreaks were being caused by heat-sensitive strains.

Even though reports had begun to increase implicating *Clostridium perfringens* as a major food-poisoning agent, its mechanism of illness remained a mystery for years. In 1967, Hauschild et al. reported the experimental production of diarrhea in lambs after ingestion of cells of *C. perfringens* type A. Until this time, no experimental animal had been found that could be used for the study of this type of food poisoning. The experiments also revealed that the intestinal mucosa was not attacked by *C. perfringens* cells during the illness, therefore suggesting the possibility of one or more toxins being produced in the intestine.

In 1968, the same group reported that they were able to produce fluid accumulation in ligated (tied off) intestinal loops of lambs when *Clostridium perfringens* was injected. In this type of experiment, a localized response can be obtained in the experimental animal by ligating the small intestine into segments, usually called loops. The loops are then injected with suspensions of vegetative cells of *C. perfringens,* which are allowed to grow for about six hours. During this time the bacteria multiply, sporulate, and cause movement of fluid from the blood stream into the cavities of the loops. As a result of the fluid accumulation and gas production, the loops become distended and sausagelike in appearance. The process is comparable to diarrhea. In 1968, Duncan et al. successfully used ligated loops of rabbit intestine to produce fluid accumulation, and the rabbit has since become a standard laboratory animal for study of this illness.

These findings in both lambs and rabbits indicated a direct action of some factor on the intestinal lining, since fluid accumulation occurred in intestinal segments injected with *Clostridium perfringens,* but not in adjacent control (not injected with *C. perfringens*) loops.

In 1969, Duncan and Strong were able, for the first time, to induce diarrhea in rabbits with cell-free preparations of various strains of *Clostridium perfringens.* This was a landmark experiment since it showed conclusively that the illness was caused by a toxin. Their study also revealed that cell extracts, prepared from cultures grown in a medium that induced sporulation, contained a heat-sensitive factor that caused fluid accumulation in rabbit intestinal loops. Cultures grown in media that did not induce sporulation failed to produce extracts that would cause fluid accumulation.

Hauschild et al. (1970) experimented with three strains of *Clostridium perfringens.* This group concluded that a heat-sensitive component of sporulating cells was likely to be the main enteropathogenic factor in *C. perfringens* food poisoning. In a later series of experiments, Hauschild et al. (1971) were able to show that *C. perfringens* grows, sporulates, and produces an erythemal (skin reddening or inflammation) factor in ligated intestinal loops of lambs. This factor was recovered from the loops and was shown to be identical to a factor that had been produced in cultures. One of the three strains of *C. perfringens* used was a strain that had caused a food-poisoning outbreak in Madison, Wisconsin. This group proposed that the term *Clostridium perfringens enterotoxin* be applied to the erythemal factor.

Purification of the enterotoxin has now been achieved in several laboratories by a variety of methods. It is a heat-sensitive protein with a molecular weight of approximately 36,000. Much of its activity is lost when heated to 140°F. (60°C) for 10 minutes.

THE BACTERIA

Clostridium perfringens is an anaerobic, spore-forming, rod-type bacterium that is nonmotile and gram-positive. Vegetative cells sporulate

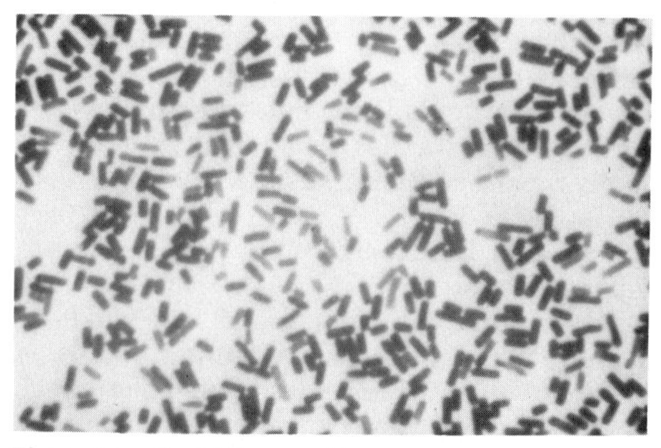

Figure 6-1 *Clostridium perfringens*

readily in the intestine, but rarely in cooked meat.

Strains of *Clostridium perfringens* are divided into five types (A to E) on the basis of four major exotoxins (alpha, beta, epsilon, iota) they may produce. Only types A and C cause foodborne illness in humans, with type A causing the greater majority of food-poisoning cases. Type A also causes gas gangrene. All types are capable of causing disease in animals, particularly in horses, sheep, goats, pigs, and cattle.

Fermentation reactions produce acid and gas from a number of carbohydrates, including maltose (malt sugar), lactose (milk sugar), glucose, and sucrose. Mannitol is not fermented and indole is not produced by this organism.

Some strains of *Clostridium perfringens* produce heat-resistant spores, others produce heat-sensitive spores. Heat-sensitive spores can be destroyed by a temperature of 212°F. (100°C) for a period of 5 to 30 minutes, whereas heat-resistant spores can survive the same temperature for one to five hours. A high proportion of English and European *C. perfringens* food-poisoning cases are caused by heat-resistant strains but most cases in the United States are caused by heat-sensitive strains. Heat-sensitive strains outnumber heat-resistant strains in most habitats.

Spores of *Clostridium perfringens* in frozen meat have been found to resist freezing and storage at 23°F. (−5°C) and 4°F. (−16°C). Canada et al. studied the spores and vegetative cells of four strains of *C. perfringens*. At 0°F. (−18°C) (normal household freezer temperature), 17 to 58 percent of the spores survived.

Strong and Canada recovered from 4.3 to 38 percent of the *Clostridium perfringens* spores after 90 days of storage and from 3.7 to 11 percent of the spores after 180 days. The storage temperature was 0°F. (−18°C) and the medium was chicken gravy. Storage temperatures in these experiments reduced the number of vegetative cells to negligible levels and even refrigerator temperatures reduced the numbers considerably over a 48-hour period of time.

The optimum temperature for growth of *Clostridium perfringens* is between 109.4° and 116.6°F. (43°–47°C). Their full growth range, however, extends from 60°–124°F. (15°–51°C). Good growth usually occurs over a temperature range of 68°–122°F. (20°–50°C), depending on the food. Barnes, for example, found no growth in beef until a temperature of 68°F. (20°C) was reached, whereas growth was observed by Hall and Angelotti in beef cubes at 65°F. (18°C). Multiplication of cells in turkey slices at 75°F. (24°C) is noticeable after four hours. Strong and Ripp observed a 400-fold increase in cell numbers after six hours.

As you can see, the temperature range for *Clostridium perfringens*, as for other food-poisoning species, is rather wide. Under favorable conditions, the number may increase nearly 1000-fold in a three-hour incubation period. Foods, though free of *C. perfringens* after cooking, can be contaminated through handling or from utensils, counters, chopping boards, or dust. Cultures can be obtained regularly from dust, soil, kitchen floors, and chopping boards.

Growth of *Clostridium perfringens* occurs within a range of pH 5.0 to 8.5. Its optimal range lies between pH 6.0 and 7.5, but fairly rapid growth occurs between pH 5.5 and 8.0.

Fischer et al. investigated the survival of four strains of *Clostridium perfringens* during 10-minute exposures to environments of varying pH values. Survival was influenced both by pH and the phase of growth cycle. Increased time of exposure to acidic conditions decreased the percent of surviving cells in most cases. The presence of spores increased the ability of the population to withstand acidic conditions. The longer the period of association between the *C. perfringens* and the acidic conditions, the greater the bacterial action.

Gough investigated the survival of five strains of *Clostridium perfringens* in salt solutions varying from 7.5 to 17 percent. Four of the five strains tested survived at least 48 days in all of the brine solutions. In other tests, the cultures survived 212°F. (100°C) for at least 30 minutes and 176°F. (80°C) for at least six hours in the presence of a combination of heat and curing salts exceeding that normally encountered in cured meat.

Arbuckle has reported a generation time as short as 8.5 minutes in broth cultures incubated at 114.8°F. (46°C) and 20 minutes at 98.6°F. (37°C). Smith (1975) has reported a generation time from 24 to 32 minutes in various poultry and meat soups. As the temperature drops from the 110° to 115°F. (43–46°C) range, the generation time of *Clostridium perfringens* increases noticeably. Thus, prevention of *C. perfringens* multiplication can be achieved by the prompt and proper use of refrigeration temperatures of 45°F. (7°C) or less and hot holding temperatures of 140°F. (60°C) or higher.

Clostridium perfringens is widespread in nature and can therefore be found in most foods. However, only certain foods, mainly meats and poultry and their gravies and dressings, supply the optimum requirements. Large numbers of cells can be produced in these foods in two to three hours. Therefore, the danger from foodborne illness lies in keeping food at temperatures that only appear hot or cold but that actually allow rapid growth of *C. perfringens*. This hazard can be present in mass feeding establishments that rely on steam tables or cooling tables to maintain foods at safe temperatures for lengthy periods of time.

THE TOXIN

It has been established that foodborne illness due to *Clostridium perfringens* is caused by an enterotoxin. It was only within recent years that this toxin was isolated and identified. It is apparently synthesized only when the vegetative cells sporulate, not when they grow or multiply. During sporulation the vegetative cells lyse (disintegrate) and the enterotoxin is released. Its production appears directly related to the ability of the strain to sporulate.

The toxin is a protein with a molecular weight of about 35,000. Much of its activity is lost when heated at 140°F. (60°C) for 10 minutes, so it is rather heat sensitive. The effect of the toxin seems to result from its direct action on the lining of the intestine.

Clostridium perfringens enterotoxin differs in two important respects from the enterotoxin of *Staphylococcus aureus:* (1) it is produced in the body rather than in food, and (2) it acts directly on the intestine rather than being regulated by the central nervous system in its intestinal action.

FOODS

Market Foods

Since *Clostridium perfringens* occurs so widely in the intestine of man and animals, and in soil and dust, foods frequently become contaminated from these sources. Foods of animal origin frequently become contaminated from direct or indirect contact with various intestinal contents during processing procedures. Meat items, irrespective of the animal from which they come, share a large number of common processing procedures. Following slaughter, the animal undergoes washing, scalding, scraping, cutting, and a number of other processes. Samples taken after all of these procedures have yielded *C. perfringens*.

Bryan and Kilpatrick, in their survey of a roast beef sandwich restaurant, reported that 23 of 83 (28 percent) samples of raw, boneless beef and 10 of 25 (40 percent) samples of raw chicken were positive for *Clostridium perfringens*.

Hall and Angelotti examined meats purchased from chain stores in the Cincinnati area. The meats were picked from the self-service counters and consisted of: unprocessed raw beef (stewing, ground, roasts, steaks, liver, kidney, and soup meat and bones); pork (chops, steaks, roasts, liver, kidney, and spare ribs); veal (chops, roasts, stew meat, liver, and kidney); lamb (stew meat, roasts, chops, and kidney); and frying chicken (leg, thigh, breast, and liver). A total of 161 specimens was examined. Of these, the veal cuts provided the highest number of isolations (82 percent), followed by beef (70 percent), chicken (58 percent), lamb (52 percent), and pork (37 percent).

Strong et al. examined 510 food samples for the presence of *Clostridium perfringens*. These samples were purchased from retail stores and were divided into five different food groups: commercially prepared frozen foods, fresh vegetables and fruits, spices, home-prepared foods, and raw meats, poultry, and fish.

The frozen foods were meat or poultry dishes with accompanying gravies or sauces, such as, TV dinners, meat and poultry pies, and similar products. Twenty types of spices were examined. The meat, poultry, and fish group included beef, pork, lamb, and veal cuts, spiced, organ, and ground meats, fish, and chicken.

Within this diverse group of food samples, a total of 31 (6.1 percent) yielded *Clostridium perfringens* isolations, with the meat, poultry, and fish group having the highest incidence (16.4 percent) and home-prepared foods the lowest (1.8 percent). Among others, 5 percent of the spices, 3.8 percent of the raw fruits and vegetables, and 2.7 percent of the commercially prepared frozen foods were positive for *C. perfringens*.

Baltzer and Wilson investigated the presence of clostridia in 12 bacon factories, six in Denmark and six in Northern Ireland. The investigation was aimed at determining the number of clostridia normally present on the rind side of pig carcasses at various points on the slaughter line. Homogenized rind is frequently used in the production of canned comminuted meat products such as luncheon meat and canned sausage. *Clostridium perfringens* was isolated from the carcasses after scalding, scraping, and inspection, and from scald tank fluid and skin sections. Dirty equipment or machinery was shown to spread the bacteria from infected to uninfected carcasses.

Ladiges et al. investigated the presence of *Clostridium perfringens* in ground beef obtained from a retail store in Denver, Colorado. A total of 95 samples were randomly selected over a period of 19 weeks. Forty-five (47.4 percent) of the samples were found to contain *C. perfringens*. Some of the samples contained high numbers of the organism, thus posing a problem of possible contamination of kitchen counters and kitchen personnel who handle the meat.

Poultry products have also been shown to be frequently contaminated with *Clostridium perfringens*. The organism has been recovered from feet, feathers, caecum, vent area, breast surface, neck skin, and scald tank water. Hagberg et al. obtained fresh skin and meat samples from a turkey eviscerating and processing plant. A total of 62 percent of the samples were positive.

Cooked Foods

The most frequently incriminated foods in *Clostridium perfringens* food-poisoning outbreaks are roast beef, turkey, and chicken, and their gravies, broths, and dressings. Foodborne illness can occur when cooked food becomes contaminated by kitchen personnel or from contaminated counters, utensils, slicers, cutting boards, and so forth. Foodborne illness can also occur when contaminated meat is cooked at a temperature low enough so that spores survive and are activated by the heat shock that results from the cooking. Several studies have shown that *Clostridium perfringens* spores can survive routine cooking procedures. These spores can then germinate and multiply.

Heat also drives off oxygen. Large items, such as beef roasts and other bulk items cooked in large containers, are cooked at lower temperatures for longer times than smaller food items. These lower temperatures drive off all or most of the oxygen present, thereby establishing anaerobic conditions. Since *Clostridium perfringens* is an anaerobic organism, its growth is encouraged in those parts of the meat lacking oxygen. After cooking, exposure to oxygen is limited to the surface.

A long, slow cooling period will then provide excellent conditions for bacterial growth and multiplication. Since the bacteria have a generation time as low as 9 to 20 minutes under optimum conditions, the required 100 million to 1 billion vegetative cells can be produced during the cooling period. The large size of some food items therefore presents a potentially hazardous situation. These foods must be handled only in the properly prescribed manner. If the food is to be retained for later use, it should be held at a temperature of 140°F. (60°C) or above, otherwise it should be cooled or refrigerated as rapidly as possible. The investigation of large numbers of *Clostridium perfringens* outbreaks has shown that the cause, in too many instances, has been large food items or masses of food cooked at low temperatures and allowed to cool slowly for several hours.

In one study, freshly boiled lamb slices were inoculated with a mixture of approximately 260,000 *Clostridium perfringens* spores and vegetative cells. This number increased to 19.5 million in three hours. When lamb slices and gravy were used, the number increased from 260,000 to 2.5 million in two hours and to 46 billion in three hours. The temperature range was 102°–120°F. (39–49°C).

Bryan and Kilpatrick (1971) reported on the presence of *Clostridium perfringens* in a roast beef sandwich restaurant. Three visits were made. Samples were taken and recordings made at each stage of the operation. *Clostridium perfringens* was recovered from 11 of 36 (31 percent) samples of cooked meat. Ninety samples were obtained from food contact surfaces of kitchen and serving equipment and from the kitchen environment. *C. perfringens* was isolated from 28 (31 percent) of these samples, including a slicer, hot pad, scale, storage pan for cooked meat, and cooking pan. Nine stool samples were obtained from seven workers. All were positive for *C. perfringens.* Four of ten (40 percent) hand-rinse cultures were also positive. Nine different serotypes were encountered during Visit 1, 22 during Visit 2, and 13 during Visit 3.

This study illustrates the close relationship between contamination of raw meat and environmental and personnel contamination. Some of the workers and equipment probably became contaminated by the raw products. These same workers and equipment then reinfected the cooked meat. Some of the cooked meat isolations were also heat-resistant strains that survived cooking and then germinated.

McKillop investigated the presence of *Clostridium perfringens* in hospital food. Cold chicken was the most frequently contaminated food. A total of 24 of 46 (52 percent) samples examined were positive. Among other foods from which isolations were made were sausage, cold roast meats, and black pudding. Cold chicken had been responsible for several outbreaks of gastroenteritis in this hospital over a two-year period.

Hall and Angelotti purchased food specimens from various chain stores in the Cincinnati area. Nineteen percent of the foods that required warming or light cooking were positive for *Clostridium perfringens.* The positive isolations came from chili, frankfurters, and garlic frankfurters. Of these ready-to-eat meats, 4.7 percent yielded *C. perfringens.*

Survival of *Clostridium perfringens* in baked ham and turkey rolls was investigated by Strong and Ripp. *C. perfringens* was recovered from each of the hams after cooking, with the center slices yielding the highest number of cells. The organism was also recovered from the turkey rolls, with the highest number of survivors again in the center slices. When roast turkey slices in broth were held under conditions of steam table storage, cells survived for six hours. *C. perfringens* was also isolated from the broth surrounding the turkeys.

THE RESERVOIR

Clostridium perfringens is probably more widespread than any other pathogenic organism. It is a common inhabitant of soil and of the intestines of man and animals.

Smith and Gardener examined eight samples of soil that had not been exposed to fecal contamination for some years. The eight samples contained from 110 to 56,700 *Clostridium perfringens* per gram of soil. Taylor and Gordon surveyed the types of *C. perfringens* normally present in farm soil and in the intestinal contents of several animal species. The soil samples were collected from fields on which horses had developed acute grass sickness. Altogether, 196 cultures from 43 samples of soil were contaminated. *C. perfringens* type A was isolated from 189 (96 percent) of the cultures.

The feces of animal origin were taken in some cases from normal animals and in other cases from animals which had died from diseases other than *Clostridium perfringens* illness. The organism was recovered from fecal samples of cats (100 percent), sheep (100 percent), guinea pigs (94 percent), pigs (93 percent), cattle (92 percent), poultry (92 percent), rabbits (87 percent), and dogs (81 percent). A total of 149 animal samples were examined with 137 (92 percent) of these yielding *C. perfringens*. Type A *C. perfringens*, which causes the most human foodborne illness, comprised 99.7 percent of the positive cultures.

Smith and Crabb examined the feces of healthy young animals, including calves, lambs, and piglets, and healthy adult animals, including cattle, sheep, horses, pigs, rabbits, guinea pigs, dogs, cats, mice, and chickens. *Clostridium perfringens* was isolated from the feces of adult chickens (80 percent), cattle (70 percent), mice (30 percent), guinea pigs (20 percent), and horses (20 percent). Even higher numbers were found in the feces of dogs and cats.

Hobbs et al. examined blowflies collected from a hospital, butcher's shop, fried-fish shop, slaughterhouse, and refuse sorting depot. *Clostridium perfringens* was recovered from every fly examined.

Hall and Hauser investigated the presence of *Clostridium perfringens* in feces obtained from food handlers in and near Baton Rouge, Louisiana. The specimens were examined at the Sanitary Engineering Center Laboratory in Cincinnati, Ohio. Of the 219 specimens collected, 171 (78 percent) were positive for *C. perfringens*.

Sutton investigated the extent of *Clostridium perfringens* in the feces of varying sections of the human population. Two hundred and fifty samples were collected from three classes of the population: primary school children, members of the general population, and aboriginal persons. Positive isolations were obtained from 175 (70 percent) of the samples.

The following cases illustrate some dangerous practices in preparing food.

CASE STUDIES

Case I: Cold Beef

This food-poisoning incident involved children who had eaten at their school cafeteria. A total of 275 of the 475 (58 percent) students who had eaten the meal of cold beef became ill with diarrhea and abdominal pain. The victims also included 18 of 41 (44 percent) of the school staff and 10 of 11 (91 percent) of the cafeteria staff.

The meat had been delivered to the school in small joints (4.5 lbs.) and kept in the pantry till midday. The joints were then boiled for 3 hours, placed in trays, covered with a tea cloth, and left in the pantry overnight. The next day they were sliced and served with freshly made gravy. Within hours, cases of food poisoning began to occur.

We see here one of the most dangerous food preparation procedures that is sometimes followed—allowing a long, slow cooling period for cooked meat. It is possible that heat resistant spores survived the boiling or that the meat became contaminated after cooking. In either case, the outbreak can be attributed to the allowance of the necessary time for an immense buildup of the bacteria. If proper refrigeration procedures had been followed, the outbreak would not have occurred.

Case 2: Roast Beef Gravy

In 1966, a food-poisoning outbreak occurred at a large state university in Wisconsin. A total of 366 students became ill. Diarrhea and abdominal cramps were the most common symptoms. The evening meal included roast beef gravy. About 27 gallons of the gravy had been held over from a meal served two days previously. This gravy had been kept warm during the serving of the first meal and then placed in nine-gallon containers and refrigerated. It was then mixed with freshly made gravy for the subsequent meal.

The origin of the contamination could not be determined but the improper and dangerous procedures were: (1) keeping a susceptible food at warm temperatures for several hours, and (2) placing this food in large containers where, even under refrigeration, the temperatures would remain within the growth range of *Clostridium perfringens* for several additional hours.

Case 3: Cold Chicken

This outbreak was one of a series that occurred in a hospital. It illustrates a little noticed but probably common source of contamination of cooked foods. Nine patients in one ward and six patients in another ward were the victims, along with a maid attached to one of the wards. Cold chicken was the incriminated food. Environmental samples taken from the kitchen revealed that the kitchen dust was heavily contaminated with *Clostridium perfringens.*

Six outbreaks of gastroenteritis had occurred within this hospital over a two-year period of time. All of the outbreaks seemed to be centered around chicken that had been cooked and then cooled in an unprotected zinc bath before being served cold. Zinc is a rough metal and was undoubtedly heavily contaminated from infected dust. The cooked chickens were not separated after removal from the cooker; they were simply transferred in bulk, with their broth, to the large open receptacle where they were allowed to remain for nearly 24 hours without refrigeration or any form of protection. In all likelihood, this is when they became infected by the dust containing *Clostridium perfringens.*

Once again, a basic tenet of food protection was violated, that is, refrigerate foods as rapidly as possible in as small portions as practicable. Most food-poisoning outbreaks can be prevented by proper refrigeration procedures.

SUMMARY

Clostridium perfringens food poisoning is a relatively new foodborne illness in the United States. Although it dates to 1895 in English reports, it was not reported in the United States until 1945. In 1976, *C. perfringens* was responsible for 6 (6.5 percent) of the 92 outbreaks and 509 (15.6 percent) of the 3,270 cases of known bacterial origin that were reported to the Communicable Disease Center Foodborne Disease Surveillance Activity.

Foods involved in outbreaks of *Clostridium perfringens* foodborne illness are usually meat or poultry, stews, sauces, gravies, dressings, or casseroles that have been cooked and held at room temperature for excessive periods of time or refrigerated in large masses.

The principal symptoms of food poisoning due to *Clostridium perfringens* are cramps, abdominal pain, diarrhea, and gas. Nausea and vomiting are not common. Recovery is usually complete within 24 hours.

Clostridium perfringens is an anaerobic, spore-forming, rod-type bacterium that is non-motile and gram-positive. Strains of *C. perfringens* are divided into five types (A to E) on the basis of four major toxins that they may produce. Only types A and C cause foodborne illness in humans, with type A causing the great majority of outbreaks. Type A also causes gas gangrene. All types are capable of causing disease in animals, particularly horses, sheep, goats, pigs, and cattle.

Fermentation reactions produce acid and gas from a number of carbohydrates, including maltose, lactose, glucose, and sucrose. Mannitol is not fermented and indole is not produced.

Some strains of *Clostridium perfringens* produce heat-sensitive spores, others produce heat-resistant spores. Both types have been responsible for foodborne illness.

The optimum temperature for growth of *Clostridium perfringens* is between 109.4° and 116.6°F. (43°–47°C). Its full growth range extends from 60°–124°F. (15°–51°C). Good growth usually occurs over a temperature range of 68°–122°F. (20°–50°C), depending on the food.

Growth of *Clostridium perfringens* occurs within a range of pH 5.0 to 8.5. Its optimal range lies between pH 6.0 and 7.5, but fairly rapid growth occurs between pH 5.5 and 8.0.

The enterotoxin produced by *Clostridium perfringens* is synthesized only during sporulation. It is a protein that acts directly on the intestinal lining, producing inflammation and diarrhea.

Clostridium perfringens is probably more widespread than any other pathogenic organism. It is a common inhabitant of soil and of the intestines of man and many animals, including most mammals.

Review Questions

1. What types of foods are usually involved in *Clostridium perfringens* outbreaks?

2. What was the major contributing factor in *Clostridium perfringens* outbreaks in 1975 and 1976?

3. Describe the typical symptoms of *C. perfringens* food poisoning.

4. a. What symptoms are common to foodborne illness caused by *S. aureus*, *Sal.* species, and *C. perfringens*? b. How do the symptoms caused by these organisms differ?

5. Name and describe three conditions other than gastroenteritis caused by salmonellae.

6. How do American strains of *C. perfringens* differ from English strains with regard to heat sensitivity?

7. List several characteristics of *C. perfringens*.

8. Which strains of *C. perfringens* cause foodborne illness in humans?

9. a. What is the effect of heat on the spores of *C. perfringens*? b. What is the effect of freezing?

10. a. Within what temperature range does *C. perfringens* grow? b. What is its optimum temperature?

11. a. Within what pH range does *C. perfringens* grow? b. What is its optimal range?

12. Describe the salt tolerance of *C. perfringens*.

13. How can prevention of *C. perfringens* food poisoning be achieved?

14. How does food of animal origin frequently become contaminated?

15. What cooked foods are most frequently incriminated in *C. perfringens* outbreaks?

16. a. What role does heat play in establishing anaerobic conditions? b. What effect does this have on the development of *C. perfringens*? c. What effect does slow cooling have?

17. Where is *C. perfringens* found in nature?

References

Andrewes, F. W. (1899). "On an Outbreak of Diarrhea in the Wards of St. Bartholomew's Hospital." *Lancet* 1:89.

Arbuckle, R. E. (1960). "The Influence of Temperature on the Growth of Clostridium Perfringens." M.S. Thesis. Library, Indiana University, Bloomington, Indiana.

Baltzer, J. and D. C. Wilson. (1965). "The Occurrence of Clostridia on Bacon Slaughter Lines." *J. Appl. Bacteriol.* 28:119–124.

Bergdoll, M. S. (1969). "Bacterial Toxins in Food." *Food Technol.* 23:532–533.

Browne, A. S., G. Lynch, A. R. Leonard, and G. Stafford. (1962). "A Clostridial or Enterococcal Food Poisoning Outbreak." *Pub. Health Rep.* 77:533–536.

Bryan, F. L. (1969). "What the Sanitarian Should Know About Clostridium Perfringens Foodborne Illness." *J. Milk Food Technol.* 32:381–389.

Bryan, F. L. (1972). "Emerging Foodborne Diseases. II. Factors that Contribute to Outbreaks and Their Control." *J. Milk Food Technol.* 35:632–638.

Bryan, F. L. and E. G. Kilpatrick. (1971). "Clostridium Perfringens Related to Roast Beef Cooking, Storage, and Contamination in a Fast Food Restaurant." *Amer. J. Pub. Health* 61:1869–1886.

Bryan, F. L., T. W. McKinley, and B. Mixon. (1971). "Use of Time-Temperature Evaluations in Detecting the Responsible Vehicle and Contributing Factors of Foodborne Disease Outbreaks." *J. Milk Food Technol.* 34:576–582.

Canada, J. C., D. H. Strong, and L. G. Scott. (1964). "Response of Clostridium Perfringens Spores and Vegetative Cells to Temperature Variation." *Appl. Microbiol.* 12:273–276.

Duncan, C. L. (1970). "Clostridium Perfringens Food Poisoning." *J. Milk Food Technol.* 33:35–41.

Duncan, C. L., and D. H. Strong. (1969). "Ileal Loop Fluid Accumulation and Production of Diarrhea in Rabbits by Cell-free Products of Clostridium Perfringens." *J. Bacteriol.* 100:86–94.

Duncan, C. L. and D. H. Strong. (1970). "Clostridium Perfringens Type A Food Poisoning. I. Response of the Rabbit Ileum as an indication of Enteropathogenicity of Strains of Clostridium Perfringens in Monkeys." *Infect. Immun.* 3:167–170.

Fischer, L. H., D. H. Strong, and C. L. Duncan. (1970). "Resistance of Clostridium Perfringens to Varying Degrees of Acidity During Growth and Sporulation." *J. Food Sci.* 35:91–95.

Fruin, J. T. (1977). "Significance of Clostridium Perfringens in Processed Foods." *J. Food Protect.* 40:330–332.

Fruin, J. T. and F. J. Babel (1977). "Changes in the Population of Clostridium Perfringens Type A Frozen in a Meat Medium." *J. Food Protect.* 40:622–625.

Gough, B. J. and J. A. Alford. (1965). "Effect of Curing Agents on the Growth and Survival of Food-poisoning Strains of Clostridium Perfringens." *J. Food Sci.* 30:1025–1028.

Guthertz, L. S., J. F. Fruin, D. Spicer, and J. L. Fowler. (1976). "Microbiology of Fresh Comminuted Turkey Meat." *J. Milk Food Technol.* 39:823–829.

Hagberg, M. M., F. F. Busta, E. A. Zottola, and E. A. Arnold. (1973). "Incidence of Potentially Pathogenic Microorganisms in Further-Processed Turkey Products." *J. Milk Food Technol.* 36:625–629.

Hall, H. E. and R. Angelotti. (1965). "Clostridium Perfringens in Meat and Meat Products." *Applied Microbiol.* 13:352–357.

Hall, H. E., R. Angelotti, K. H. Lewis, and M. J. Foster. (1963). "Characteristics of Clostridium Perfringens Strains Associated with Food and Food-borne Disease." *J. Bacteriol.* 85:1094–1103.

Hall, H. E. and G. H. Hauser. (1966). "Examination of Feces from Food Handlers for Salmonella, Shigellae, Enteropathogenic, Escherichia Coli, and Clostridium Perfringens." *Appl. Microbiol.* 14:928–933.

Hart, C. J., W. W. Sherwood, and E. Wilson. (1960). "A Food Poisoning Outbreak Aboard a Common Carrier." *Pub. Health Rep.* 75:527–531.

Hauschild, A. H. W. (1971). "Clostridium Perfringens Enterotoxin." *J. Milk Food Technol.* 34:596–599.

Hauschild, A. H. W., L. Nillo, and W. J. Dorward. (1967). "Experimental Enteritis with Food Poisoning and Classical Strains of Clostridium Perfringens Type A in Lambs." *J. Infect. Dis.* 117:379–386.

Hauschild, A. H. W., L. Nillo, and W. J. Dorward. (1968). "Clostridium Perfringens Type A Infection of Ligated Intestinal Loops in Lambs." *Appl. Microbiol.* 16:1235–1239.

Hauschild, A. H. W., L. Nillo, and W. J. Dorward. (1970). "Enteropathogenic Factors of Food-Poisoning Clostridium Perfringens Type A." *Can. J. Microbiol.* 16:331–338.

Hauschild, A. H. W., L. Nillo, and W. J. Dorward. (1968). "The Role of Enterotoxin in Clostridium Perfringens Type A Enteritis." *Can. J. Microbiol.* 17:987–991.

Hobbs, B. C. (1965). "Clostridium Welchii as a Food Poisoning Organism." *J. Appl. Bacteriol.* 28:74–82.

Hobbs, B. C., M. E. Smith, C. L. Oakley, G. H. Warrock, and J. C. Cruickshank. (1953). "Clostridium Welchii Food Poisoning." *J. Hyg.* 51:74–101.

Kemp, G. E., R. Proctor, and A. S. Browne. (1962). "Foodborne Disease in California with Special Reference to Clostridium Perfringens (Welchii)." *Pub. Health Rep.* 77:910–914.

Knox, R. and E. K. McDonald. (1943). "Outbreaks of Food Poisoning in Certain Leicester Institutions." *Med. Offr.* 69:21–22.

Ladiges, W. C., J. F. Foster, and W. Ganz. (1974). "Incidence and Viability of Clostridium Perfringens in Ground Beef." *J. Milk Food Technol.* 37:622–623.

McClung, L. S. (1945). "Human Food Poisoning Due to Growth of Clostridium Perfringens (C. Welchii) in Freshly Cooked Chicken: Preliminary Note." *J. Bacteriol.* 50: 229–231.

McKillop, E. J. (1959). "Bacterial Contamination of Hospital Food with Special Reference to Clostridium Welchii Food Poisoning." *J. Hyg.* 57:31–46.

Pasch, J. H. (1974). "Food and Other Sources of Pathogenic Microorganisms in Hospitals. A Review." *J. Milk Food Technol.* 37:487–493.

Smith, L. DS. (1963). "Clostridium Perfringens Food Poisoning." In *Microbiological Quality of Foods*, ed. L. W. Slanetz. Academic Press, Inc. New York.

Smith, L. DS. (1975). *The Pathogenic Anaerobic Bacteria.* Charles C. Thomas. Springfield, Illinois.

Smith, L. DS. and M. V. Gardner. (1949). "The Occurrence of Vegetative Cells of Clostridium Perfringens in Soil." *J. Bacteriol.* 58:407–408.

Smith, H. W. and W. E. Crabb. (1961). "The Faecal Bacterial Flora of Animals and Man: Its Development in the Young." *J. Pathol. Bacteriol.* 82:53–66.

Sterne, M. and G. H. Warrack. (1964). "The Types of Clostridium Perfringens." *J. Pathol. Bacteriol.* 88:279–283.

Strong, D. H. and J. C. Canada. (1964). "Survival of Clostridium Perfringens in Frozen Chicken Gravy." *J. Food Sci.* 29:479–482.

Strong, D. H. and J. C. Canada, and B. B. Griffiths. (1963). "Incidence of Clostridium Perfringens in American Foods." *Appl. Microbiol.* 11:42–44.

Strong, D. H. and N. M. Ripp. (1967). "Effect of Cookery and Holding on Ham and Turkey Rolls Contaminated with Clostridium Perfringens." *Appl. Microbiol.* 15:1172–1177.

Sutton, R. G. A. (1966). "Enumeration of Clostridium Welchii in the Faeces of Varying Sections of the Human Population." *J. Hyg. Camb.* 64:367–374.

Sutton, R. G. A. and B. C. Hobbs. (1968). "Food Poisoning Caused by Heat-Sensitive Clostridium Welchii. A Report of Five Recent Outbreaks." *J. Hyg. Camb.* 66:135–145.

Sylvester, P. K. and J. Green. (1961). "The Effect of Different Types of Cooking on Artificially Infected Meat." *Med. Offr.* 105:231–235.

Taylor, A. W. and W. S. Gordon. (1940). "A Survey of the Types of Cl. Welchii Present in Soil and in the Intestinal Contents of Animals and Man." *J. Pathol. Bacteriol.* 50:271–277.

U.S. Department of Health, Education and Welfare. Communicable Disease Center. (1966). *Morbidity, Mortality Weekly Reports* 15:103.

U.S. Department of Health, Education and Welfare. Communicable Disease Center. (1968). *Morbidity, Mortality Weekly Reports* 17:415–416.

Walker, H. W. (1975). "Food Borne Illness From Clostridium Perfringens." Crit. Rev. in *Food Sci. and Nutrition.* 7:71–104.

Woodburn, M. and C. H. Kim (1966). "Survival of Clostridium Perfringens During Baking and Holding of Turkey Stuffing." *Appl. Microbiol.* 14:914–920.

7 Botulism

On April 13, 1978, a 35-year-old enlisted man was flown from Cannon Air Force Base in New Mexico to El Paso, Texas, because of a progressing neurological condition typical of the serious disease, botulism. Cannon Air Force Base is located outside Clovis, New Mexico, a town of 40,000 population. The enlisted man was also a part-time employee in a country club restaurant in Clovis.

On April 15, two more cases appeared. They were transferred from Clovis to Amarillo, Texas. Both patients were women and, at admittance, exhibited drooping eyelids, eye muscle tremors, decreased gag reflexes, and generalized weakness. Within hours both women required breathing tubes and mechanical respiratory assistance. One woman had eaten at the country club restaurant in Clovis on April 9 and the other on April 12.

During the next 48 hours, 29 additional patients with similar symptoms were admitted to hospitals in Amarillo and Lubbock, Texas, and Clovis, Albuquerque, and Santa Fe, New Mexico. Eleven of these people eventually required breathing tubes and mechanical respiration to aid their breathing muscles.

All 32 patients had eaten at the country club in Clovis on April 9, 12, or 13. Two foods served at the salad bar, potato salad and three-bean salad, were eventually incriminated in the outbreak, the second largest outbreak of botulism reported in the United States since recording began in 1899. The Clovis outbreak was caused by type A *Clostridium botulinum*.

Botulism outbreaks, although infrequent, are extremely serious because of the high fatality rate associated with the disease. Respiratory paralysis is not uncommon. The disease can have far-reaching and disastrous economic consequences as the following two cases illustrate.

On the morning of June 30, 1971, an elderly man residing in Westchester County, New York, began to experience double vision. That afternoon he was hospitalized and by late evening he was dead as a result of respiratory failure. The following morning his 63-year-old wife was also admitted to the hospital having difficulty in swallowing, forming words, and speaking. Botulism was again diagnosed.

On June 29, one day previous to the husband's illness, the couple had consumed part of a can of vichyssoise at their evening meal. They had eaten very little because the contents tasted bad. Examination of a second can of vichyssoise from the home and of the blood serum of the patients revealed the presence of type A botulinal toxin. Toxin was also found in four other cans of the same manufacturer's lot. All four cans were swollen. In addition to the vichyssoise, which was distributed nationally under 22 brand names, including the canner's own name, the company canned 89 other products. Examination of several of these products revealed a high incidence of swollen cans.

At this point, the company voluntarily recalled all of its products. The impact of the recall on the company was disastrous. The resultant publicity and expense associated with the recall drove the company out of business.

On February 17, 1973, a canning company in Ohio issued a recall of 11 lots of canned mushrooms after botulinal toxin type B was detected in cans from one lot. Within several days, 20 lots of these canned mushrooms had been recalled. Botulinal toxin was found in cans from five lots. The mushrooms were distributed under 11 labels to restaurants and wholesale distributors in Illinois, Ohio, New York, and Pennsylvania.

Because the possibly contaminated mushrooms were used in their frozen food products, two other companies also felt compelled to recall their products, thus spreading the economic hardship and damaged reputations to even greater limits. One company in Ohio recalled four of its products that contained the suspected mushrooms—Tuna Noodle Casserole, Escalloped Chicken and Noodles Casserole, Green Beans and Mushroom Casserole, and Cream of Mushroom Soup; the other company, in Michigan, recalled its frozen Mushroom Pizza.

No reported cases developed from this mishap, probably because the products involved required cooking at a temperature high enough to destroy the toxin. So fearsome is the reputation of this disease that any recall because of the possibility of its existence can cause serious economic damage to the company involved, not only because of the large sums of money that must be spent to recall the product, but also because of the loss of income due to public wariness about the company's products.

THE ILLNESS

Incidence

In 1976, there were 438 reported outbreaks of foodborne illness in the United States. These outbreaks affected 12,463 persons. The cause was determined in 132 (30 percent) of the outbreaks. Ninety-two (70 percent) of these outbreaks, affecting 3,270 people, were of bacterial origin. *Clostridium botulinum* was responsible for 23 (25 percent) of the 92 outbreaks, placing it third for that year, just slightly behind salmonella and staphylococcus outbreaks. The 23 outbreaks affected 40 persons and caused 5 deaths, a fatality rate of 12.5 percent.

During the period 1899 through 1969, there were 659 reported outbreaks and 1,696 cases of botulism in the United States. A total of 957 persons died from the illness, a fatality rate of 56 percent. In only 201 cases was the type of *Clostridium botulinum* identified. Of these, 144 (71.6 percent) were due to type A strains, 37 (18.4 percent) to type B, 17 (8.5 percent) to type E, and 1 (0.5 percent) to type F.

During the years 1970 through 1976, 88 outbreaks were recorded, an average of 12.5 outbreaks per year. A total of 182 cases resulted, an average of 2.07 cases per outbreak. Thirty-five deaths occurred, a fatality rate of 19 percent. In 72 of the 88 outbreaks, the strain of *Clostridium botulinum* involved was typed. Forty-two (58.3 percent) were caused by type A, 21 (29.2 percent) by type B, 8 (11.1 percent) by type E, and 1 (1.3 percent) by type F.

More cases of botulism have probably occurred than have been diagnosed, since isolated cases are difficult to recognize. Some have been diagnosed initially as gastroenteritis, intestinal obstruction, muscular dystrophy, poliomyelitis, tick paralysis, chemical intoxication, and trichinosis.

Four hundred and thirty-four outbreaks, with 1,294 cases and a fatality rate of 13.8 percent were recorded in Germany during the first half of this century. Most of the outbreaks were caused by type B strains, which appear to be less lethal than the other strains. More than 500 outbreaks were recorded in France during World War II, 98.5 percent of which were caused by type B strains. Between 1932 and 1964, 50 outbreaks were reported from Japan. A total of 307 cases and 73 deaths (23.8 percent) resulted from the outbreaks. Japan currently has the highest incidence of type E botulism.

Vegetables have been the most commonly implicated food in type A and type B outbreaks. Peppers have been implicated more frequently than any other vegetable. String beans have also caused many outbreaks. Almost all of the type E outbreaks have been traced to fish or fish products. Table 7-1 lists some of the larger outbreaks of botulism in the United States and other countries.

Type A is the most frequent cause of botulism in North America while type B is more prevalent in Europe. Type E has come to be of paramount significance in countries of northern latitudes, probably because it can grow and produce toxin at lower temperatures than types A and B. In the twentieth century, type E has been responsible for 46 percent of the out-

Table 7-1
Some large outbreaks of botulism.

Food	Year	Number Sick	Deaths	Type	Country
Stuffed egg plant	1933	230	94	A	Russia
Peppers	1977	59	0	B	United States
Kirikomi	1962	55	1	E	Japan
Potato salad and/or three-bean salad	1978	32	0	A	United States
Beets	1919	23	12	—	United States
Spinach	1921	21	3	A	United States
Whitefish	1963	17	8	E	Japan
Izushi	1961	16	16	E	Japan
Peas	1939	16	5	B	United States
Izushi	1951	14	4	E	Japan
Chili peppers	1937	14	9	A	United States
String beans	1931	13	13	A	United States

breaks of botulism in Japan, Canada, and Scandinavia.

Although outbreaks in the United States have been reported from almost all states, five western states (California, Oregon, Washington, Colorado, and New Mexico) have accounted for well over half of all reported outbreaks in the past. About 90 percent of reported type A outbreaks have occurred west of the Mississippi River.

Most type B outbreaks (about 70 percent) have been reported from states east of the Mississippi River. Type E has been associated primarily with Alaska and with the Great Lakes area, particularly Lake Michigan.

These figures indicate that there is a regional distribution of the types of *Clostridium botulinum* in the United States. Type A strains are the predominant form found in the soils of the western states, while type B predominates in the east. There are apparently pockets of type E strains occurring in the soils near the Great Lakes that have been washed into the lakes in certain areas and are responsible for the contamination of many fish in these lakes.

Symptoms

Botulism is a serious foodborne intoxication that usually occurs as small-scale scattered outbreaks in the human population but often as large-scale outbreaks in animals. Although the fatality rate has dropped since the development of antitoxins, it can still be highly lethal in any individual outbreak. In 1963, eight of the 17 persons affected died from eating contaminated fish caught in Lake Michigan. A 1961 outbreak in Japan resulted in the death of all 16 persons affected.

The illness is caused by the toxin of *Clostridium botulinum*. This toxin is formed in the food before it is consumed. In this respect it is similar to staphylococcal foodborne intoxication. Most cases result from the ingestion of contaminated foods from jars and cans.

The typical symptoms of botulism usually appear within 12 to 36 hours, but may vary from 4 hours to 8 days in extreme cases. These symptoms typically involve the nervous system, but may be preceded by nausea, vomiting, diarrhea, and other digestive disturbances, usually within

12 to 24 hours of consumption of the contaminated food. The presence of gastrointestinal symptoms seems to depend on the degree of spoilage and the amount of food consumed. Constipation often develops in the later stages of the disease. There is usually no fever.

The gastrointestinal upset, if present, is usually followed by feelings of weakness and dizziness, often accompanied by dryness of the mouth and tongue due to decreased saliva production. Sore throat sometimes occurs. The pulse is normal or slow and most patients are responsive and oriented.

Typical neurological symptoms are blurred and double vision resulting from paralysis of eye muscles. The pupils may be dilated. Difficulty in swallowing and speaking results from the increasing paralysis of the muscles of the head and neck. As the paralysis proceeds downward, the respiratory muscles are affected and death results. There is no impairment of the patient's mental abilities during the course of the disease.

Administration of antitoxin, intensive respiratory supportive care, and the elimination of toxin from the digestive tract by using cathartics, enemas, and gastric lavage are the most important measures in the treatment of botulism. Generally, patients with short incubation periods (less than 24 hours) are more seriously affected, have longer acute and recovery stages, and are more likely to die. Antitoxin treatment seems to be most effective in the type E cases.

The death rate in adults is higher than in children. Improvements in diagnosis, the availability of antitoxin, and intensive care have contributed to lowering the death rate.

Botulinal toxin interferes with the release of acetylcholine at the neuromuscular junction. This leads to a paralysis of the motor nerves of the autonomic nervous system, which regulates the involuntary muscles. Among the involuntary muscles affected are those involved in swallowing, some eye muscles, and the breathing muscles.

Infant botulism is also known. It was detected first in 1976 in six infants, five in California and one in New Jersey, and since then it has become recognized as a serious infant disease. By 1978, there were 63 recorded instances of the illness in the United States. Two deaths have been reported. Great Britain's first recorded case came in February 1978.

The following is taken from the case history of one infant.

> A male baby born March 22, 1976, in Los Angeles had normal development until the onset of constipation occurred when he was 20 weeks of age. Two days later, his mother noted that he sucked poorly from her breast. By the following day, drooping eyelids, generalized muscle weakness (particularly in the neck), a weak cry, and shallow respiration developed. He appeared lethargic, irritable, and floppy. After two days with no improvement on a regimen of oral ampicillin sodium, the infant was admitted to a local hospital for examination. . . . Although muscle tone remained poor, the infant's condition improved during his nine-day stay, with partial return of head control and reaching ability. . . . Four days after discharge, the infant's condition deteriorated, with onset of increased weakness, shallow respiration, and a poor cry.
>
> He was then admitted to a metropolitan hospital. . . . He appeared drowsy and had a weak, constant cry. . . . The infant had generalized decreased muscle tone and diminished spontaneous movements. Cranial nerve function was intact except for . . . sluggishly reactive pupils; absent upward gaze; and a weak gag reflex.
>
> He gradually recovered neurologic function and was discharged after a total hospitalization of 60 days (Greenberg, 1978).

On the basis of cases documented thus far, it is estimated that there are about 250 infant botulism attacks a year in the United States. There is no solid evidence to date to indicate how these cases originate. The infants seem to become ill without having ingested contaminated food.

Cases of botulism from wound sources have also been reported. The first case occurred in 1943, and by 1973 nine cases had been recorded. All nine patients complained of difficulty in swallowing and breathing; all but two had double vision and difficulty in speaking; and all of the patients were severely ill and required mechanical respiratory assistance. Four of the nine cases resulted in death. Six of the patients yielded either type A toxin or *Clostridium botulinum* from the wound.

Botulism also occurs in animals and is known under many different names, depending on the geographic location of the disease and the type of animal afflicted. "Lamziekte," "forage poisoning," "dying disease," "coast disease," "spinal typhus," and "loin disease" are some of the names for botulism in livestock. It is called "limberneck," "alkali disease," and "western duck disease" in waterfowl.

Lamziekte occurs in cattle grazing in fields that are highly deficient in phosphorus. These cattle tend to eat the carcasses of decaying small animals. The carcasses sometimes are infected with *Clostridium botulinum* types C or D and can become highly toxic to cattle. It is estimated that in the Union of South Africa, 100,000 cattle have died in some years from this disease. In 1932 and 1933, 100,000 sheep died in western Australia from the same cause. The disease has also been reported from Cuba, Brazil, Senegal, and the United States. In Senegal, some cases of botulism in cattle have resulted from decomposing animals making the drinking ponds toxic.

Forage poisioning occurs in cattle and horses in many different and widely separated countries. It results from ingesting toxic hay or silage that contains the decaying bodies of small animals that have become infected with *Clostridium botulinum*. The toxin, usually type C, will sometimes diffuse from the animal into the hay or silage.

Serious outbreaks on mink farms have been reported. In 1961, approximately 200 minks on three different fur farms in Japan died suddenly. One food was common to all three farms—meat from a whale that had been caught off the coast and frozen. The whale meat was mixed with other ingredients and fed to the minks in the evening. The next morning the minks began showing cyanosis of the lips, bleeding from the nostrils, diarrhea, salivation, and paralysis. Death occurred within one to two days. *Clostridium botulinum* type C was isolated from the whale meat.

An outbreak in 1964 in Japan killed 1,249 minks within five days, 55 percent of the minks on the farm. These minks had been fed with the meat from a horse that had died after exhibiting coliclike signs. The minks exhibited such symptoms as general weakness, paralysis of lower extremities, and difficulty in breathing. *Clostridium botulinum* type C was isolated from the horse meat.

Botulism in chickens has been known for more than 70 years. It is frequently caused by feeding spoiled food containing toxin to the fowl. Types A, B, and C have been implicated in outbreaks of botulism in chickens and barnyard ducks, turkey, and geese. One outbreak caused the death of 643 chickens.

Type C toxin has been responsible for major outbreaks of botulism in aquatic and shore birds. Thousands of birds may be involved in a single outbreak. Wild ducks are affected most frequently, although outbreaks have occurred among gulls, loons, and sandpipers. A total of 250,000 aquatic wild birds died in 1932 on the north side of the Great Salt Lake. Botulism was also responsible for the death of several thousand herring gulls, ring-billed gulls, common loons, and other aquatic birds on Lake Michigan in 1963 and 1964.

Outbreaks of botulism occur almost annually on certain lakes and mud flats in the United States, Canada, Argentina, Mexico, Uruguay, and Australia. Outbreaks have also been reported from Great Britain, Denmark, Sweden, and Holland.

History

Clostridium botulinum was first isolated by van Ermengen in 1896 during the course of an investigation into a food-poisoning outbreak in Ellezeles, Belgium. Thirty-four members of a music club had gathered for a banquet that included a raw salted ham. Twenty-three of the party-goers became ill within 20 to 36 hours and three of them died within a week. Van Ermengen isolated an anaerobic, spore-forming, rod-shaped bacterium from the ham and from the spleen of one of the victims. This organism produced an extremely potent toxin.

The symptoms (widely dilated pupils and progressive muscular paralysis) of the people involved in the Ellezelles outbreak were similar to a sausage-poisoning illness, known as *wurstvergiftung,* which had been recognized in Germany for many years. It was named botulism in 1870, when Mueller gave it that title from the latin *botulus* meaning sausage. For this reason, van Ermengen named the organism that he isolated *Bacillus botulinus.* This name was later changed to *Clostridium botulinum* as the classification of bacteria became more organized.

Sausage poisoning had been a recognized disease since the early 1800s when the German, Justinus Kerner, published two papers resulting from a study of 230 cases of the illness. As a result of the papers, warnings were issued against consumption of spoiled sausages. The warnings were not very effective, however, since some 2,000 cases of botulism with a mortality rate of 30 percent occurred in Europe from the mid-nineteenth century to the beginning of the twentieth century.

Botulism had been known in Russia, even before the time of Kerner's papers, as *ichthyosismus* (fish poisoning) because of its frequent association with the ingestion of smoked or pickled fish. Although the illness was primarily associated with fish such as sturgeon and herring, the largest outbreak ever recorded occurred in Dniepropetrovsk, where 230 cases were caused by stuffed egg plant, commonly used as a relish. Ninety-four deaths resulted from the outbreak.

In 1904, an outbreak in Darmstadt, Germany, resulted in the isolation by Landman of a somewhat different type of *Clostridium botulinum* than had been identified on other occasions. This new strain was named *C. botulinum* type B. Both of the type C strains were discovered in 1922, C alpha by Bengston in fly larvae and C beta by Seddon from the bodies of cattle that had died of botulism in Australia.

Type D was isolated by Theiler in 1927. The organism was isolated from cattle that had died of lamziekte in South Africa. Type E was identified in 1936 by Gunnison of the California Medical School. He had received two cultures of bacteria that had been isolated from a sturgeon and sent to him from Dniepropetrovsk, Russia.

Type F appeared on the scene as a recognized disease agent in 1958, when Moeller and Scheibel isolated this strain in liver paste that was involved in an outbreak of botulism on the Danish island of Langeland. Type G was isolated by Gimoz and Ciccarelli in 1970 during a soil survey in Argentina. The organism was recovered from a cornfield.

Until 1904, virtually all botulism outbreaks were caused by foods with a high animal protein content, such as, ham, sausage, and fish. However, the Darmstadt outbreak in 1904 was caused by wax bean salad, and a 1913 outbreak in California (involving 12 persons) was caused by home-canned beans. One person died in this outbreak. From this time on botulism became associated with home-canned foods, particularly vegetables. From 1918 to 1922, 83 outbreaks involving 297 persons and 185 deaths (62.3 percent) occurred in California, alone. Underprocessed vegetables were implicated in most of the outbreaks.

This is not to say that home-preserved foods were the only cause of botulism at that time. During the period 1918 to 1926, commercially canned foods in the United States caused 31 known outbreaks with 160 cases and 74 deaths (46 percent). It was during these years, however, that the canning industry developed processes that virtually eliminated the disease from commercially canned products in the United States for 40 years.

An outbreak in 1922 in Loch Maree, Scotland, resulted in eight deaths. The outbreak occurred at a resort hotel and was caused by wild duck paste used in making a sandwich spread. Type A toxin was the agent of death on this occasion. Type A toxin was implicated again in a 1924 outbreak that killed a family of 12 in Albany, Oregon, who had eaten home-canned string beans.

A tragedy similar to that in Albany took place in Japan in 1970. A grandmother had prepared izushi (rice patties containing diced vegetables and raw fish compressed together in a tub and allowed to ferment for about three weeks). She sampled one, became sick within a few hours and was dead by the next day. Relatives and friends gathered for her funeral and consumed the rest of the cakes. Eleven more people died as a result. Type E botulism, probably from the fish, was the cause.

An outbreak of type E botulism worthy of noting occurred in the United States, also in 1960. Seven cases and two deaths resulted from a dinner at which vacuum-packed smoked ciscoes were served. The ciscoes had been caught in the Great Lakes, normally processed, and kept in the refrigerator of the hostess for about a week or ten days. The fish had been boned and served without further cooking.

Research resulting from this incident showed that, while vacuum packaging did not increase the rate of toxin production by cells resulting from spores that survived the smoking process, it did reduce spoilage and made the product appear safe. This condition occurs because type E botulinal cells do not produce noticeable spoilage effects and the vegetative cells are capable of growth and toxin formation at refrigerator temperatures.

In 1963, botulism in commercially prepared foods reappeared with a flourish. Four outbreaks from these foods were reported. One outbreak, type E, was traced to canned tuna fish and caused two deaths in the Detroit area; a second outbreak, type E, was caused by vacuum-packed smoked whitefish chubs and resulted in seven deaths among 17 cases in Tennessee, Alabama, and Kentucky; a third outbreak, type E, caused two deaths in Kalamazoo, Michigan, and was traced to whitefish; the fourth outbreak, in New York City, was attributed to canned liver paste and caused no deaths. Types A and B toxin were found in the cans of liver paste.

Even though *Clostridium botulinum* type E was first isolated from an inland Russian sturgeon, it was not until many years later that American inland waters were suspected of harboring this organism. The first recorded outbreak in the United States of botulism caused by inland fish occurred in 1960 and resulted from vacuum-packaged smoked ciscoes caught in the Great Lakes. Two of the 1963 outbreaks were caused by Lake Michigan whitefish. Other outbreaks have occurred in North America since these, some resulting from Great Lakes fish and others from fish caught in the inland waters of Alaska and Canada.

THE BACTERIA

Clostridium botulinum is an anaerobic, spore-forming, motile rod (bacillus) that occurs singly, in pairs, or in short chains, and is commonly found in soils throughout the world. It also occurs on vegetables, fruits, human and animal feces, and decomposing or decaying animal life. The organism is gram-positive. All strains ferment glucose and fructose with the subsequent production of acid and gas. The ability to ferment other carbohydrates, such as, sucrose, maltose, and dextrin, depends on the

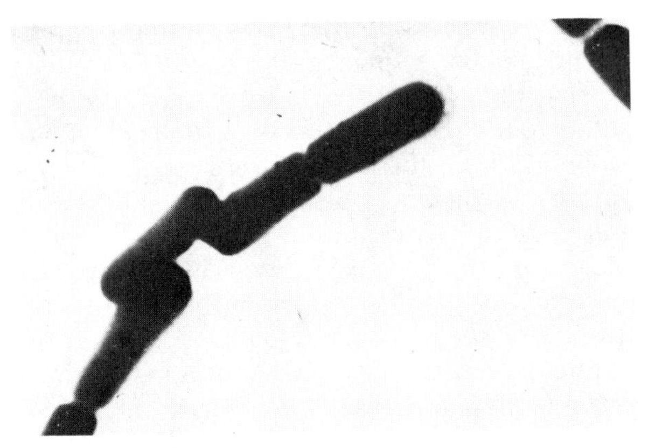

Figure 7-1 *Clostridium botulinum*

strain. The high mortality rate of the intoxication produced by this organism has made it the most feared of the food-poisoning bacteria and the danger of botulism has been a major factor in formulating food processing techniques.

Clostridium botulinum is divided into eight toxigenic types on the basis of the neurotoxins that they produce. These types are known as A, B, C alpha, C beta, D, E, F, and G. Three of the types, A, B, and E, are responsible for better than 99 percent of the botulism cases in humans. Types C and D are important in animal botulism. Types D and F have caused only occasional outbreaks in man, while type G has not yet been associated with any outbreak of botulism.

All serotypes form spores and grow rapidly under anaerobic conditions in various kinds of organic matter. Types C and D can be found in decaying animals. The powerful toxins are formed during the growth stage and cause intoxication when ingested by man.

An important property of *Clostridium botulinum* is its ability to form spores that are highly resistant, not only to heat, but also to irradiation, drying, freezing, and exposure to many chemical agents. Spores of types A and B are the most resistant, requiring temperatures of over 200°F. (93°C) to destroy them. Type E spores are destroyed at 176°F. (80°C). A temperature of 250°F. (80°C) for 3 minutes will destroy all spores.

The temperature range for growth and toxin production by types A and B extends from 50°F. (10°C) to 122°F. (50°C). However, the minimum growth temperature for type E is much lower. Schmidt reported that growth and toxin formation of four type E strains occurred at 38°F. (3.3°C) in 31 days. Normal refrigeration would thus restrict growth and toxin production of types A and B but would be ineffective in preventing type E.

The presence of *Clostridium botulinum* types A and B can sometimes be recognized by the swelling produced in cans or by the spoiled taste of the product. Type E, however, does not produce obvious spoilage, particularly in the fish products in which it occurs most often.

In controlled experiments, no toxin formation has been detected in foods with pH lower than 4.6. The minimum pH seems to vary from 4.6 to 5.3, depending on the type of food. Toxin formation seems to be independent of pH within the growth range of the organism, but the toxin is more stable at low pH values.

Salt and sugar can have an inhibitory effect on the growth of *Clostridium botulinum*. Experiments indicate that 50 percent sugar or 10 percent salt completely inhibits the growth of spores of types A and B. Type E spores are considerably less salt tolerant than A and B and concentrations of 5 to 6 percent will prevent toxin formation.

Among the factors, then, that determine the growth and toxin production of *Clostridium botulinum* are temperature, pH, oxygen, the presence of growth factors such as essential amino acids and fatty acids, and the presence of inhibitory chemicals such as sugar and salt. Since the limits of tolerance for these factors are rather narrow, the actual occurrence of botulism in man is rare, even though the organism may be present in or on many foods.

THE TOXIN

The botulism toxins are the most poisonous substances known. Once in the bloodstream, the equivalent of one drop of botulinal toxin per 8.7 million gallons of water can be lethal to an adult. Less than .001 microgram causes the death of a mouse. It has been estimated that one gram of pure toxin will kill ten billion mice (220,000 tons). Fortunately, the toxin of all types is heat-sensitive.

Since these toxins act on the nervous system rather than on the digestive system, they are called neurotoxins (*neuro*—nerve). They are absorbed into the blood from the digestive system and are then carried to their site of action. The action is primarily on the autonomic nervous system, that part of the nervous system that regulates involuntary activities such as eye movements and breathing. The peripheral motor nerves (those nerves leading from the brain to the muscles) are primarily affected, with the major action being at the nerve-muscle junctions. Here, the toxin prevents the passage of nerve messages from the motor nerves to the muscle fibers, and the muscles involved totally or partially cease to function.

Most of the neurotoxins are synthesized by the cells as nontoxic prototoxins or pretoxins. These may then be activated into the toxic form by cell enzymes or proteins such as trypsin. Only a few strains of *Clostridium botulinum* synthesize a fully toxic molecule that needs no activation. Type A toxin causes most fatalities in the United States.

Five types of *Clostridium botulinum* toxin have been obtained in more or less purified form. Type A crystals are white with a molecular weight of about 900,000. They are readily denatured by heat, strong alkali, and many oxidizing agents. All of the purified types are composed entirely of amino acids and are therefore simple proteins. They are capable of dissociating into smaller toxic molecules in the digestive system.

FOODS

Human botulism is almost invariably caused by food that has been inadequately preserved, stored for some time, and then consumed cold or without sufficient heating. It may take something only slightly unusual to produce an outbreak of botulism, such as a defect in can-seaming operations or a change in packaging technology.

In general, the types of food that have caused botulism outbreaks are related to national, regional, or local eating habits. Many types have been involved over the years. Almost any food with pH above 4.5, which exists for a period of time under anaerobic conditions, can support growth and toxin formation.

Historically, most outbreaks have been caused by preserved vegetables. Home-canned string beans, corn, beans, spinach, and asparagus have accounted for many outbreaks in the United States and Australia. Various meat products have been implicated frequently in England, Belgium, and Germany. Fish have been the primary botulism source in the U.S.S.R. and Japan, with home-processed fish accounting for 57 percent of Russian outbreaks from 1958 to 1964.

The growth of types A and B *Clostridium botulinum* in low-acid foods having a bland taste will usually produce a musty or rancid odor and a bad taste that can be recognized and make the food unpalatable. The presence of the organism in low protein or highly spiced or fermented foods is more difficult to detect, however, as the tell-tale odor and taste may be masked by the odor of the food, itself. Spoilage odors are also more difficult to detect in cold dishes, and many botulism outbreaks result from food that is consumed cold.

During the period 1960 to 1969, the majority (59 percent) of outbreaks in the United States were caused by vegetables. Thirteen percent were traced to preserved fruits and 12 percent to fish or fish products.

Sixty-five percent of the type A outbreaks and 61 percent of the type B outbreaks were caused by preserved vegetables. Only one type E outbreak resulted from a vegetable (mushroom) during this time. Fifteen percent of the type A and 11 percent of the type B outbreaks were caused by preserved fruits. Fish or fish products were responsible for 4 percent of the type A outbreaks, 6 percent of the type B outbreaks, and 94 percent of the type E outbreaks.

Three of the 30 outbreaks from 1970 through 1973 were caused by commercially prepared foods. One (type A) was caused by a nationally distributed vichyssoise soup; the second (type B) involved a pepper product distributed over a three-state region; the third (type A) involved canned meatballs and spaghetti sauce.

Although only three of the reported outbreaks during this time span involved commercially prepared foods, there were nine occasions when *Clostridium botulinum* or its toxin was found in commercial food products. One of the products, once again, was a nationally distributed soup. The other eight were mushroom products.

Among the foods incriminated in recent years have been: (type A) peppers, potato salad, pickles, three-bean salad, vichyssoise, fish eggs, beef pot pie, carrots, eel, beaver tail, chicken pot pie, beets, green beans, salmon, beef stew, tomato juice, and corn; (type B) cabbage, green beans, salted salmon, corn, beets, figs, mushrooms, and dried whitefish; and (type E) whitefish, fish eggs, mullet, ciscoe, and seal.

The major factors that influence the production of botulinal toxin in food are:

1. The presence of spores in suitable numbers. The smaller the number of spores present, the shorter the heating time necessary to kill them.
2. Insufficient heating. Botulinal spores are highly heat-resistant. Temperatures of 250°F. (121°C) or higher are used by the canning industry.
3. pH. Most foods of pH 4.5 or above can support growth of *Clostridium botulinum*.
4. Anaerobic conditions. Lack of oxygen favors germination, growth, and toxin production of this organism.

THE RESERVOIR

Clostridium botulinum, like *Clostridium perfringens*, is commonly found in soil. Approximately 1 to 25 percent, or more, of thousands of soil samples taken in North America, Australia, Hawaii, China, Belgium, Sweden, Denmark, Holland, Switzerland, and Germany have yielded *C. botulinum*. A total of 10.5 percent of 4,000 soil samples in the Soviet Union yielded the organism. *C. botulinum* is also found in fresh and marine waters. Types C and D are found as saprophytes in decomposing animal carcasses.

In general, the organism is found in the northern hemisphere, mostly between 35° and 55° north latitude. In the United States, the types causing human botulism are also distributed on an east-west basis. Generally, type A is most prevalent from the Rocky Mountains (100th meridian) westward. In one soil survey of this area, 23 percent of the samples contained *Clostridium botulinum* and 84 percent of these were type A.

Type B predominates east of the 100th meridian. The same survey obtained *Clostridium botulinum* from 14.5 percent of the eastern samples and 81 percent of the isolates were type B. Soil from the Mississippi River Valley and the Great Plains, however, contained few cells.

Type E predominates in the Great Lakes, with Green Bay of Lake Michigan showing the highest incidence of contamination. Bott et al. (1966) reported that more than half of 72 fish taken from the southern part of the bay harbored this strain. Type E was also found in 59 percent of alewives examined. The organism has also been found in yellow perch, whitefish, lake trout, brown trout, carp, catfish, and suckers from the Great Lakes.

Type E is also common in the Pacific Northwest. The organism has been recovered in high numbers from sediments of Lake Washington. Every sediment sample taken along the Columbia River two hundred miles above the mouth contained type A or type E cells.

Many marine habitats have yielded *Clostridium botulinum*. Twenty-seven samples taken from shallow bottoms along the British Columbia coast yielded 7 (26 percent) type E strains. Forty-seven percent of soil samples from fish markets, canals, and harbors in Copenhagen yielded type E toxin. Another survey showed type E toxin present in every sample taken from the Baltic Sea, the Sound Sea, and the straits of Kattegat and Skagerrak. Marine sediments and shore samples from the coasts of Japan, British Columbia, Alaska, the Soviet Union, France, Denmark, and Greenland regularly yield type E *C. botulinum*. The waters surrounding Scandinavia, particularly the Baltic Sea, are heavily contaminated. The organism has also been found in sediments of the Gulf of Mexico off the United States coast. The locations and types of some *C. botulinum* isolations are:

Location	Type
Coastal waters north of Long Island	E
Gulf of Mexico	all types—63% type E
Pacific coast—northern U.S.A.	E
Pacific coast—southern U.S.A.	A, B, F
Sediments of coastal waters—Sweden	E
Baltic Sea	E
Great Britain—soil	A, B
Sweden—soil	A, B, E
Calcutta	A
South Africa	B
China	A, B
Japan	A, E

SUMMARY

Clostridium botulinum is an anaerobic, spore-forming, motile rod that occurs singly, in pairs, or in short chains, and is commonly found in soils throughout the world. It also occurs on vegetables, fruits, human and animal feces, and decomposing or decaying animal life.

The organism is divided into eight toxigenic types, based on the neurotoxin produced. The types are known as A, B, C alpha, C beta, D, E, F, and G. Three of these types A, B, and E, account for better than 99 percent of the human botulism cases. Types C and D are important causes of botulism in animals. All serotypes form spores and grow rapidly under anaerobic conditions in various kinds of organic matter. The spores are highly heat-resistant and temperatures of 250°F. (121°C) or higher are commonly utilized in killing them.

The temperature range for growth and toxin production by types A and B extends from 50°F. (10°C) to 122°F. (50°C). Type E can grow and produce toxin at 38°F. (3°C). In controlled experiments no toxin formation has been detected in foods with pH lower than 4.6. Fifty percent sugar and 10 percent salt have inhibitory effects on the growth of spores.

The botulism toxins are the most poisonous substances known. They act on the nervous system and hence are called neurotoxins. They affect the autonomic nervous system at the neuro-muscular junction of the peripheral motor nerves. Death results from paralysis of the breathing muscles. The toxins are simple proteins.

Human botulism is almost invariably caused by food that has been inadequately preserved, stored for some time, and then consumed cold or without sufficient heating. Vegetables have been the most commonly implicated food in type A and type B outbreaks. Almost all of the type E outbreaks have been traced to fish or fish products.

Clostridium botulinum is commonly found in soils and waters, both fresh and marine, in northern latitude countries all over the world. In the United States, type A predominates west of the 100th meridian, type B predominates east of the 100th meridian, and type E predominates in the Great Lakes.

Review Questions

1. a. How does the death rate from botulism compare with the death rate from staphylococcal, salmonella, and *Clostridium perfringens* food poisoning? b. How does the incidence of botulism compare with that of the other three major kinds of food poisoning?

2. How does the occurrence of strains A, B, and E vary in the United States, Europe, and Japan?

3. a. What type of food has been most commonly implicated in types A and B outbreaks? b. What type of food has been most commonly implicated in type E outbreaks?

4. Why is type E botulism the prevalent type in northern latitude countries?

5. How do types A and B differ in regional distribution in the United States?

6. Given the four major foodborne illnesses—staphylococcal, salmonellosis, *Clostridium perfringens,* and botulism—which type(s) are intoxications and which are infections?

7. a. What are the typical symptoms of botulism? b. How do these symptoms differ from those of the other three foodborne illnesses that have been discussed?

8. How does the incubation period relate to the seriousness of the disease?

9. What are the symptoms of wound botulism?

10. a. What danger is involved in livestock eating the carcasses of dead animals? b. How do fowl occasionally contract botulism?

11. What are the characteristics of *Clostridium botulinum?*

12. a. What are the *Clostridium botulinum* types? b. Which types cause human illness? c. What types are important in animal botulism?

13. Why are *Clostridium botulinum* spores difficult to destroy?

14. How does the temperature range for types A and B differ from that of type E?

15. How can the presence of *Clostridium botulinum* types A and B sometimes be recognized in canned goods?

16. What two common preservatives have an inhibitory effect on *C. botulinum?*

17. a. Why is botulinal toxin called a neurotoxin? b. How does this toxin act on the nervous system?

18. What conditions in food can lead to extensive growth of *C. botulinum?*

19. What major factors influence the production of botulinal toxin in food?

20. What is the natural habitat of *C. botulinum*?

References

Angelotti, R. (1970). "The Heat Resistance of Clostridium botulinum Type E in Food." In *Proceedings First U.S.-Japan Conference on Toxic Microorganisms*, ed. M. Herzeberg. Pp. 404–409. U.S. Government Printing Office. Washington, D.C.

Baltzer, J. and D. C. Wilson. (1965). "The Occurrence of Clostridia on Bacon Slaughter Lines," *J. Appl. Bacteriol.* 28:119–124.

Blandford, T. B. and T. A. Roberts. (1970). "An Outbreak of Botulism in Broiler Chickens." *Vet. Rec.* 87:258–261.

Bott, T. L., J. S. Deffner, and E. M. Foster. (1967). "Occurrence of C. Botulinum Type E in Fish from the Great Lakes, with Special Reference to Certain Large Bays." In *Botulism 1966: Proceedings 5th International Symposium Food Microbiology*, ed. M.

Ingram and T. H. Roberts. Moscow. Pp 25–33. Chapman and Hall, Ltd. London, England.

Bott, T. L., J. S. Deffner, E. M. Foster, and E. McCoy. (1964). "Ecology of C. Botulinum in the Great Lakes." In *Botulism: Proceedings of a Symposium,* ed. K. H. Lewis and K. Cassal, Jr. U.S. Public Health Service Publication 999-FP-1. Cincinnati, Ohio.

Bott, T. L., J. S. Deffner, E. McCoy, and E. M. Foster. (1966). "Clostridium botulinum type E in Fish from the Great Lakes." *J. Bacteriol.* 91:919–923.

Cann, D. C., B. B. Wilson, and G. Hobbs. (1968). "Incidence of Clostridium botulinum in Bottom Deposits in British Coastal Waters." *J. Appl. Bacteriol.* 31:511–514.

Crisley, F. D., J. T. Peeler, R. Angelotti, and H. E. Hall. (1968). "Thermal Resistance of Spores of Five Strains of Clostridium botulinum Type E in Ground Whitefish Chubs." *J. Food Sci.* 33:411–416.

Dolman, C. E. and H. Iida. (1963). "Type E Botulism: Its Epidemiology, Prevention and Specific Treatment." *Can. J. Pub. Health.* 54: 293–308.

Donadio, J. A., E. J. Gangarosa, and G. A. Faich. (1971). "Diagnosis and Treatment of Botulism." *J. Infect. Dis.* 124:108–112.

Dubovsky, B. J. and K. F. Meyer. (1922). "The Distribution of the Spores of C. Botulinum in the Territory of Alaska and the Dominion of Canada." *J. Infect. Dis.* 31:595–599.

Eklund, M. W. and F. Poysky. (1966). "Incidence of Clostridium botulinum Type E from the Pacific Coast of the U.S." In *Botulism 1966: Proceedings 5th International Symposium Food Microbiology,* ed. M. Ingram and T. H. Roberts. Moscow. Chapman and Hall, Ltd. London, England.

Eklund, M. W., and F. T. Poysky. (1970). "Distribution of Clostridium botulinum on the Pacific Coast of the United States." In: M. Herzberg (Ed.). Proceedings First U.S.-Japan Conference Toxic Microorganisms. Pp. 304–308. U.S. Government Printing Office. Washington, D.C.

Eklund, M. W., D. I. Weiler, and F. T. Poysky. (1967). "Outgrowth and Toxin Production of Nonproteolytic Type B Clostridium botulinum at 3.3° to 5.6°C." *J. Bacteriol.* 93:1461–1462.

Foster, E. M., J. S. Deffner, T. L. Bott, and E. McCoy. (1965). "Clostridium botulinum Food Poisoning," *J. Milk Food Technol.* 28:86–91.

Gangarosa, E. J., J. A. Donadio, R. W. Armstrong, K. F. Meyer, P. S. Brachman, and V. R. Dowell (1971). "Botulism in the United States, 1899–1969." *American J. Epidemiol.* 93:93–101.

Graikoski, J. T., E. W. Bowman, R. A. Robohm, and R. A. Koch. (1970). "Distribution of Clostridium botulinum in the Ecosystem of the Great Lakes." In *Proceedings First U.S.-Japan Conference Toxic Microorganisms.* ed. M. Herzberg. Pp. 271–277. U.S. Government Printing Office. Washington, D.C.

Greenberg, J. (1978). "Infant Botulism." *Sci. News* 113:297, 302.

Gunnison, J. B., J. R. Cummings, and K. F. Meyer. (1936). "Clostridium botulinum Type E." *Proc. Soc. Expl. Med.* 35:278–280.

Johannsen, A. (1963). "Clostridium botulinum in Sweden and the Adjacent Waters." *J. Appl. Bacteriol.* 26:43–47.

Johannsen, A. (1965). "Clostridium botulinum Type E in Foods and the Environment Generally." *J. Appl. Bacteriol.* 28:90–94.

Kautter, D. A. (1964). "Clostridium botulinum Type E in Smoked Fish." *J. Food Sci.* 29:843–849.

Koenig, M. G., D. J. Drutz, A. I Mushlin, W. Schaffner, and D. E. Rogers. (1967). "Type B Botulism in Man." *American J. Med.* 42:208–219.

Merson, M. H. and V. R. Dowell, Jr. (1973). "Epidemiologic, Clinical, and Laboratory Aspects of Wound Botulism." *New England J. Med.* 289:1005–1010.

Merson, M. H., J. M. Hughes, V. R. Dowell, A. Taylor, W. H. Barker, and E. J. Gangarosa. (1974). "Current Trends in Botulism in the United States." *J. American Med. Assoc.* 229:1305–1308.

Meyer, K. F. (1956). "The Status of Botulism as a World Health Problem." *Bull. Wld. Health Org.* 15:281–298.

Meyer K. F. and B. Eddie. (1965). *Sixty-five Years of Human Botulism in the United States and Canada*. University of California Printing Department. San Francisco, California.

Osheroff, B. J., G. G. Slocum, and W. M. Decker. (1964). "Status of Botulism in the United States." *Public Health Rep.* 79:871–878.

Pace, P. J. and E. R. Krumbiegel. (1973). "Clostridium botulinum and Smoked Fish Production: 1963–1972." *J. Milk Food Technol.* 36:42–49.

Pivnick, H. and H. Barnett. (1965). "Effect of Salt and Temperature on Toxinogensis by Clostridium Botulinum in Perishable Cooked Meats Vaccum-Packed in Air-Impermeable Plastic Pouches." *Food Technol.* 19:140–147.

Riemann, H. ed. (1969). *Food-borne Infections and Intoxications*. Academic Press, Inc. New York.

Roberts, T. A., and M. Ingram. (1965). "The Resistance of Spores of Clostridium botulinum Type E to Heat and Radiation." *J. Appl. Bacteriol.* 28:125–141.

Rogers, D. E. (1963). "Botulism, Vintage 1963." *Ann. Internal. Med.* 61:581–588.

Ryan, D. W. and M. Cherington. (1971). "Human Type A Botulism." *J. American Med. Assoc.* 216:513–514.

Schmidt, C. F., R. V. Lechowich, and J. E. Folinazzo. (1961). "Growth and Toxin Production by Type E Clostridium Botulinum Below 40°F." *J. Food Sci.* 26:626–630.

Smith, L. DS. (1975). *The Pathogenic Anaerobic Bacteria*, 2nd ed. Charles C Thomas, Pub. Springfield, Illinois.

Sugiyami, H., T. L. Bott, and E. M. Foster. (1970). "Clostridium botulinum Type E in an Inland Bay." In *Proceedings First U.S.-Japan Conference on Toxic Microorganisms*, ed. M. Herzberg. Pp. 287–291. U.S. Government Printing Office. Washington, D.C.

Thatcher, F. S. (1966). "Food-Borne Bacterial Toxins." *Can. Med. Assoc. J.* 94:582–590.

U.S. Department of Health, Education and Welfare. (1976). *Foodborne and Waterborne Disease Outbreaks Annual Summary. 1974*. Center for Disease Control. Atlanta, Georgia.

U.S. Department of Health, Education and Welfare. (1976). *Foodborne and Waterborne Disease Outbreaks Annual Summary. 1975*. Center for Disease Control. Atlanta, Georgia.

U.S. Department of Health, Education and Welfare. (1977). *Foodborne and Waterborne Disease Outbreaks Annual Summary. 1976*. Center for Disease Control. Atlanta, Georgia.

U.S. Department of Health, Education and Welfare. Communicable Disease Center. (1973). *Morbidity, Mortality Weekly Reports* 12:336–340.

U.S. Department of Health, Education and Welfare. Communicable Disease Center. (1971). *Morbidity, Mortality Weekly Reports* 20:242.

U.S. Department of Health, Education and Welfare. Communicable Disease Center. (1973). *Morbidity, Mortality Weekly Reports* 22:57–58.

U.S. Department of Health, Education and Welfare. Communicable Disease Center. (1978). *Morbidity, Mortality Weekly Reports* 23:2.

U.S. Department of Health, Education and Welfare. Communicable Disease Center. (1974). *Morbidity, Mortality Weekly Reports* 23:12.

U.S. Department of Health, Education and Welfare. Communicable Disease Center. (1977). *Morbidity, Mortality Weekly Reports* 26:117.

U.S. Department of Health, Education and Welfare. Communicable Disease Center. (1978). *Morbidity, Mortality Weekly Reports* 27:16.

Zacks, S. I. and M. F. Sheff. (1970). "Studies on Botulinus Toxin." In *Proceedings First U.S.-Japan Conference on Toxic Microorganisms*, ed. M. Herzberg. Pp. 348–350. U.S. Government Printing Office. Washington, D.C.

8 Other Foodborne Illnesses

During the three-year period 1974 to 1976, a total of 20,096 cases of foodborne illness of known origin were reported to the Communicable Disease Center. These cases represented 524 outbreaks, about 38 percent of the total outbreaks reported (1,391 outbreaks—46,212 cases). In the remaining outbreaks (62 percent), the agent causing the foodborne illness could not be or was not determined.

Of the 524 outbreaks and 20,096 cases of known origin, 310 outbreaks (59 percent) and 16,685 cases (83 percent) were attributed to bacteria discussed in the previous four chapters (see Table 8-1). Although these bacteria account for the majority of the cases, they do not, by any means, account for all. Other agents, both biological and chemical, have been implicated in food-poisoning outbreaks. Many of these, while infrequent, can be serious.

INFECTIOUS HEPATITIS

Infectious hepatitis is the most common of the virus-caused foodborne illnesses. During the period 1974 to 1976, 11 food-associated outbreaks of this disease, involving 492 persons, were reported to the Communicable Disease Center. Ten states reported outbreaks (Table 8-2). Eight of the 11 outbreaks involved foodservice establishments. In one other case, a food handler in a foodservice establishment tested positive for hepatitis virus but nothing additional could be shown. In seven of the out-

Table 8-1
Summary of outbreaks and cases involving *Staphylococcus aureus, Clostridium botulinum, Clostridium perfringens,* and *Salmonella,* 1974–1976. CDC. Foodborne and Waterborne Disease Outbreaks. Annual Summaries. 1974–1976.

	Year	Outbreaks	Cases
Staphylococcus aureus	1974	43	1565
	1975	45	4067
	1976	26	930
Total		114	6592
Salmonella	1974	35	5499
	1975	38	1573
	1976	38	1169
Total		111	8241
Clostridium botulinum	1974	21	32
	1975	14	19
	1976	23	40
Total		58	91
Clostridium perfringens	1974	15	863
	1975	16	419
	1976	6	509
Total		37	1791

Table 8-2
Foodborne infectious hepatitis outbreaks, 1974–1976. CDC. Foodborne and Waterborne Disease Outbreaks. Annual Summaries. 1974–1976.

State	Cases	Food	Place where Food was Mishandled
		1974	
California	132	Salad	Foodservice establishment
Illinois	12	Unknown	Unknown but food handler pos.
Michigan	15	Ham	Home
Minnesota	107	Cold sandwiches	Foodservice establishment
Ohio	9	Ice cream	Unknown
Ohio	7	Unknown	Foodservice establishment
		1975	
Oklahoma	116	Sandwiches, salads	Foodservice establishment
New York	34	Glazed donuts	Foodservice establishment
Oregon	23	Sandwiches	Foodservice establishment
		1976	
Georgia	26	Unknown	Foodservice establishment
Pennsylvania	11	Unknown	Foodservice establishment

breaks, poor personal hygiene on the part of food-handling persons was a factor in the outbreak.

The disease was first reported waterborne in 1895 when 34 persons in an English village contracted the illness from drinking water obtained from a stream. In subsequent years, waterborne infectious hepatitis struck the United States, German, Canadian, and British armies, villages and cities, schools, summer camps, and a college, sanitarium, resort hotel, and bowling alley.

One of the largest outbreaks of infectious hepatitis ever recorded occurred in New Delhi, India, in 1955–56, when more than 50,000 cases of the disease resulted from contaminated drinking water. The water supposedly had been chlorinated.

Foods responsible for hepatitis outbreaks include salads, cold cuts, ice cream, glazed doughnuts, milk and dairy products, shellfish, potato salad, roast pork, orange juice, and frozen strawberries. Although the hepatitis virus cannot withstand thorough heating, cooked foods can be infected by a food handler after cooking. Viruses excreted from the body in the stool may contaminate the hands. Hands can then contaminate foods directly or indirectly by contaminating utensils or equipment. Food handlers working while ill with the disease have been implicated as the source in several outbreaks. Food must therefore be regarded as an important link in the transmission of this disease.

In the summer of 1962, an epidemic of infectious hepatitis occurred at a hospital in St. Louis. There were 24 cases in the outbreak, primarily affecting hospital employees. No cases of clinical or subclinical hepatitis could be discovered in the staff.

The investigation determined from several factors that the most likely date of exposure was July 1. On that date, 20 of the 24 persons affected were in the hospital at the same time. It seemed likely that the outbreak was foodborne. All 24 victims had eaten in the employee cafeteria on July 1 or July 2. More of them ate in the cafeteria at breakfast than at any other meal. Of the 20 patients who were in the hospital July 1, 19 ate breakfast in the cafeteria. The four who were not in the hospital July 1 ate breakfast in the cafeteria July 2. Orange juice was singled out as the most likely vehicle of transmission since all of the victims consumed this drink.

The kitchen worker who prepared the orange juice had not been ill at the time, but her husband had the symptoms of a mild case of infectious hepatitis. Most likely, the kitchen worker was the source of the virus and contaminated the orange juice during preparation.

In 1956, the first epidemic of infectious hepatitis caused by the ingestion of raw shellfish was reported in Sweden. In 1961, two outbreaks of shellfish hepatitis were reported in the United States. The first outbreak was traced to raw oysters and resulted in 84 cases, mostly in Alabama and Mississippi. The second outbreak, which resulted from the ingestion of raw clams, consisted of 459 cases in New Jersey and an undetermined number in other states.

In the first outbreak, the source of the oysters was Pascagoula Bay at the mouth of the Pascagoula River in Mississippi. The city of Pascagoula is on the river, just above the bay. During the 1950s, the city nearly doubled in size, taxing its sewage treatment plant. Raw sewage began to be discharged into the river. In 1960, sewage treatment was totally stopped to allow for construction of a larger plant. A shipyard at the mouth of the river employing 2,500 workers was also feeding raw sewage into the river.

During the last three months of 1960, infectious hepatitis cases were reported from the area of raw sewage discharges into the Pascagoula River. This virus-contaminated sewage was undoubtedly the source of the viruses that contaminated the oyster bed in Pascagoula Bay. All of the cases were related to the ingestion of oysters from this bed. The oysters were eaten either raw or minimally cooked.

This outbreak stresses the importance of continuous monitoring of waters surrounding shellfish beds and the prevention of shellfish harvesting in contaminated waters.

A six-month outbreak, which occurred in Connecticut in 1963–64, resulted in 123 cases of infectious hepatitis. The source of the outbreak was thought to be clams taken from closed areas by one or more Rhode Island harvestors, who then sold a portion of their catch to primary shippers in Connecticut. An outbreak in Rhode Island occurred concurrently with the Connecticut outbreak.

In another outbreak, cold cuts were contaminated by a food handler in a delicatessen. About 2½ weeks before the start of the outbreak, a sewer backup occurred at the delicatessen. One of the food handlers who cleaned the sewage-contaminated floors apparently contaminated his hands or clothing and then dispensed cold cuts, contaminating them also. More than 80 people who purchased these cold cuts contracted hepatitis. In another outbreak involving a delicatessen, the owner continued to work and handle food while ill, with the result that 66 customers also became ill from contaminated salads and cold cuts.

SHIGELLOSIS

Shigellosis is an illness, sometimes food- or waterborne, that is characterized by fever, nausea, diarrhea, and abdominal pain. Death rarely results. The incubation period is usually less than four days and frequently requires only 7 to 36 hours. The illness is caused by gram-negative, nonmotile, rod-shaped bacteria of the genus *Shigella*. It is also known as bacillary dysentery.

During the period 1974 to 1976, twelve food-associated outbreaks of this disease, involving 398 persons and one death, were reported to the Communicable Disease Center. Nine states reported outbreaks (Table 8-3). Eight of the 12 outbreaks involved foodservice establishments. Poor personal hygiene on the part of food handlers was a contributing factor in many cases.

Although the shigellae accounted for only 2.2 percent of the outbreaks during 1974 to 1976, they accounted for 4.4 percent of the cases of known origin, having an average of almost 75 cases per outbreak. In 1973, eight foodborne outbreaks of shigellosis resulted in 1,388 cases, an average of 174 cases per outbreak.

Donadio reported on a five-year period during which 21 foodborne or waterborne outbreaks were reported to the Communicable Disease Center. In each of the outbreaks involving food, the source of contamination was a human carrier who infected it through poor personal hygiene. Among the incriminated foods were potato salad, shrimp salad, and chicken salad, foods that are usually mixed by hand and that require extensive handling.

Inadequate hand washing after using the rest room can leave the hands contaminated with fecal residue. Microorganisms present can be transferred to food or utensils. If the food is not promptly and properly refrigerated, extensive bacterial growth can occur. Such foods as potato salad, tuna salad, shrimp salad, and turkey salad have been implicated on many occasions in foodborne shigellosis outbreaks.

Almost all foodborne shigellosis outbreaks can be traced to this type of pattern. Investigations of these outbreaks usually reveal that one

Table 8-3
Foodborne shigellosis outbreaks, 1974–1976. CDC. Foodborne and Waterborne Disease outbreaks. Annual Summaries. 1974–1976.

State	Cases	Food	Place where Food was Mishandled
		1974	
Washington	100	Unknown	Unknown
Iowa	15	Potato salad	Unknown
Arizona	7	Unknown	Unknown
		1975	
Oregon	150	Unknown	Unknown
Montana	144	Unknown	Foodservice establishment
Texas	119	Unknown	Foodservice establishment
		1976	
Texas	176	Spaghetti	Foodservice establishment
California	46	Unknown	Foodservice establishment
Washington	21	Tossed salad	Home
Washington	13	Polynesian food	Foodservice establishment
Colorado	12	Fruit compote	Foodservice establishment
Massachusetts	5	Chopped liver	Foodservice establishment

or more food handlers had diarrheal illness about the time of the outbreak. *Shigella flexneri* and *Shigella sonnei* cause 90 percent of the foodborne shigellosis outbreaks in the United States.

Watt and Lindsay, and Lindsay et al. have shown the importance of the housefly as a vector in the spread of shigellosis. Their studies revealed that in areas of high and moderate incidence of shigellosis, a decrease in the fly population has led to a decrease in shigellosis.

BACILLUS CEREUS

Bacillus cereus food poisoning is an illness characterized by diarrhea, abdominal pain, and nausea. Vomiting and fever occur infrequently. The symptoms closely resemble those of salmonellosis, and some outbreaks attributed in the past to salmonellae have probably been caused by *B. cereus*. The incubation period for the illness averages about 8 to 16 hours.

The organism is a large, gram-positive, rod-shaped, motile bacterium that forms spores. The spores are fairly resistant to heat. It is aerobic but capable of growing under anaerobic conditions. *Bacillus cereus* grows at temperatures between 50°F. (10°C) and 122°F. (50°C), with the optimum temperature around 86°F. (30°C). It grows best within a pH range of 4.9 to 9.3.

These bacteria are widely distributed in nature. They are common soil saprophytes and are also found in air, water, milk, and dust.

The first well-documented food-poisoning outbreak caused by *B. cereus* was reported by Hauge. Four outbreaks caused by this organism, involving 600 persons, occurred in a hospital in Oslo, Norway, during 1947, 1948, and 1949.

A 1948 outbreak, typical of the series, resulted in 80 of 99 persons who ate a Sunday dinner becoming ill. The dessert, which was chocolate pudding and vanilla sauce, was determined to be the vehicle of transmission. Both items of the dessert had been prepared on the morning of the previous day and stored in a large container at room temperature.

Laboratory examination revealed the presence of spores of *Bacillus cereus* in the cornstarch, one ingredient in the vanilla sauce. Vanilla sauce was the vehicle in all four of the outbreaks that occurred. In each outbreak, the sauce was prepared in the same manner.

Once again, we see the danger of placing warm food in large containers in which they cannot cool rapidly. In this case, spores of *Bacillus cereus* survived the heat used to cook the sauce, and germinated and multiplied profusely in the warm sauce for about 24 hours. The resulting bacterial population (110 million per gram) was large enough to cause foodborne illness.

The first well-documented *Bacillus cereus* foodborne outbreak in the United States was described by Midura et al. in 1970. The illness struck 15 college students in a fraternity house where they had eaten a dinner that included meat loaf. Laboratory examination of the remaining meat loaf resulted in isolations of up to 70 million *B. cereus* per gram.

During the period 1974 to 1976, six foodborne outbreaks caused by *B. cereus* were reported to the Communicable Disease Center (Table 8-4). Four outbreaks involved foodservice establishments. Improper holding temperatures were implicated in all cases.

Kim and Goepfert investigated 170 samples of dried food products. The samples were purchased from retail stores in the Madison, Wisconsin area. Seasoning mixes (55 percent) were contaminated most frequently, followed by spices (40 percent), dry potatoes (40 percent), and milk powder (37.5 percent). Soup mixes, gravy mixes, and flour and starch samples were also contaminated, but far less frequently.

Table 8-4
Bacillus cereus Foodborne Outbreaks, 1974–1976. CDC. Foodborne and Waterborne Disease Outbreaks. Annual Summaries. 1974–1976.

State	Cases	Food	Place where Food was Mishandled
		1974	
California	11	Fried rice	Foodservice establishment
		1975	
Wisconsin	25	Beef	Unknown
California	18	Fried rice	Foodservice establishment
Wisconsin	2	Mashed potatoes	Home
		1976	
Wisconsin	55	Chicken stew	Foodservice establishment
New York	8	Pork-fried rice	Foodservice establishment

VIBRIO PARAHAEMOLYTICUS

Vibrio parahaemolyticus is a gram-negative, curved, rod-shaped bacterium that occurs primarily in marine coastal waters and sediments. It thrives in a medium containing 2 to 4 percent salt and does not occur in fresh water. It is primarily anaerobic but can exist where oxygen is present. The organism grows best in an alkaline environment and multiplies rapidly at 99°F. (37°C), having a generation time as short as 10 to 12 minutes. It may survive in seafood cooked at temperatures of up to 176°F. (80°C). *V. parahaemolyticus* is the major cause of bacterial foodborne illness in Japan, causing about 14,000 to 20,000 cases per year or approximately 40 to 70 percent of the yearly food-poisoning cases.

The first recorded food-poisoning outbreak positively attributed to this organism occurred in Osaka, Japan, on October 21–22, 1950. A total of 272 persons became ill, 22 of whom died, after eating *shirasuboshi* (boiled sardine larvae). Other *Vibrio parahaemolyticus* outbreaks soon followed.

Most of the food-poisoning outbreaks in Japan that are caused by this organism result from eating raw seafoods, and usually occur in coastal areas. Symptoms appear about 14 to 20 hours after eating contaminated food. Severe abdominal pains, cramps, nausea, vomiting, and diarrhea mark the usual course of the disease. Fever, headache, and chills occur less frequently. Recovery usually occurs within a day or two.

Vibrio parahaemolyticus has been implicated in pathogenic conditions other than food poisoning. Occasionally, it has infected cuts, sores, and bruises of swimmers and clam diggers in marine waters. In one such case, the loss of a leg resulted. The organism has also been responsible for deaths of crabs, shrimp, and oysters.

Vibrio parahaemolyticus has been recovered from marine waters, particularly coastal areas, and from sediments in many parts of the world. It has been recovered from marine waters, sediments, fish, shellfish, and crabs from Germany, Korea, Hong Kong, Taiwan, Singapore, the Philippines, the Netherlands, Hawaii, and the United States.

There was no evidence that the organism would be a problem in the United States until 1968, when Baross and Liston reported the isolation of *Vibrio parahaemolyticus* from the waters and sediments and from fish, shellfish, and crustaceans of Puget Sound, Washington. Since then, it has been found in samples taken from almost all major bodies of water surrounding the United States.

Although *Vibrio parahaemolyticus* had been isolated in several unconfirmed foodborne disease outbreaks in the United States between 1967 and 1970, the first confirmed outbreak occurred in 1971.

On August 14, 1971, approximately 550 persons attended a picnic in Bainbridge, Maryland, at which steamed crabs were served. Some hours later, about 320 of them were sick with severe abdominal pains, diarrhea, nausea, vomiting, headache, fever, and chills. Two persons required hospitalization. Some of the persons attending the picnic brought clams home for friends and relatives, with the result that 43 other known cases occurred, six of which required hospitalization.

The crabs served at the picnic had been harvested on August 11 and 12 from Chesapeake Bay. They were steamed by the wholesaler, placed in bushel baskets, and transported to the retailer in an unrefrigerated truck, which also contained live crabs. The steamed and live crabs were kept in a walk-in cooler (50°–55°F.) [10–13°C] through August 13. On the morning of August 14, 80 of the 85 bushels of steamed crabs were taken to the picnic, where they were kept without refrigeration and dispensed throughout the day.

Three of the five remaining bushels of steamed crabs were purchased for another picnic, also held on August 14. Fifteen of the 21 persons attending this picnic also became sick. The other bushels were sold by the retailer to the public and five reported cases of food poisoning resulted from these.

By 1972, thirteen outbreaks of *V. parahaemolyticus* had been reported from eight states. Six of the outbreaks occurred in Maryland; the others in Massachusetts, Louisiana, New Jersey, Washington, Texas, Hawaii, and Florida (all coastal states). Eight of the outbreaks were caused by contaminated crabs. Lack of refrigeration and failure to cook food at adequate temperatures were contributing factors in most cases. In many instances, cooked seafood was contaminated after cooking by placing it in containers that had held raw seafood.

Two large outbreaks of *V. parahaemolyticus* foodborne illness occurred in 1975 on cruise ships sailing between Florida and the Caribbean Sea. The first outbreak downed 252 (36 percent) of 703 passengers and resulted from using seawater containing *V. parahaemolyticus* to wash the galley decks. Splash from the wash water fell on food preparation surfaces which, in turn, contaminated shrimp that were prepared on them. The shrimp were responsible for the outbreak.

The second outbreak struck 445 (61 percent) of 734 passengers on another pleasure ship. In this outbreak, precooked, frozen lobsters were thawed at room temperature for about 8 hours. While thawing, they were periodically washed with seawater from the ship's fire system. The seawater probably contained *Vibrio parahaemolyticus.*

THE ARIZONA GROUP

The Arizona group comprises a number of motile, gram-negative, rod-type bacteria that occasionally cause foodborne illness in man. They are closely related to the salmonella, and their infection is every bit as severe as that group. The illness is characterized by fever, vomiting, diarrhea, and sometimes prostration.

The first culture was isolated in 1939 from a disease of gila monsters and chuckwallas. The organisms are widely distributed in reptiles, fowls, and lower mammals, causing a specific disease in each group.

The organism was isolated in 1944 from a woman afflicted with fever, vomiting, and diarrhea, and from the feces of an 11-month-old baby with acute colitis. The first reported large-scale outbreak involving this organism occurred in 1945 in Australia, when 26 members of a hospital staff were taken ill. In 1946, an outbreak in Washington, D.C., felled 51 student nurses. Many of those affected required hospitalization, some for more than a week. Custard was the food involved.

During the period 1974 to 1976, no outbreaks due to these organisms were reported to the Communicable Disease Center.

CHEMICAL FOODBORNE ILLNESS

Chemical foodborne intoxication results from the ingestion of a number of toxic chemicals. The chemicals may be man-made (insecticides, rodenticides, or food preservatives) or produced by plants such as certain mushrooms. They may be produced by animals, such as fish, or they may be the salts of such heavy metals as zinc, copper, or tin.

During the period 1974 to 1976, 524 foodborne outbreaks of known origin were reported to the Communicable Disease Center. Of these outbreaks, a total of 128 (24 percent) were caused by chemicals. Five hundred eighty-six cases resulted from the outbreaks. The majority of these were caused by fish toxins (70 outbreaks), heavy metals (14 outbreaks), and mushroom poisons (12 outbreaks).

Toxins from Fish

More than 520 species of fish have been incriminated in fish poisonings. Ciguatoxin, a poison of small molecular weight that is not destroyed by heat or gastric juice, is occasionally produced by certain fish, such as grouper, red snapper, kingfish, amberjack, po'ou fish, sea bass, barracuda, parrot fish, and sturgeon. More than 300 species of fish have been involved in ciguatoxin poisoning. The toxin is probably associated with the ingestion of certain blue green algae by the fish.

United States servicemen stationed in the Pacific were the victims of ciguatoxin poisoning on many occasions during World War II. The illness has a 2 to 3 percent fatality rate and recovery sometimes takes weeks to months.

In July of 1972, 24 of the 25 crew members of a Brazilian cargo ship were struck by cigautoxin poisoning. The illness resulted from consumption of barracuda. Within 30 minutes to 6 hours after eating a lunch of barracuda stew, symptoms of ciguatoxin poisoning began to appear. Abdominal pain, vomiting, diarrhea, dizziness, and severe prostration characterized the outbreak. Several of the victims experienced temporary blindness or blurred vision. The day following the outbreak, the ship reached Mobile, Alabama, where 20 of the crew members required hospitalization.

Ciguatoxin was responsible for 51 (73 percent) of the 70 outbreaks involving fish toxin during the period 1974 to 1976. Scombrotoxin, produced in tuna, skipjack, and Spanish mackeral, accounted for 18 (26 percent) of the remaining outbreaks.

Heavy Metals

Many metals required by the human body in trace amounts are highly toxic when ingested in large amounts. Copper, zinc, tin, cadmium, and iron are the most frequent offenders. Highly acidic food and drink can react with these metals and become contaminated. The resulting intoxication can produce severe gastrointestinal symptoms.

During the period 1974 to 1976, a total of 14 metal-poisoning outbreaks due to food or drink were reported to the Communicable Disease Center. One hundred thirty-three cases resulted from the outbreaks, most of which were due to copper contamination of carbonated soft drinks.

1. Copper. A total of 11 (78 percent) of the 14 metal-poisoning outbreaks during the period 1974 to 1976 were due to copper. This metal readily reacts with various acids, sometimes present in food and drink, to produce copper salts. When these copper salts are ingested in large quantities, the resultant intestinal reaction is normally rapid and severe. Vomiting is usually so quick, however, that little copper is absorbed into the system.

Most of the copper-poisoning cases that occurred during the period 1974 to 1976 were due to the consumption of soft drinks that had become contaminated from carbonated water reacting with copper tubing. This reaction can place large quantities of copper salts in the carbonated water that is mixed with the soft drink syrup. Copper intoxication can then result.

2. Zinc. Zinc intoxication can result from reactions between acidic food and galvanized (zinc-lined) containers. On one such occasion, during an India Independence Day Celebration in California, around 300 to 350 persons became ill with severe diarrhea and abdominal cramping. The symptoms appeared from 3 to 10 hours after eating. A total of nine persons required hospitalization.

The food had been prepared the day before the picnic. Some had been placed in galvanized tubs for holding until the afternoon of the picnic. The food was then dispensed to about 400 people.

During the course of the investigation it was learned that two other outbreaks, three and five years previous, had occurred at India Independence Day celebrations conducted by the same group of people. In each case, the vessels that were used to cook and hold the food had been borrowed from a Moslem temple in another city in California. A similar outbreak had also occurred at this temple in the past.

Examination of the containers that were used at all of these events showed that two galvanized tubs and a galvanized bucket were releasing large quantities of zinc into acidic foods. These vessels were undoubtedly responsible for every one of the outbreaks. They had remained in use because none of the past outbreaks had been reported to the public health authorities.

On another occasion, at the opening of a gift shop, a fruit punch was served that had been held in galvanized containers for 55 hours after preparation. The punch contained a mixture of orange, lemon, and lime juices. Approximately 90 persons drank the punch and almost all reported nausea and vomiting, some within 20 minutes. The acute symptoms were hot taste and dryness in the mouth, nausea, vomiting, and diarrhea.

3. Tin. High levels of tin in food can cause gastrointestinal upset characterized by bloating, nausea, vomiting, abdominal cramps, and diarrhea. Tin intoxication results from detinning of the inner surface of a container, usually a can, that holds the food.

A widespread outbreak of tin intoxication was reported in Washington and Oregon in 1972. A total of 32 separate outbreaks involving 113 cases resulted from commercially canned tomato juice produced by a company in central Washington. Two of the outbreaks occurred at banquets.

Investigation of the company's production procedures revealed that one farmer had delivered tomatoes that had been raised in excessively nitrated soil. It appeared likely that nitrates in the tomatoes caused the detinning. Other canned products involved in tin intoxication have included cherries, salmon, rhubarb, fruit salad, asparagus, herring, and apricots.

Another incident involved the members and guests of a ladies bowling league who attended a banquet. Thirty-one of the 38 ladies became ill after drinking a fruit punch made of pineapple and grapefruit juices. The punch had been held in a detinned five-gallon milk can that showed obvious evidence of corrosion.

SUMMARY

Infectious hepatitis is the most common of the viral foodborne illnesses. Viruses excreted from the body in the stool can contaminate the hands. Hands then contaminate foods directly or indirectly by contaminating utensils or equipment. Infected shellfish taken from contaminated waters have frequently been implicated in transmission of the disease.

Shigellosis is an illness, sometimes foodborne or waterborne, that is characterized by fever, nausea, diarrhea, and abdominal pain. The illness is caused by gram-negative, nonmotile, rod-shaped bacteria of the genus *Shigella*. Shigellosis outbreaks tend to involve large numbers of people. The bacteria are frequently transmitted by foods such as potato and chicken salad. These foods require extensive handling during preparation. The housefly is an important vector of transmission in contaminating food and food contact surfaces.

Bacillus cereus food poisoning is an illness characterized by diarrhea, abdominal pain, and nausea. The organism is a large, gram-positive, rod-shaped, motile bacterium that forms spores. The spores are fairly resistant to heat. These bacteria are widely distributed soil saprophytes.

Vibrio parahaemolyticus is a gram-negative, curved, rod-shaped bacterium that occurs primarily in marine coastal waters and sediments. It is primarily anaerobic but can exist where oxygen is present. Symptoms usually occur about 14 to 20 hours after eating contaminated food and are characterized by severe abdominal pain, cramps, nausea, vomiting, and diarrhea. Recovery usually occurs within two days.

The Arizona group comprises a number of motile, gram-negative rod-type bacteria that occasionally cause foodborne illness in man. They are closely related to the salmonella. The illness is characterized by fever, vomiting, diarrhea, and sometimes prostration.

Chemical foodborne intoxication results from the ingestion of a number of toxic chemicals. The chemicals may be man-made (insecticides, rodenticides, or food preservatives), or produced by plants such as certain mushrooms. They may be produced by animals such as fish, or they may be the salts of heavy metals (zinc, copper, or tin).

Review Questions

1. a. Which type of bacterium was responsible for the most outbreaks during the years 1974 to 1976? b. Which type was responsible for the most cases?

2. What is the most common of the virus-caused foodborne illnesses?

3. How can contaminated hands frequently transmit foodborne viral hepatitis?

4. Refer to the hospital outbreak of infectious hepatitis. a. What led the investigators to suspect that July 1 was the exposure date? b. Why did the investigators suspect that the illness was foodborne? c. Why was orange juice singled out as the suspected food? d. Why was a healthy kitchen worker thought to have transmitted the virus?

5. How can water contaminated with sewage be responsible for foodborne infectious hepatitis?

6. a. Describe the symptoms of shigellosis. b. Describe the organism that causes the illness. c. Do shigellosis outbreaks usually involve large or small numbers of people? Upon what facts do you base your answer? d. What is characteristic about shigellosis outbreaks?

7. a. Describe the symptoms of *Bacillus cereus* food poisoning. b. Describe the organism that causes the illness. c. Where can the organism be found in nature? d. What faulty procedure has been implicated in most cases of *Bacillus cereus* foodborne illness?

8. a. Describe the organism that causes *Vibrio parahaemolyticus* foodborne illness. b. In what type of environment is it found? c. In what country is it the major cause of foodborne illness? d. Describe the symptoms of the illness.

9. a. Describe the bacteria that comprise the Arizona group. b. What are the symptoms of the foodborne illness caused by this group?

10. List four types of chemicals that can cause foodborne illness.

References

Barker, W. H., Jr. (1974). "Vibrio parahaemolyticus Outbreaks in the United States." *Lancet* 1:551–554.

Barker, W. H., Jr. and V. Runte. (1969). "Tomato Juice-Associated Gastroenteritis, Washington and Oregon, 1969." *Amer. J. Epidemiol.* 96:219–226.

Barkin, R. M. (1974). "Ciguatera Poisoning: A Common Source Outbreak." *Southern Med. J.* 67:13–16.

Baross, J. and J. Liston. (1968). "Isolation of Vibrio Parahaemolyticus from the Northwest Pacific." *Nature* (London) 217:1263–1264.

Baross, J. and J. Liston. (1970). "Occurrence of Vibrio parahaemolyticus and Related Hemolytic Vibrios in Marine Environments of Washington State." *Appl. Microbiol.* 20:179–186.

Becker, M. E. (1966). "Water-Borne and Food-Borne Viruses." *J. Milk Food Technol.* 29:243–245.

Brown, M., J. V. Thom, G. L. Orth, P. Cova, and J. Juarez. (1964). "Food Poisoning Involving Zinc Contamination." *Archives Environ. Health* 8:657–660.

Caldwell, M. E. and D. L. Ryerson. (1939). "Salmonellosis in Certain Reptiles." *J. Infect. Dis.* 65:242–245.

Cliver, D. O. (1966). "Implications of Foodborne Infectious Hepatitis." *U.S. Publ. Hlth. Rep.* 81:159–165.

Dadisman, T. A., Jr., R. Nelson, J. R. Molenda, and H. J. Garber. (1973). "Vibrio parahaemolyticus Gastroenteritis in Maryland. I. Clinical and Epidemiological Aspects." *Amer. J. Epidemiol.* 96:414–426.

Donadio, J. A. (1969). "Foodborne Shigellosis," *J. Infect. Dis.* 119:666–668.

Dougherty, W. J. and R. Altman. (1962). "Viral Hepatitis in New Jersey." *Amer. J. Med.* 32:704–716.

Edwards, P. R., A. C. McWhorter, and M. A. Fife. (1956). "The Arizona Group of Enterobacteriaceae in Animals and Man." *Bull. Wld. Hlth. Org.* 14:511–528.

Edwards, P. R., A. C. McWhorter, and M. A. Fife. (1956). "The Occurrence of Bacteria of the Arizona Group in Man." *Can. J. Microbiol.* 2:281–287.

Eichner, E. R., E. J. Gangarosa, and J. B. Goldsby. (1968). "The Current Status of Shigellosis in the United States." *Amer. J. Pub. Hlth.* 58:753–763.

Eisenstein, A. B., R. D. Aach, W. Jacobsohn, and A. Goldman. (1963). "An Epidemic of Infectious Hepatitis in a General Hospital." *JAMA* 185:171–174.

Fishbein, M. (1971). "Vibrio parahaemolyticus: A Real Foodbourne Disease Problem." *FDA Papers* 5:16–22.

Goepfert, J. M., W. M. Spira, and H. V. Kim. (1972). "Bacillus cereus: Food Poisoning Organism. A Review." *J. Milk Food Technol.* 35:213–227.

Hauge, S. (1955). "Food Poisoning Caused by Aerobic Spore-forming Bacilli." *J. Appl. Bact.* 18:591–595.

Hiscock, I. V. and D. F. Rogers. (1922). "Outbreak of Epidemic Jaundice among College Students." *J. Amer. Med. Assoc.* 78:488–490.

Hopper, S. H. and H. S. Adams. (1958). "Copper Poisoning from Vending Machines." *Pub. Health Rep.* 3:910–914.

Johnson, H. C., J. A. Baross, and J. Liston. (1971). "Vibrio parahaemolyticus and its Importance in Seafood Hygiene." *J. Amer. Vet. Med. Assoc.* 159:1470–1473.

Kim, H. V. and J. M. Goepfert. (1971). "Occurrence of Bacillus cereus in Selected Dry Food Products." *J. Milk Food Technol.* 34:12–15.

Lindsay, D. R., W. H. Stewart, and J. Watt. (1953). "Effect of Fly Control on Diarrheal Disease in an Area of Moderate Morbidity." *Pub. Health Rep.* 68:361–367.

Longree, K. (1972). *Quantity Food Sanitation.* John Wiley and Sons, Inc. New York.

Mason, J. O. and W. R. McLean. (1962). "Infectious Hepatitis Traced to the Consumption of Raw Oysters." *Amer. J. Hyg.* 75:90–111.

Midura, T., M. Greber, R. Wood, and A. R. Leonard. (1970). "Outbreak of Food Poisoning Caused by Bacillus cereus." *Publ. Health Rep.* 85:45–48.

Mossel, D. A., M. J. Koopman, and E. Jongerius. (1967). "Enumeration of Bacillus cereus in Foods." *Appl. Microbiol.* 15:650–653.

Murphy, W. J. and J. F. Morris. (1950). "Two Outbreaks of Gastroenteritis Apparently Caused by a Paracolon of the Arizona Group." *J. Infect. Dis.* 86:253–259.

Plowright, C. B. (1896). "On Epidemic of Jaundice in King's Lynn." *Brit. Med. J.* 1:1321.

Read, M. R., H. Brancroft, J. A. Doull, and R. F. Parker. (1946). "Infectious Hepatitis—Presumably Food-Borne Outbreak." *Amer. J. Public Health.* 36:367–370.

Riemann, H. ed. (1969). *Food-borne Infections and Intoxications.* Academic Press, Inc. New York.

Roland, F. P. (1970). "Leg Gangrene and Endotoxin Shock Due to Vibrio parahaemolyticus—An Infection Acquired in New England Coastal Waters." *New England J. Med.* 282:1306.

Ruddy, S. J., R. F. Johnson, J. W. Mosley, J. B. Atwater, M. A. Rosetti, and J. C. Hart. (1969). "An Epidemic of Clam-Associated Hepatitis." *JAMA* 208:649–655.

U.S. Department of Health, Education, and Welfare. Center for Disease Control. (1975). "Vibrio parahaemolyticus Gastroenteritis on Cruise Ships." *Morbidity, Mortality Weekly Reports* 24:109, 110, 115.

U.S. Department of Health, Education, and Welfare. (1976). *Foodborne and Waterborne Disease Outbreaks Annual Summary.* 1975. Center for Disease Control. Atlanta, Georgia.

U.S. Department of Health, Education, and Welfare. (1976). *Foodborne and Waterborne Disease Outbreaks Annual Summary.* 1974. Center for Disease Control. Atlanta, Georgia.

U.S. Department of Health, Education, and Welfare. (1977). *Foodborne and Waterborne Disease Outbreaks Annual Summary.* 1976. Center for Disease Control. Atlanta, Georgia.

Verder, E. (1961). "Bacteriological and Serological Studies of Organisms of the Arizona Group Associated with a Food-Borne Outbreak of Gastroenteritis." *J. Food Sci.* 26:618–621.

Warburtan, S., W. Udler, R. M. Ewert, and W. S. Haynes. (1962). "Outbreak of Foodborne Illness Attributed to Tin." *Pub. Health Rep.* 77:798–800.

Watt, J. and D. R. Lindsay, (1948). "Diarrheal Disease Control Studies. I. Effect of Fly Control in a High Morbidity Area." *Public Hlth. Rep.* 63:1319–1334.

Zen-Yoji, H., S. Sakai, T. Terayama, Y. Kudo, T. Ito, M. Benoki, and M. Nagasaki. (1965). "Epidemiology, Enteropathogenicity, and Classification of Vibrio parahaemolyticus." *J. Infect. Dis.* 115:436–444.

Pest Control

- I. Rodents
- II. The House Mouse
  - A. Origin
  - B. Life History
  - C. Senses
  - D. Nesting and Harborage
  - E. Food Habits
- III. Rats
  - A. Origin
  - B. Life History
  - C. Senses
  - D. Nesting and Harborage
- IV. Signs of Rodents
  - A. Droppings
  - B. Runways and Rubmarks
- V. Control
  - A. Sanitation
  - B. Food Storage
  - C. Ratproofing
  - D. Rodenticides
- VI. Insects
- VII. The Housefly
  - A. Life Cycle
  - B. Physical Characteristics
  - C. Eating Habits
  - D. Disease
- VIII. The Cockroach
  - A. Life Cycle
  - B. Physical Characteristics
  - C. Disease
- IX. Insect Control
  - A. Sanitation
  - B. Insecticides
- X. Summary

For many centuries, man has been waging war against the harmful insects and rodents that destroy crops, eat and contaminate foods, spread various diseases, destroy property, and disturb the economy. Because of their infestations of human dwellings, there are legal requirements at federal, state, and local levels for the control of these pests. Nearly every health code contains provisions that require building owners and managers to eliminate conditions conducive to the development of pest infestations.

Included among the relatively few types directly involved in the cost and sanitation phases of foodservice operations are, unfortunately, some of the most intelligent examples of insect and rodent life. Flies and cockroaches are our major insect pests; the rodents of most concern are rats and mice.

Sanitation is the most important principle in the control of insects and rodents that infest the foodservice establishment. Sanitation has been defined as "a modification of the environment in such a way that a maximum of health, comfort, safety, and well-being occurs to man" (Pratt and Johnson, 1975). It includes, therefore, the control of insects and rodents, for in large measure the control of these pests depends on maintaining the building and its premises in a sanitary manner.

Pests can be controlled by modern sanitary practices. Sanitation and good housekeeping call for the prompt removal of trash, garbage, and all waste materials. The elimination of pest harborages and breeding areas is also necessary to prevent and control infestations. The maintenance of the structural integrity of the building to eliminate all cracks and unnecessary openings is of vital importance in keeping insects and rodents from within, where they can multiply and spread throughout the building. It is actually easier and cheaper to keep pests out than it is to eliminate them once they gain entrance. Careful attention to a vigorous cleaning program and to the constant maintenance of the building and property will, in the long run, be the least costly and most effective pest control program. Research and community demonstration programs have shown that the application of the basic principles of sanitation result in substantial reductions in fly and rodent populations. In a number of communities, it has been estimated that refuse sanitation will accomplish 90 percent of the job in fly control and 65 percent in rat control.

For years after World War II, pest control programs were based primarily on the use of insecticides and rodenticides. In recent years, however, the emphasis has shifted to a more balanced program of chemical pesticides and physical techniques such as sanitation and exclusion. This greater emphasis on sanitation has been necessitated by the increasing resistance of flies to insecticides and of rats to anticoagulant rodenticides. Reducing the capability of the environment to support a large number of rodents not only decreases the population but also increases competition between remaining individuals. This results in a lower rate of reproduction and a higher mortality.

Insect and rodent infestations result from neglect of basic principles of cleanliness. Insects and rodents will always abound where there is food and shelter. Their numbers increase rapidly as standards of cleanliness and tidiness decrease. Lack of knowledge, carelessness, and indifference are usually the basic reasons for the lowering of these standards.

RODENTS

Rodents make up a large order of mammals that have specially adapted teeth for gnawing. These teeth are the large, curved, and deeply rooted incisors, or front teeth (Fig. 9-1). Like

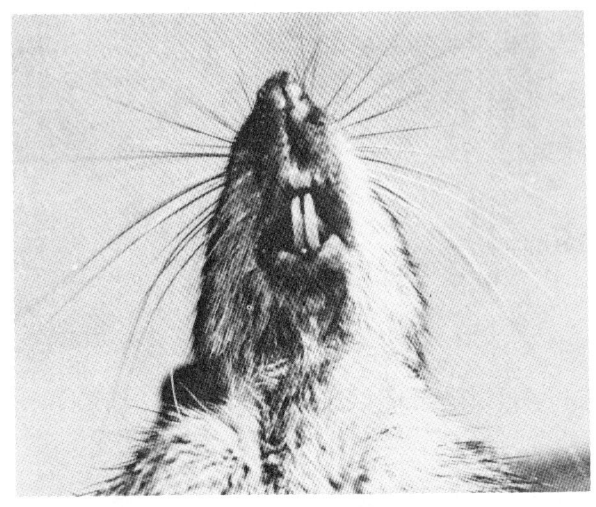

Figure 9-1 Incisor teeth of a Norway rat. These conspicuous teeth are found in all rodents. From *Biological Factors in Domestic Rodent Control.* U.S. Dept. Health, Education and Welfare.

all mammals, they are warm-blooded and bear their young alive. They make up the largest order of mammals.

North American rodents include such native animals as field and wood mice, wood rats, squirrels, rabbits, woodchucks, gophers, muskrats, porcupines, and beavers. They also include three species that reached the United States from other countries: the house mouse (*Mus musculus*), the roof rat (*Rattus rattus*), and the Norway rat (*Rattus norvegicus*). These three species are far more destructive to people and property than any of the native rodents. They are found almost anywhere that man lives. They inhabit buildings and eat or damage food.

THE HOUSE MOUSE

Origin

The house mouse probably originated in the grassy borderland of Iran and Russia and followed man as trade routes to Mediterranean port cities became established. From here, it traveled by boat to Western Europe, where it became firmly established. It reached the New World with the early explorers. It is now distributed over the entire United States, both in fields and in cities.

Life History

The life span of the house mouse runs somewhat less than one year. The young are born 19 days after mating and mature rapidly. They reach sexual maturity 1½ to 2 months after birth. Female mice can mate 48 hours after the birth of their young.

The female can bear as many as eight litters, averaging five or six mice per litter. Many of the young do not survive, however. Some die at, or shortly after, birth and some are killed by other animals. Some die if the female is forced to move her nest.

Mice that survive the hazards of birth and the first few days of life grow very rapidly. At first, however, they are virtually helpless. Their eyes and ears are not open, they are hairless, and their legs are small and uncoordinated. Hair appears on the body in about a week. At the age of 12 to 14 days, they open their eyes. The young depend on the mother for food for about three weeks. After this time, they are able to survive on their own.

Mice are usually nocturnal in their habits. They will generally be found indoors, particularly in the wintertime, in any convenient space between walls, in cabinets, in furniture such as upholstered chairs or couches, or in stored goods.

Senses

The success of the rodent group is due, in no small measure, to their highly developed senses.

1. Touch. Well developed. Mice have highly sensitive whiskers and guard hairs. They prefer to

run along sidewalls, between objects, or in runways where they can keep their whiskers in contact with surrounding surfaces. Their repeated use of the same routes leads to the formation of trails or runways. This behavior is so well known that rodent control workers use it as a guide for setting traps or baits so rodents will find them.

2. Vision. Not as well developed as in human beings. The eye of the mouse is adapted to dim light. It can detect motion and recognize simple shapes up to 45 feet away. Mice are apparently color-blind. This can be helpful when using poisons for control purposes. Cornmeal bait, colored green as a warning to humans, remains quite attractive to mice.
3. Smell. The sense of smell is keen. There is little need to worry about the odor of man on traps or baits, however. Mice are well adjusted to the human odor.
4. Taste. Mice eat most foods that man does. They will eat strychnine-poisoned grain but reject baits containing red squill, which has an extremely bitter taste.
5. Hearing. Their hearing is keen. They recognize noises quickly and locate them to within six inches of their source. Since most mouse activities are carried on in the dark, this sensitivity to noise is highly advantageous to them. Loud noises will usually drive them from their hiding places.
6. Balance. Mice have an excellent sense of balance. When falling, they invariably land on their feet. Mice can climb the vertical walls of most brick buildings. They can climb any vertical surface on which they can get a claw hold. Stucco is often rough enough to allow climbing. Vine-covered walls make perfect runways since vines offer footholds and concealment. Mice can also travel along wires.

Nesting and Harborage

Mice will nest in any safe place if food and water are near. In buildings they are generally found between walls, in the spaces around counters, or in any place hidden from view. The more rubbish that is littered around, and the more empty boxes and crates that are stacked, the more nesting and hiding places available.

Mice build their nests in hiding places that are relatively quiet. They gather any type of soft material to build and line the bowl-shaped nest. Mouse nests are about five inches in diameter.

Food Habits

The original habitat of the mouse was a grain-producing area. Since then, because of its association with man, it has adapted to a wide variety of foods. Its choice of food is determined largely by its environment. Mice consume about one-tenth of an ounce of dry food per day.

Mice have regular eating habits, usually at night. They normally carry their food to a hiding place before eating it. Mice are nibblers when they eat, eating a little here and a little there. Therefore, when using a single-dose poison, many poison baits should be used and placed close together in secluded spots so that the mice will nibble enough to obtain a lethal dose.

RATS

Rats are indicators. Their presence in a community or area is firm evidence of poor sanitary practices, lack of property maintenance, or inadequate refuse collection. The roof rat and the Norway rat are the two species that most affect man.

Origin

The roof rat is a native of southeast Asia (Fig. 9-2). It followed the caravan routes across India to the Mediterranean ports and probably reached Europe aboard trading vessels. Carrying infected rat fleas, it was responsible for catastrophic outbreaks of bubonic plague. It was the first of the two types of rats to be carried to the United States, where it was well documented in the early colonies. There are

Figure 9-2 The roof rat. Note the slender body, pointed nose, prominent ears, and long tail. From *Biological Factors in Domestic Rodent Control.* U.S. Dept. Health, Education and Welfare.

Figure 9-3 The Norway rat. Note the heavy body, blunt nose, small ears, and tail shorter than the combined head and body. From *Biological Factors in Domestic Rodent Control.* U.S. Dept. Health, Education and Welfare.

three varieties or color phases of this rat: the black rat, the Alexandrine rat, and the fruit rat.

The Norway rat is predominantly a burrowing rodent (Fig. 9-3). It is the most common and the largest of the two types of rat. It probably originated in the dry grassy plains of Central Asia and, being an aggressive animal, spread rapidly. This rat reached Europe in the 1700s and was soon carried by ship to the New World, spreading throughout the country just like the roof rat some years earlier. The Norway rat is also known as the brown rat, house rat, barn rat, sewer rat, and wharf rat. Table 9-1 shows a comparison of the physical characteristics of the two types of rat.

Table 9-1
Physical characteristics of the Norway rat and the roof rat.

Characteristic	Norway Rat	Roof Rat
Body	Heavy, stocky	Slender, graceful
Body weight	About 1 pound	About $\frac{2}{3}$ pound
Length of head and body	7 to 10 inches	$6\frac{1}{2}$ to 8 inches
Length of tail	6 to $8\frac{1}{2}$ inches	$7\frac{1}{2}$ to 10 inches
Total length	13 to $18\frac{1}{2}$ inches	14 to 18 inches
Fur	Coarse	Fine
Nose	Blunt	Pointed
Ears	Small	Large
Eyes	Small	Large

The Norway and roof rats do not usually inhabit the same area. When they compete for the same territory, the aggressive Norway rat frequently becomes dominant and the roof rat soon disappears. Only under special conditions do both rats inhabit the same area. Roof rats, because of their better climbing ability, will sometimes occupy the upper portions of a structure that supports a Norway rat population in the lower portion.

The roof rat is presently confined mostly to the warm southern states and along the Pacific Coast into western Canada. It is found only sporadically in the northern part of the continent. The Norway rat is found throughout the United States and southern Canada. On the West Coast, its range extends into Alaska.

Life History

The life span of the Norway rat and the roof rat is about one year. The young are born about 22 days after mating. They reach sexual maturity three to five months after birth. There may be six to seven litters during the year, averaging eight to twelve per litter in the Norway rat and six to eight per litter in the roof rat. Many of the young die at birth, or shortly after, through accident or the actions of other animals. Young rats are just as helpless as young mice. Their eyes and ears are closed, they are hairless, and their legs are small and poorly developed. Hair appears on the body in about a week. Their eyes open at about 12 to 14 days of age. After three weeks, they are capable of surviving on their own. Rats are usually nocturnal in their habits and are seldom seen.

Senses

The senses of the rat are developed to the same high degree as those of the mouse.

1. Touch. The sense of touch is highly developed in the whiskers and guard hairs.
2. Vision. Vision is adapted to dim light and motion can be recognized at distances of up to 30 feet.
3. Smell. The sense of smell is keen. The odor of man on baits and traps does not disturb them.
4. Taste. Rats will eat most of the foods that man does. Rats seem to have an aversion to strychnine (mice do not). Hulbert and Krumbiegel reported that the addition of four food flavors to standard warfarin-cornmeal bait improved its acceptance. The four flavors were butter vanilla, roast beef, coconut, and maple.
5. Hearing. Rats recognize noises readily and can locate them to within six inches. Loud noises cause them to panic and attempt to escape.
6. Balance. Rats have an excellent sense of balance. They can fall several stories without being injured. Rats have been known to enter buildings by dropping through open skylights. They are notoriously good climbers and have even been observed crossing city streets by walking along telephone wires.

Nesting and Harborage

The rat, like the mouse, originated in a grain-producing area. It has also adapted to most of man's foods. Schein and Orgain reported that Norway rats feeding on garbage preferred meats, grain and grain products, cooked eggs, and potatoes. They were not attracted to raw beets, peaches, onions, celery, cauliflower, and green peppers. They seemed to be repelled by highly spiced foods.

The choice of food is sometimes dictated by circumstances, and rats have proven to be adaptable. For example, neither rats nor mice prefer citrus fruits but, in Florida, the roof rat is a serious pest in citrus groves. There seems to be a scarcity of other foods in this area.

The average adult rat eats about one ounce of dry food a day. This fact must be taken into account in control programs using single-dose poisons. Since rats consume a rather small amount of food, enough poison must be used so that a normal feeding will provide a lethal dosage. Otherwise, the poison will only make the rat sick and it will then avoid that bait.

Rats usually begin searching for food after sunset. They carry it to a hiding place, if possible, sometimes a bit at a time. They rarely eat in the open. This habit should be considered when setting traps or baits.

SIGNS OF RODENTS

Rats and mice are usually nocturnal, secretive, and quick, and thus, are rarely seen. It is necessary, therefore, to be able to identify signs that indicate their presence. These signs are found along walls, under piles of rubbish, and behind or under boxes and boards. Proper interpretation of these signs can tell whether the number of rats is large or small and whether the signs are old or new.

Droppings

Presence of rat or mouse feces is one of the best indications of their presence. Fresh droppings are usually moist, soft, shiny, and dark, but in a few days they become dry and hard. Old droppings are dull and grayish and crumble when pressed.

Of the three rodents under discussion, the Norway rat releases the largest droppings (Fig. 9-4). They range up to $\frac{3}{4}$ of an inch in length and $\frac{1}{4}$ inch in diameter and vary in shape from having blunt ends to being spindle-shaped in appearance. Roof rat droppings are usually smaller with blunt ends. House mouse droppings are very small and pointed at both ends.

Fresh droppings mean that at least one rat or mouse is present. The presence of several

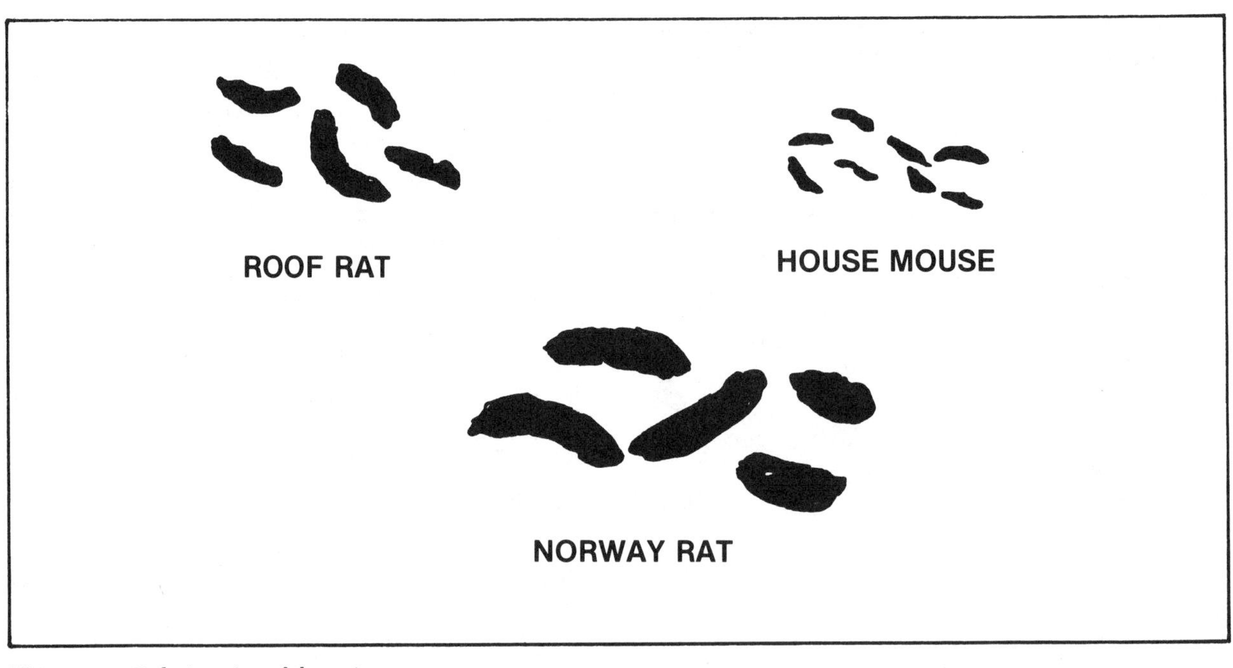

Figure 9-4 Relative size of droppings of the house mouse, roof rat, and Norway rat. From *Biological Factors in Domestic Rodent Control*. U.S. Dept. Health, Education and Welfare

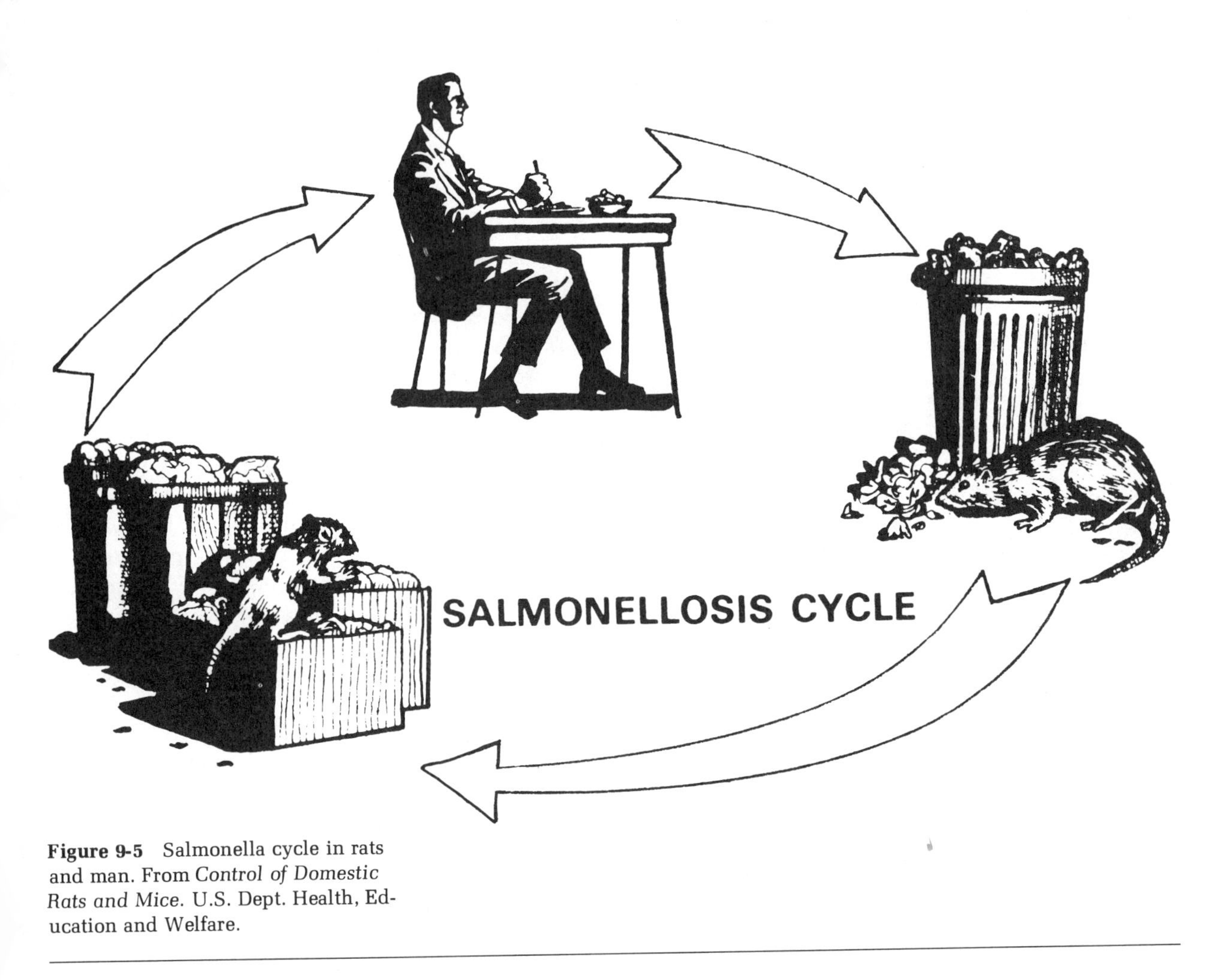

Figure 9-5 Salmonella cycle in rats and man. From *Control of Domestic Rats and Mice.* U.S. Dept. Health, Education and Welfare.

sizes of droppings usually means that several ages are represented in the population and is a good indication that the infestation has been present long enough for breeding to occur. Droppings are most numerous along runways, near harborage, in secluded corners, and near food supplies.

Rats are known carriers of salmonella (Fig. 9-5). Various studies have indicated that from 0.7 to 13 percent of these animals harbor food poisoning varieties of salmonella. Ludlam found, at one time in his investigation, that 40 percent of the rats examined were harboring these organisms.

Since these bacteria occur in the intestine, they are released in the droppings. Welch et al. found that *Salmonella enteritidis* can remain viable in rat droppings for as long as 148 days. Salthe and Krumwiede describe a food-poisoning outbreak involving 59 persons resulting from infected rat droppings that contaminated cream filler used in cakes and eclairs.

Runways and Rubmarks

Rats and mice generally use the same runway as long as it provides a safe, convenient route to their food supply. Their runways in or on buildings are frequently marked by rubmarks (Fig. 9-6) since they prefer to maintain body or whisker contact with a vertical surface. Fresh rubmarks are soft and will smear if rubbed. Rubmarks may be found around gnawed holes,

Figure 9-6 Roof rat "swing marks" under floor joints on an overhead beam indicator runway. From *Biological Factors in Domestic Rodent Control*. U.S. Dept. Health, Education and Welfare.

along pipes and beams, on the edges of stairs, or along walls. The rubmarks of the Norway rat are most commonly found along runways near ground or floor level, whereas those of the roof rat are usually seen as swing marks beneath beams or rafters at the place where they connect to the walls.

Small vertical pipes and columns should be periodically checked for rubmarks since they are commonly used by rats and mice to move up or down in buildings. Rubmarks of mice are usually faint and difficult to detect.

CONTROL

Sanitation

Rats thrive wherever ample food and harborage are available, conditions that may exist in and around homes, restaurants, or other food businesses. In addition, community sewage systems are often heavily infested with Norway and roof rats. The movement of rats from sewers to above-ground harborages can cause public health problems.

To achieve successful rodent control, sanitation measures, such as prompt garbage disposal, harborage elimination, proper food storage, and ratproofing must be followed diligently. It is useless to try to reduce the rat population in an area where garbage and rubbish are available to them. Li and Davis reported finding that 4.2 percent of the rats in Baltimore harbored salmonella. In blocks that were kept in an unsanitary condition, however, the rate was 24 percent. The number of rats in any area is directly proportional to the amount of food and harborage available. If these factors are decreased, then competition between rats for the remaining food and harborage results in increased mortality and decreased population.

Proper sanitation consists chiefly of adequate refuse storage, collection, and disposal practices. The proper storage of refuse or garbage on the premises is probably the most important of the three practices; it also tends to be the most poorly accomplished. Garbage containers should be:

1. Rust resistant
2. Watertight
3. Easy to clean
4. Provided with a close-fitting lid
5. Rat and damage resistant
6. Of adequate capacity.

Elevation of the container on racks and stands (Fig. 9-7) prevents corrosion, permits

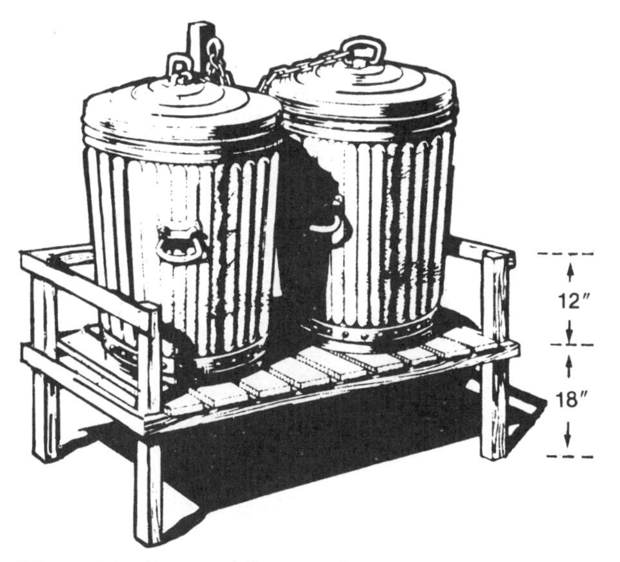

Figure 9-7 Proper placement of garbage cans on stand. From *Control of Domestic Rats and Mice.* U.S. Dept. Health, Education and Welfare.

regular cleaning underneath, reduces rodent harborage under the stand, and lessens the possibility of cans being overturned. Lids should always be closed tightly (Fig. 9-8).

Refuse collections should be thorough and frequent enough to prevent the accumulation of garbage outside the container and the buildup of boxes and cartons. These two poor practices provide food and harborage for rats and mice.

Proper refuse disposal is the final aspect of rodent control through refuse sanitation. Unsatisfactory disposal can result in an infestation of rats and mice that can spread to homes and business establishments in the area. Incinerators, sanitary landfills, and garbage grinders are the most satisfactory methods of disposal.

Food Storage

Rats and mice must be kept out of any area that is concerned with food. This includes the kitchen, bakeshop, pantry, dining room, and all storage rooms. Rats and mice consume and damage stored foods, making them unusable because of droppings and hairs. They are dirty animals because of their association with garbage, sewers, and filth. Rats have even been observed feeding on manure.

In addition to these undesirable qualities, rats are carriers of several species of salmonella. *Salmonella typhimurium, enteritidis,* and *newport* have been isolated from rats. Stored food (such as flour) can be contaminated with these bacteria if rats can gnaw into it. Rat droppings containing salmonellae can become incorporated in the dust of the environment

Figure 9-8 Garbage can with properly chained lid. From *Control of Domestic Rats and Mice.* U.S. Dept. Health, Education and Welfare.

and thus contaminate food, utensils, clothing, and hands.

Stored foods must be protected against pests of all sorts. Floors, walls, and windows of storage rooms should be rodent-proof. Places in walls and ceilings where pipes enter and leave must be sealed. Rats can squeeze through an opening no larger than a quarter. Mice need only a nickel-sized opening.

Foods should never be placed directly on the floor. Wooden pallets that elevate boxed or sealed materials 12 to 18 inches off the floor are highly recommended (Fig. 9-9). The open space under the pallet is easily cleaned and inspected. The material on the pallet should be tightly stacked, if possible, to prevent accessibility.

Aisles at least two feet wide should be provided along all walls and at intervals throughout the storage room. A six-inch stripe painted on the floor along the walls will make rodent signs more visible. All spillage should be cleaned up immediately and thoroughly. Rodents are not attracted to areas where there is no food.

Ratproofing

The ratproofing of buildings in which food is stored, prepared, or distributed, such as restaurants, cafeterias, and fast-food establishments, is highly desirable. Rats carry diseases that can be transmitted to both patrons and employees. Ratproofing is one of the most effective methods of insuring that rats and their diseases do not affect our food.

The objectives in ratproofing are threefold: (1) to prevent rats from entering buildings, (2)

Figure 9-9 Proper storage of dry goods for protection against rats and mice. From *Control of Domestic Rats and Mice*. U.S. Dept. Health, Education and Welfare.

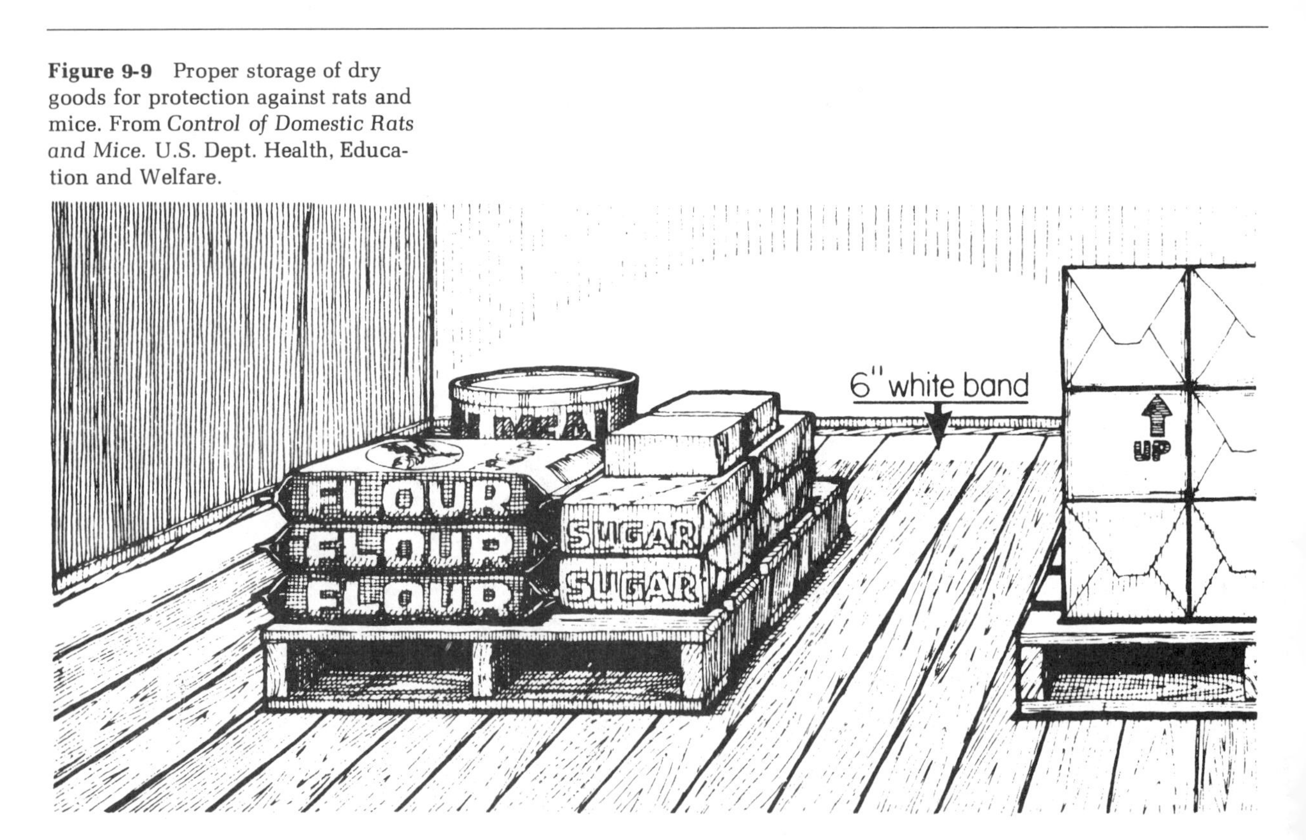

to facilitate the eradication of rodents present in the building, and (3) to make future rodent control programs easier and less costly. The following practices are recommended in rat-proofing a building:

1. All holes should be closed with mortar or openings reduced with hardware cloth. Sheet metal should be used where practicable.
2. Doors and windows should fit snugly. Doors should be provided with automatic closing devices. Windows and ventilators should be screened or covered with gratings. Floor drains should have covers that fit tightly. Rats have entered many buildings through sewer drains.
3. All openings in floors, ceilings, and walls for the passage of pipes and wires should be closed with cement or some material through which rats cannot gnaw.

Rodenticides

Rodenticides are supplemental to and not a substitute for proper sanitation and exclusion. If the latter two methods are not employed, the rodent infestation will recur. Used under proper circumstances, however, rodenticides can be extremely useful in combating rodents that already inhabit the premises.

General-use pesticides are those that can be used safely by the general public if they follow the directions on the container. Restricted-use pesticides are highly toxic to human beings (sodium fluoroacetate) or have an adverse effect on the environment (DDT). These pesticides require a certified or trained applicator.

There are two types of rodenticides: the single-dose type, which is fatal to the rodent through a single feeding (zinc phosphide, red squill); and the multiple-dose type, which requires many feedings to be effective (warfarin, diphacinone). The latter are more popular because they are acceptable to rodents in small quantities, do not cause bait shyness, and are easy to use, economical, and safe around human beings and animals.

Multiple-dose chemicals interfere with the ability of the blood to clot, and the rodent bleeds to death internally. These poisons are known as anticoagulants. Their successful action depends on consumption of at least a small amount of poison almost every day for five days or more. To achieve effective control, anticoagulant baits must be available to the rats for at least two weeks. Establishment of permanent bait stations in places subject to continual reinfestation gives good control. The anticoagulant rodenticides are available as solid or liquid concentrates, ready-to-use baits, bait blocks, water baits, pellets, toss packs, and tracking powders. Warfarin, fumarin, diphacinone, chlorophacinone, phenylacetyl, PMP, and Pival are examples of multiple-dose poisons.

The single-dose rodenticides are especially useful for rapid reduction of large numbers of rats and mice. Red squill, ANTU, zinc phosphide, and strychnine have been used effectively and safely in rodent control programs for many years. Also included in this group are sodium fluoroacetate (1080), fluoroacetamide (1081), norbormide, and vacor. A comparison of single-dose rodenticides follows:

Red squill is effective against Norway rats but mice will not consume it. It is only effective against roof rats in very high dosages. Reacceptance of red squill bait is poor by Norway rats that survive the first poisoning. It should not be used again for a period of about six months.

ANTU is toxic to Norway rats but ineffective against roof rats and mice. Rats are not inclined to accept a second dose of this poison, either, and a six-month interval between use is recommended. It has a good safety record where human beings are concerned but is highly toxic to dogs, cats, and pigs.

Zinc phosphide is effective against rats and mice and has been used extensively as a rodenticide. It is toxic, however, to human beings and domestic animals.

Strychnine is effective against mice but is rejected by rats. It is highly toxic to human beings and pets.

Norbormide is extremely effective against the Norway rat, moderately effective against the roof rat, and ineffective against mice. It is one of the safest rodenticides.

Sodium fluoroacetate and fluoroacetamide are fast-acting, highly toxic to man and animals, and extremely dangerous to use. Their application is restricted almost entirely to certified professional exterminators.

Bait mixtures can be put into paper, metal, or plastic pie plates or into permanent bait stations. The baits should be inspected at regular intervals to ensure that they remain attractive. They should be replaced when necessary. Treatment with anticoagulant baits is most effective if continued for two weeks or more.

Insects

Man has had to compete for food and shelter with many other life forms since his appearance on planet Earth. His major rival in this battle undoubtedly has been a small foe called the insect. This foe eats his food, destroys his crops, annoys and bites him, transmits disease, and lives in his dwellings, sometimes even eating those. Insects and other household pests destroy well over one billion dollars worth of stored food annually in the United States.

Of about one million varieties of insects thus far identified, only a few thousand affect man enough to force him to wage war against them. Of these few thousand, two are of major public health importance to the operator of a foodservice establishment—the fly and the cockroach.

THE HOUSEFLY

The housefly (*musca domestica*) is one of the most common domestic insect pests (Fig. 9-10). It is also one of the dirtiest. Flies are a major pest in every industry. They spread disease and contaminate products and equipment around dairies, bottling plants, and food-processing and foodservice establishments. The tiny hairs on their legs carry bacteria and other microorganisms that cause disease, such as salmonellosis, typhoid fever, shigellosis, and a host of other diarrheal diseases. Open dumps and garbage cans are dangerous, therefore, because they offer breeding places and a possible supply of germs for flies to carry into foodservice establishments.

The known habits of flies make them important vectors of all kinds of bacteria with which they come in contact. They are an extremely potent source of spoilage organisms, particularly in those food plants where little attention is paid to sanitation and where the food is prepared for the consumer without a final treatment to destroy these organisms. Flies should not be tolerated in or around any business that deals in food.

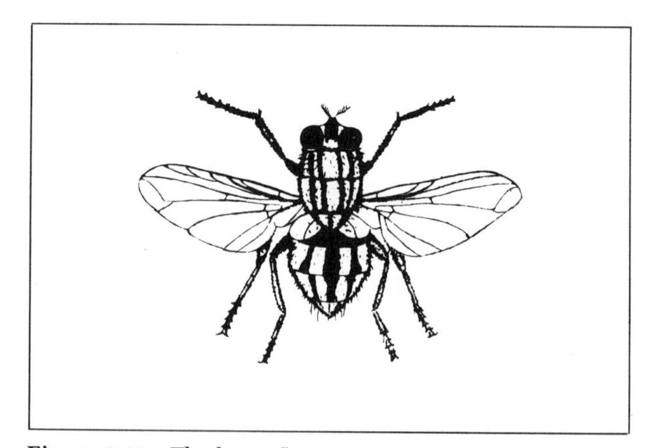

Figure 9-10 The housefly

Life Cycle

The female fly lays from 75 to 150 small, white, oval eggs on manure or some other rotting material. Five or six batches are laid during the life of the average female. In one or two days, these eggs hatch into maggots, the tiny, white, wormlike larvae of the insect. Unlike the adult they have no legs. The maggots burrow into the breeding material, where they remain and feed for periods extending from three to seven days. After the feeding state, the larvae look for some dry, sheltered place and enter the next stage of development, the pupal stage.

Activity ceases for three or four days while the larva inside the pupal case changes into the adult fly. When the change is complete, the fly pushes out of its pupal case, makes its way to

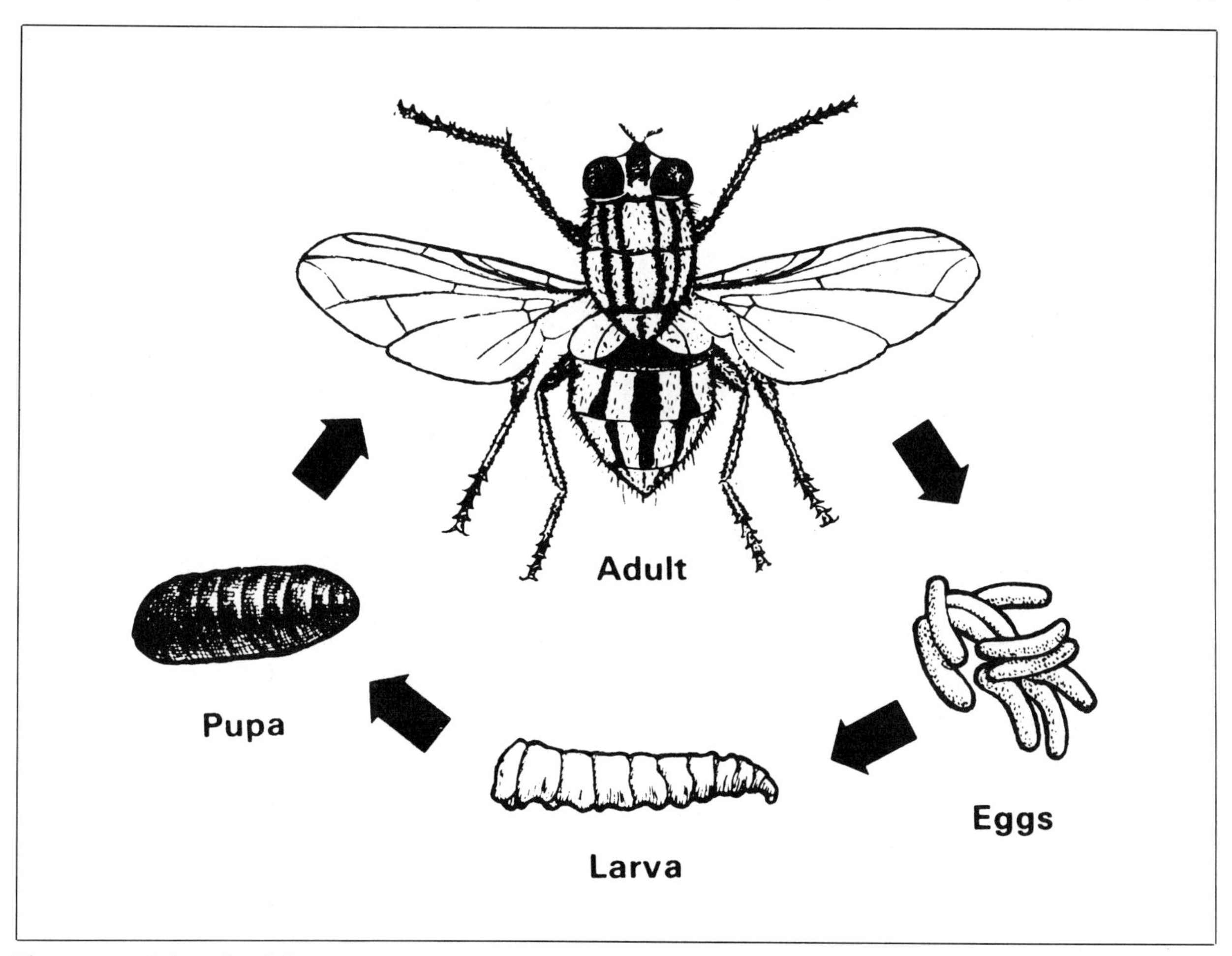

Figure 9-11 Life cycle of the housefly. From *Flies of Public Health Importance and Their Control*. U.S. Dept. Health, Education and Welfare.

the surface, spreads its wings, and waits for its body to dry and harden. It is now ready to begin an existence of one to two months as an adult housefly (Fig. 9-11).

Physical Characteristics

The adult fly has a hairy body, with feet that resemble delicate, transparent, webbed leaves. Two flattened pads on each foot allow a sticky material to ooze out. This sticky substance enables the fly to walk on walls and ceilings, or cling to thin objects like wires, hanging cords, or strings. This same substance collects filth and germs that are transferred to food or any other surface that the fly contacts.

The fly has no teeth. It has a sucking mouth tube called a proboscis, which is shaped like an upside-down funnel (Fig. 9-12). It uses the proboscis to suck up food, which it moistens and dissolves first with stomach fluid that it regurgitates.

Its eyes are compound, that is, they are composed of thousands of smaller eyes or facets. They enable the fly to see in many directions at once and make it quickly aware of movement. The fly obtains air through tiny pores on the sides of its body. Insecticides enter these pores rapidly, causing flies that are sprayed to die rather quickly.

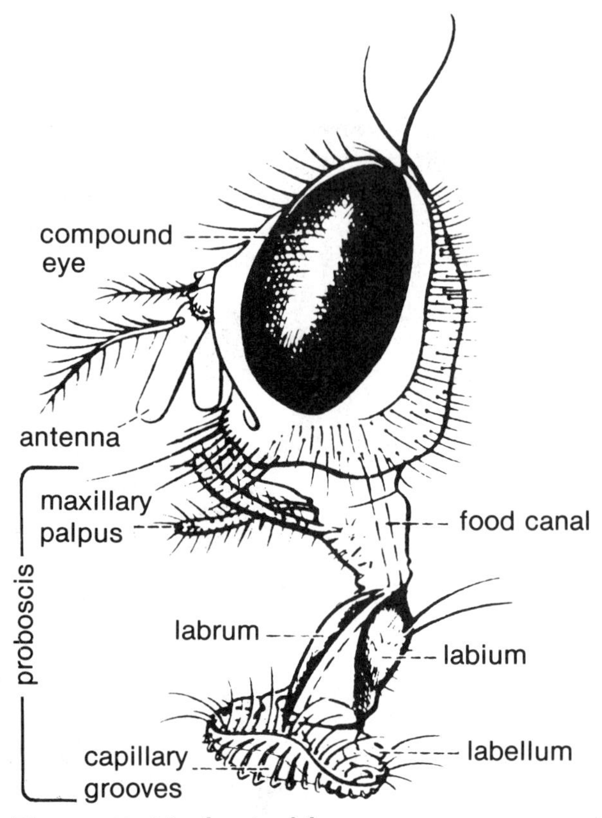

Figure 9-12 Mouthparts of the housefly. From *Flies of Public Health Importance and Their Control*. U.S. Dept. Health, Education and Welfare.

Eating Habits

When the fly feeds, it sucks liquids into its tube-like proboscis. It also likes solid foods, particularly sweets like sugar, candy, and cake frosting. In order to eat solids, it regurgitates a fluid from its stomach. This fluid is deposited on the food to moisten, soften, and dissolve it. It then sucks up the resulting fluid.

The fly often eats and eliminates at the same time. The light-colored spots seen on feeding areas are the regurgitated stomach fluids; the darker spots are fecal material. Both are called fly specks and both contain germs. Laboratory examinations of the fly's digestive tract have revealed as many as 28 million bacteria.

Disease

The habits of the housefly make it an efficient vector of disease (Fig. 9-13). It is probably the most dangerous insect closely associated with man. Flies can be as effective in spreading intestinal diseases as fingers, dirty eating utensils, and contaminated food. The organism frequently emerges in manure and here it has its first feeding. It will probably walk and feed on manure and garbage before coming indoors. It

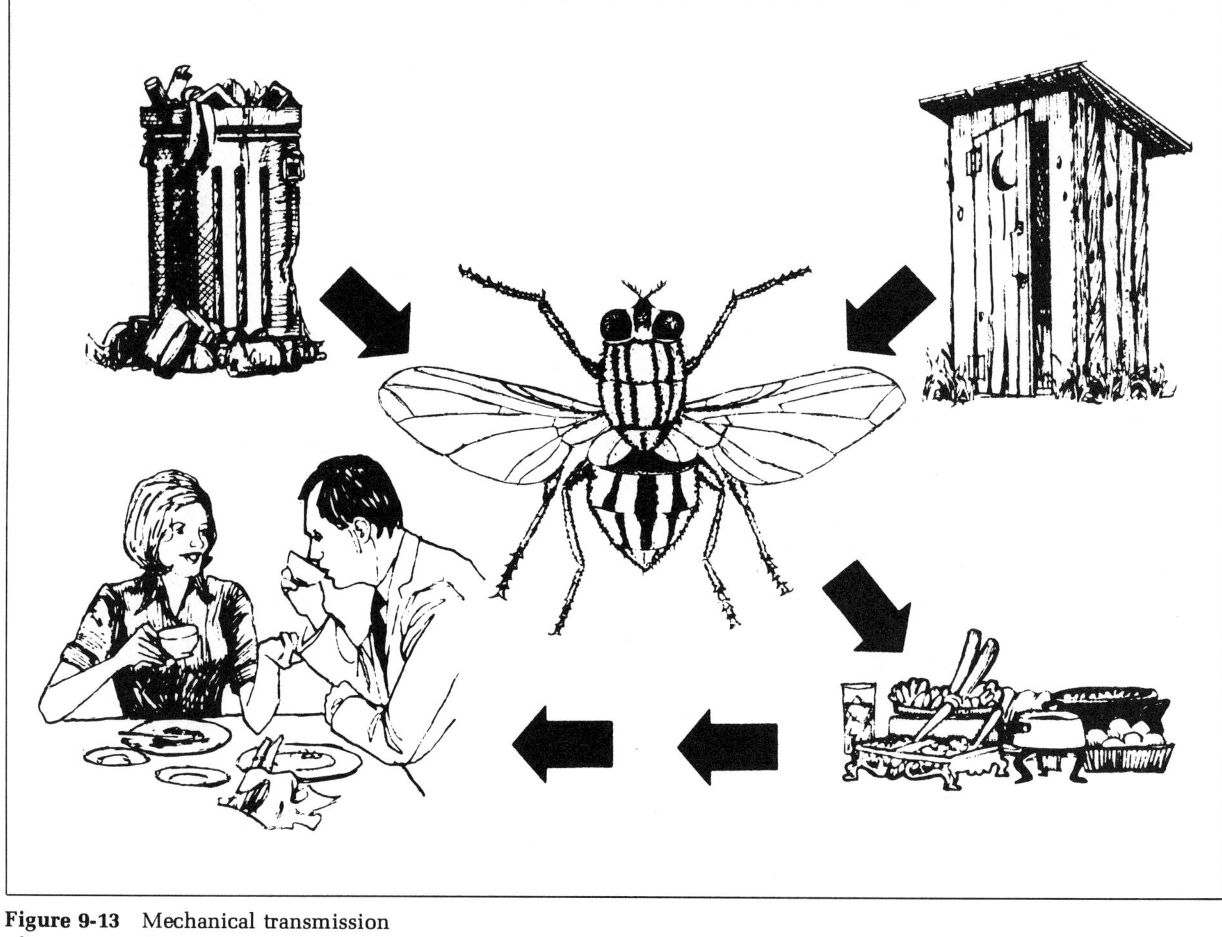

Figure 9-13 Mechanical transmission of organisms that cause disease. From *Household and Stored-Food Insects of Public Health Importance and Their Control.* U.S. Dept. Health, Education and Welfare.

will distribute the bacteria carried on its feet and body and in its intestine to any surface on which it walks. If the surface is food, the bacteria multiply and illness results.

Man has long been aware of the disease-spreading ability of the housefly. Early writing on the subject dates back to the Greeks and Romans. It was shown in the mid-1800s that flies could carry anthrax bacteria, both on their bodies and in their intestines. By 1885, the housefly had been incriminated as a carrier of cholera. Identification of other diseases that could be transmitted soon followed. Among them were tuberculosis, opthalmia, typhoid fever, polio, and shigellosis. Large-scale epidemics of foodborne diseases, such as typhoid fever and dysentery, occurred among soldiers during the Boer and Spanish-American wars, many involving more than 20,000 cases. Moorehead and Weiser later reported that the housefly could carry *Staphylococcus aureus*.

Ostrolenk and Welch established that *Salmonella enteritidis* could be transmitted between flies and that an infected fly could be a lifetime carrier, shedding the organism in its droppings and vomitus. It was also shown that salmonella could be transferred from infected flies to healthy mice and vice versa.

The ease of transfer of food-poisoning organisms from fly to fly lends added significance to their potentialities as sources of food-poisoning outbreaks. The transmission of food-poisoning bacteria by flies to mice and the later retransfer from infected mice to flies mark these insects as potential health hazards wherever food that is to be eaten raw is prepared.

Watt and Lindsay studied the relationship between flies and shigellosis in Hidalgo County, Texas, which had a high rate of *Shigella* infection. In those towns treated with insecticides, the incidence of the disease dropped dramatically; in untreated towns, the incidence remained high. Lindsay et al. repeated the experiment in Thomas County, Georgia, with similar results.

THE COCKROACH

Cockroaches are found in every industrial and domestic habitat. They infest processing and storage areas, washrooms, and basements. They can live in any place that is warm and that provides food and water. Their eating habits include almost any type of food. While feeding, they regurgitate and eliminate on it like the fly. The two substances are laden with bacteria. These insects, like the fly, carry many intestinal diseases. They are associated with sewers, garbage, and filth, and from these sources they pick up and carry many pathogenic microorganisms. They rank second only to the housefly as a domestic insect pest of public health importance.

Roaches are generally nocturnal insects. During the day, they hide in cracks in walls, furniture, under and behind cabinets, stoves, refrigerators and sinks, and in basements and sewers. A high level of sanitation will reduce their numbers but not eliminate them. A periodic residual insecticide program is usually necessary.

Life Cycle

Cockroaches have three stages in their life cycle—egg, nymph, and adult. The female drops or attaches egg capsules behind store windows, under shelves and counters, behind furniture, in folds of drapes, behind and under drawers, and in any cracks and crevices.

When the tiny nymphs emerge, they do not have wings nor are they sexually mature. They look like minature cockroaches. Growth occurs by molting, as few as 5 or as many as 13 times, depending on the sex and species of roach. Males mature more rapidly than females. When cockroaches of various sizes of the same species are observed, it indicates a well-established breeding colony and action should be taken immediately.

Physical Characteristics

Cockroaches are rather flat insects, and even large roaches can squeeze into small cracks. Domestic roaches vary in color from tan to chestnut brown to black. The head bears one pair of long, thin antennae. The eyes are compound and the mouthparts are adapted for chewing.

There are four common species of cockroaches in the United States: the German cockroach, the brown-banded cockroach, the oriental cockroach, and the American cockroach.

The German cockroach (*Blatella germanica*) is the most common type of roach in homes and restaurants (Fig. 9-14). It is about one-half inch in length, and is grayish in color with two

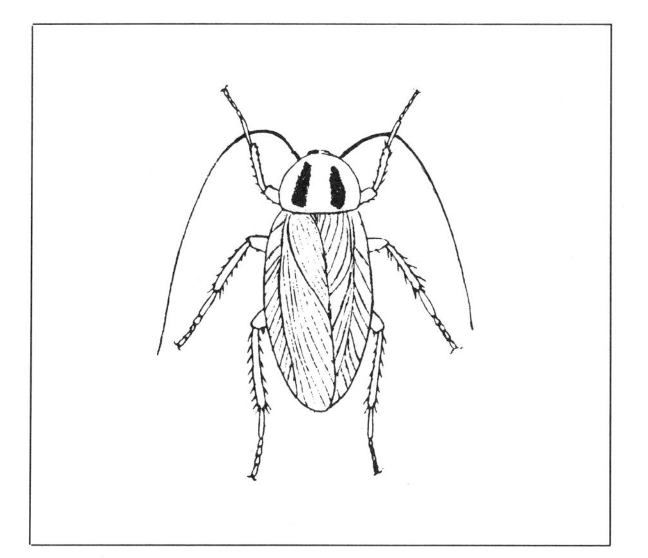

Figure 9-14 The German cockroach

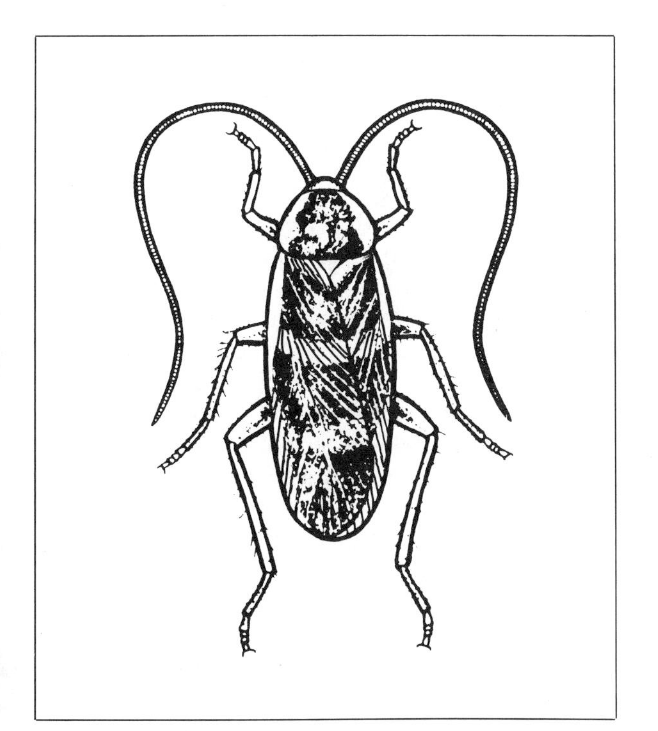

Figure 9-15 The brown-banded cockroach

Figure 9-16 The Oriental cockroach

black bars on the head. The German roach is found in buildings throughout the world, particularly in kitchens and pantries. It commonly enters a building in cartons of bottled drinks or bags of potatoes or onions that come from an infested warehouse.

The brown-banded cockroach (*Supella longipalpa*) is about one-half inch in length with two brownish and yellowish cross-bands on the wings (Fig. 9-15). There is a dark area on the center of the head. This roach can be found throughout the house or building, often in electric clocks, radios, and television sets, which are warm and dark. It frequently arrives in crates of furniture.

The oriental cockroach (*Blatta orientalis*) is about one inch in length and black in color (Fig. 9-16). It is commonly found in trashy yards, filthy outbuildings, sewers, cool, damp basements, and under sinks and refrigerators. It enters buildings with food or laundry, or crawls under doors, through basement windows, or cracks in the structure.

The American cockroach (*Periplaneta americana*) is the largest of the domestic roaches (Fig. 9-17). It is about $1\frac{1}{2}$ inches in length, reddish or dark brown in color, and prefers the warm, humid environment of boiler rooms, basements, and kitchens. The roaches of this

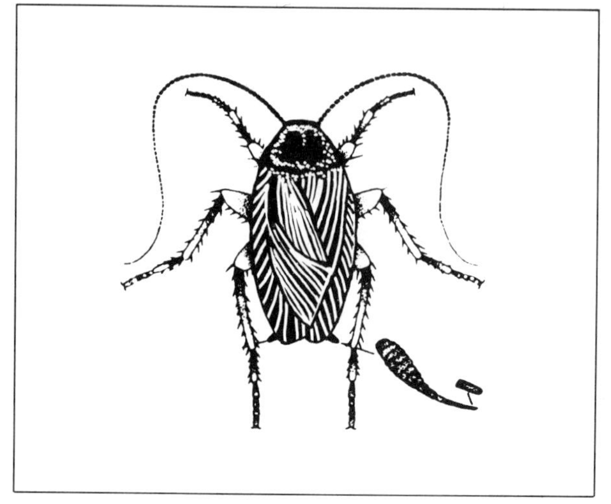

Figure 9-17 The American cockroach

species usually enter a building by crawling under doors or through cracks in the foundation or frame. They are sometimes called "waterbugs" because they frequent sewers and moist places.

Disease

The cockroach ranks second only to the fly as a household pest and disease carrier. It may carry the organisms causing typhoid fever, cholera, rickettsial pox, shigellosis, salmonellosis, and many other diarrheal diseases. Reuger and Olson studied American cockroaches in 19 cities in the United States. They found that these roaches carried many strains of salmonella and staphylococcus.

Olson and Reuger conducted experiments with roaches to determine how long they would excrete salmonella in their feces after becoming infected. The type used was *Salmonella oranienburg*, a food-poisoning species. These researchers found viable salmonella in the feces of the American roach for 10 days, the German roach for 12 days, and the oriental roach for 20 days. The organism was able to survive in droppings of the American roach for 199 days.

Insect Control

Sanitation

The basic principles of sanitation—cleanliness, proper food storage, sanitary refuse storage, maintenance of buildings and premises—are necessary to keep insect populations at a low level. Sanitation prevents development of large populations by depriving insects of food and shelter.

Doors should fit tightly and be equipped with automatic closing devices so that they are open for as short a time as possible. Daily and careful washing and vacuum cleaning of floors will remove available food. Counters, serving tables, and cabinets, if unclean, can serve as feeding stations for flies and roaches. Basements and storerooms should be checked at night for roaches.

Insecticides

There are two types of insecticides in common use—stomach poisons and contact poisons (Fig. 9-18).

Stomach poisons must be swallowed to cause death. They are primarily effective against insects with chewing mouthparts, which consume the poison with their food. Stomach poisons are used regularly to protect crops. They are ineffective against such insects as mosquitoes, lice, fleas, and true bugs that penetrate into the food supply with their sucking mouthparts, thereby missing the poison spread on the surface. Stomach poisons can be incorporated into bait form, however, such as sugar-water baits with dichlorvos for fly control and peanut butter baits with chlordecone for cockroach control.

> The arsenic group. Arsenic trioxide is the base for a number of stomach poisons including Paris green, lead arsenate, calcium arsenate, and sodium arsenite. These chemicals cause deterioration of the intestinal lining and subsequent death. They are also toxic to man and are not generally used against domestic in-

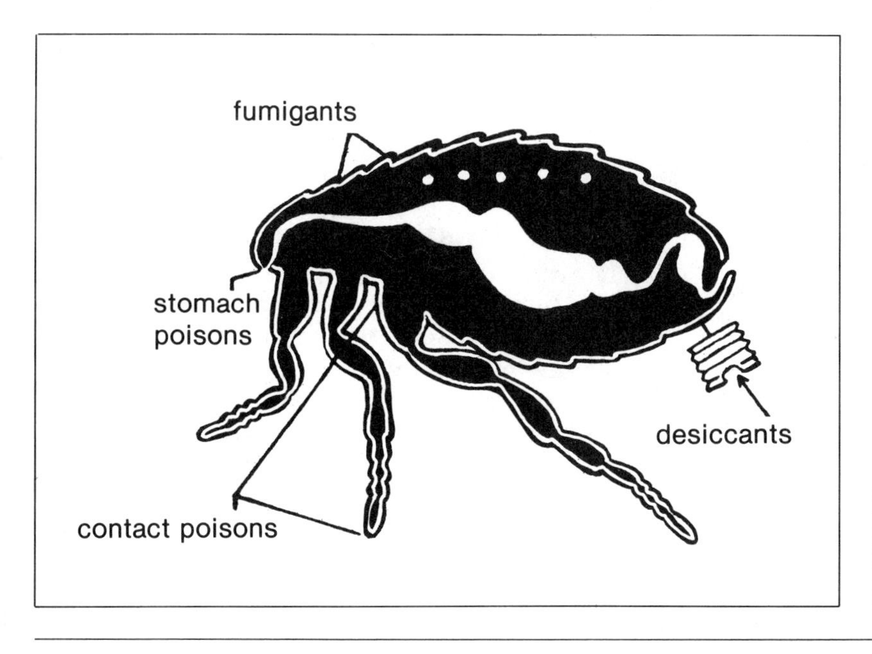

Figure 9-18 Types of insecticides. From *Insecticides For the Control of Insects of Public Health Importance.* U.S. Dept. Health, Education and Welfare.

sects, although sodium arsenite solution is sometimes used in ant baits.

The fluorine group. Sodium fluoride will successfully control roaches. It is usually used in a mixture with other chemicals. Since sodium fluoride is highly toxic to human beings, it should never be stored with foods. Other insecticides in this group are cryolite and sodium fluorosilicate.

Borax. Sodium tetraborate (borax) has been an effective stomach poison for many years. It is usually purchased in tablet form and can be used successfully against roaches.

Contact poisons pass through the body wall or tarsi of the insect. Most of these poisons affect the nervous system. They are available as space sprays, which pass into the insect as it crawls, flies, or rests, and as residual sprays, which are applied to areas like baseboards, cracks, and the space under cabinets, refrigerators, and stoves. As the insect crawls over the poison it absorbs it until a lethal dose is obtained.

Insecticides of plant origin. Chemicals from plants furnish some of the most widely known and safest insecticides. The two most common of these chemicals are pyrethrum and rotenone. Pyrethrum is one of the few insecticides permitted in kitchens and dining rooms. It is the primary insecticide in many insect sprays and aerosols because of its quick action. It is frequently combined with other insecticides for maximum killing action. Allethrin is a synthetic chemical that is similar to one of the active ingredients of pyrethrum. It is effective and safe for space spraying.

Chlorinated hydrocarbons contain chlorine, hydrogen, and carbon. Some also contain other elements. They are primarily central nervous system poisons. They were highly effective when they first appeared in the 1940s, but many insects have become resistant to them and they tend to persist in the environment for long periods of time. One of them, DDT, has been banned in the United States for almost all uses. Lindane and Chlordane act as a stomach poison against roaches and flies, but their general use indoors is not recommended because their residue vaporizes readily into toxic fumes. Roaches and flies have become increasingly resistant to them. Dieldrin has been approved for indoor treatment for cockroaches.

Organophosphates are derived basically from phosphoric acid. They have tended to replace the chlorinated hydrocarbons because they are effective against some insects, such as roaches and flies, that have become resistant to other insecticides. They also remain in the environment for a short period of time, that is, they are biodegradable. Malathion, one of the safest insecticides, is effective against flies and roaches. Dichlorvos is effective in different forms against flies and roaches. It is highly effective when incorporated in resin strips or pellets and placed in enclosed areas. Dichlorvos is not approved, however, for use in kitchens, restaurants, or other areas where food is prepared or served. Diazinon, Fenthion, and Ronnel are effective against cockroaches and flies.

Carbamates are derivatives of carbamic acid. Most contain nitrogen rather than chlorine (chlorinated hydrocarbons) or phosphorus (organophosphates). The majority are contact poisons. Proposur acts as a stomach and contact poison. It is used in bait form for cockroach control and in spray form for fly eradication. It is a quick-acting poison and must be used with care.

SUMMARY

The major rodent pests in foodservice establishments are the house mouse, roof rat, and Norway rat. The major insect pests are flies and cockroaches. Sanitation is the most important weapon in the control of these pests.

Mice dwell in any safe place near food and water. In buildings, they are generally found between walls, in spaces and corners, or in any place hidden from view that is relatively undisturbed. Their choice of food depends on what is available, but they eat most of what man eats.

Fresh mouse droppings are usually soft, moist, shiny, and dark. They are very small, pointed at both ends, and are most numerous along runways, near harborage, in secluded corners, and near food supplies.

Rats are indicators. Their presence in a community or area is firm evidence that sanitary practices have broken down. In buildings they live between walls, underneath floors, and in undisturbed rubbish and stored materials. The Norway rat and the roof rat do not usually inhabit the same area due to the aggressive habits of the Norway rat.

In general, rats eat the same food as man. Fresh rat droppings are usually moist, soft, shiny, and dark, and are most numerous along runways and near harborage and food supplies. Norway rat droppings vary in shape from having blunt ends to being spindle-shaped in appearance. Roof rat droppings are usually smaller with blunt ends.

Rats and mice thrive wherever food and harborage are available. To achieve successful rodent control, such sanitation measures as prompt garbage removal, harborage elimination, proper food storage, and ratproofing must be followed diligently. Proper sanitation consists chiefly of adequate refuse storage and maintenance of buildings and premises.

Rodenticides are supplemental to, but not a substitute for, proper sanitation and exclusion. There are two types of rodenticides: the single-dose type, which is fatal to the rodent through a single feeding, and the multiple-dose type, which requires many feedings to be effective.

The known habits of flies make them important vectors of all bacteria with which they come in contact. Their sticky feet and hairy bodies collect and transmit microorganisms as they walk and feed on manure and garbage. The fly often eats and eliminates at the same time. Their saliva and feces are both laden with bacteria.

Flies can be as effective in spreading intestinal diseases as are fingers, dirty eating utensils, and contaminated food. They transmit anthrax, cholera, ophthalmia, typhoid fever, polio, shigellosis, and several food-poisoning illnesses.

Cockroaches are found in every industrial and domestic habitat. Their eating habits include almost any type of food. While feeding, they regurgitate and eliminate on the food. They are associated with sewers, garbage, and filth, and from these sources they collect and carry many pathogenic organisms. Among the diseases transmitted are typhoid fever, cholera, rickettsial pox, shigellosis, salmonellosis, and many other diarrheal diseases.

Basic principles of sanitation—cleanliness, proper food storage, sanitary refuse storage, maintenance of buildings and premises—are necessary to keep insect populations at a low level. Basements and storerooms should be checked periodically at night for roaches.

There are two types of insecticides in common use—stomach poisons and contact poisons. Stomach poisons must be swallowed to cause death. Contact poisons pass through the breathing pores on the body of the insect. They are available as space sprays and residual sprays.

Review Questions

1. a. What rodents are of major importance? b. What insects are of major importance to the foodservice operator?

2. What is the most important principle in the control of those rodent and insect pests that infest the foodservice establishment?

3. What sanitary practices are required to control rodent and insect pests?

4. What two environmental conditions will always attract rodents and insects?

5. What are rodents?

6. a. How many litters can the house mouse bear during its lifetime? b. How many mice are in one litter?

7. Where can house mice be found?

8. Describe the six senses of the house mouse.

9. Where will mice nest?

10. Describe the feeding habits of the mouse.

11. a. What two species of rat most affect man? b. Which is most common? c. Which is largest? d. Which is the most aggressive?

12. a. How many litters can the rat bear during its lifetime? b. How many young are in a litter?

13. Describe the six senses of the rat.

14. What does the rat prefer for food?

15. Describe the feeding habits of the rat.

16. What two major signs can indicate the presence of rats and mice?

17. List four characteristics of a good garbage container.

18. How can rats and mice make stored food unusable?

19. a. How do single-dose rodenticides differ from multiple-dose rodenticides? b. List two examples of each type of rodenticide.

20. In what ways is man affected by insects?

21. Why are flies dangerous insects to allow around food?

22. Why do open dumps and piles of rotting materials make fly control difficult?

23. How does a feeding fly transmit pathogenic microorganisms?

24. a. Where are cockroaches found? b. In what respect are their feeding habits like those of the fly? c. Why are they of public health importance?

25. List the four common species of cockroach that occur in the United States and briefly describe each.

26. Describe the two types of insecticides and list three of each type.

References

Barnett, S. A. (1963). *The Rat, A Study in Behavior*. Aldine Publishing Co. Chicago, Illinois.

Bentley, E. W. (1972). "A Review of Anticoagulant Rodenticides in Current Use." *Bull. Wld. Hlth. Org.* 47:275–280.

Brooks, J. E. (1973). "A Review of Commensal Rodents and Their Control." *Critical Rev. Environmental Control* 3:405–453.

Brooks, J. E. and A. M. Bowerman. (1975). "Anticoagulant Resistance in Rodents in the United States and Europe." *J. Environ. Health* 37:537–542.

Center for Disease Control. (1968). "Public Health Pesticides." *Pest Control*. March: 1–16.

Center for Disease Control. (1972). "Public Health Pesticides." *Pest Control*. April: 1–21.

Davis, D. E. (1972). "Rodent Control Strategy." In *Pest Control Strategies for the Future*. National Academy of Sciences. Washington, D.C.

Ebeling, W. (1975). *Urban Entomology*. University of California, Division of Agricultural Sciences. Los Angeles, California.

Gratz, N. G. (1973). "A Critical Review of Currently Used Single-dose Rodenticides." *Bull. Wld. Hlth. Org.* 48:469–477.

Greensberg, B. (1964). "Experimental Transmission of Salmonella typhimurium by Houseflies to Man." *Amer. J. Hyg.* 80:149–155.

Guthrie, R. K. (1972). *Food Sanitation*. Avi Publishing Co. Westport, Connecticut.

Hayes, W. J., Jr. (1969). "Pesticides and Human Toxicity." *Ann. N.Y. Acad. Sci.* 160:40–54.

Hopkins, M. (1953). "Distance Perception in Mus musculus." *J. Mammology* 34:393.

Hulbert, R. H. and E. R. Krumbiegel. (1972). "Synthetic Flavors Improve Anticoagulant Type Rodenticides." *J. Environ. Health* 34:407.

Li, H. and D. E. Davis. (1952). "The Prevalence of Carriers of Leptospira and Salmonella in Norway Rats of Baltimore." *Amer. J. Hyg.* 56:90–99.

Lindsay, D. R., W. H. Stewart, and J. Watt. (1953). "Effect of Fly Control on Diarrheal Disease in an Area of Moderate Morbidity." *Pub. Health Rep.* 68:361–367.

Longree, K. (1972). *Quantity Food Sanitation*. John Wiley and Sons, Inc. New York.

Ludlam, G. B. (1954). "Salmonella in Rats with Special Reference to Findings in a Butcher's By-Products Factory." *Mon. Bull. Min. Health* 13:196–203.

Moorehead, S. and H. H. Weiser. (1946). "The Survival of Staphylococcol Food Poisoning Strain in the Gut and Excreta of the House Fly." *J. Milk Food Technol.* 9:253–259.

Olson, T. A. and M. E. Reuger. (1950). "Experimental Transmission of Salmonella oranienburg Through Cockroaches." *Pub. Health Rep.* 65:531–540.

Ostrolenk, M. and H. Welch. (1942). "The House Fly as a Vector of Food Poisoning Organisms in Food-Producing Establishments." *Amer. J. Pub. Health.* 32:487–494.

Pratt, H. D. and K. S. Littig. (1974). "Insecticides for the Control of Insects of Public Health Importance." U.S. Department of Health, Education and Welfare. Publication no. (CDC) 76-8229. Center for Disease Control. Atlanta, Georgia.

Pratt, H. D. and W. H. Johnson. (1975). "Sanitation in the Control of Insects and Rodents of Public Health Importance." U.S. Department of Health, Education and Welfare. Publication no. (CDC) 77-8138. Center for Disease Control. Atlanta, Georgia.

Pratt, H. D. and R. Z. Brown. (1976). "Biological Factors in Domestic Rodent Control." U.S. Department of Health, Education and Welfare. Publication no. (CDC) 76-8144. Center for Disease Control. Atlanta, Georgia.

Pratt, H. D., K. S. Littig, and H. G. Scott. (1975). "Flies of Public Health Importance and Their Control." U.S. Department of Health, Education and Welfare. Publication no. (CDC) 75-8218. Center for Disease Control. Atlanta, Georgia.

Pratt, H. D., B. F. Bjornson, and K. S. Littig. (1976). "Control of Domestic Rats and Mice." U.S. Department of Health, Education and Welfare. Publication no. (CDC) 76-8141. Center for Disease Control. Atlanta, Georgia.

Pratt, H. D., K. S. Littig, and H. G. Scott. (1976). "Household and Stored-Food Insects of Public Health Importance and Their Control." U.S. Department of Health, Education and Welfare. Publication no. (CDC) 76-8122. Center for Disease Control. Atlanta, Georgia.

Rueger, M. E. and T. A. Olson. (1969). "Cockroaches (Blattaria) as Vectors of Food Poisoning and Food Infection Organisms." *J. Med. Ent.* 6:185–189.

Salthe, O. and C. Krumwiede. (1924). "Studies on the Paratyphoid-Enteritidis Group." *Amer. J. Hyg.* 4:23–32.

Schein, M. W. and H. Orgain. (1953). "A Preliminary Analysis of Garbage as Food for the Norway Rat." *Amer. J. Trop. Med. Hyg.* 2:1117–1130.

Southwick, C. H. (1955). "Regulatory Mechanisms of House-Mouse Populations. Social Behavior Affecting Litter Survival." *Ecology* 36:627–634.

Staff, E. J. and M. L. Grover. (1936). "An Outbreak of Salmonella Food Infection Caused by Filled Bakery Products." *Food Research* 1:5.

Watt, J. and D. R. Lindsay. (1948). "Diarrheal Disease Control Studies. I. Effect of Fly Control in a High Morbidity Area." *Pub. Health Rep.* 63:1319–1334.

Welch. H., M. Ostrolenk, and M. T. Bartram. (1941). "Role of Rats in the Spread of Food Poisoning Bacteria of the Salmonella Group." *Amer. J. Pub. Health* 31:332–340.

Worth, C. B. (1950). "Field and Laboratory Observations on Roof Rats, Rattus rattus Linnaeus, in Florida." *J. Mammalogy* 31:293–304.

I. Location
II. Manual Dishwashing
  A. Regulations
  B. Washing Operations
  C. Glasses and Silverware
  D. Pots and Pans
  E. Types of Soil
  F. Water Temperatures
III. Mechanical Dishwashing
  A. Types of Dishwashing Machines
  B. Regulations
  C. Hot Water Temperatures
  D. Machine Dishwashing
  E. Pot- and Panwashing Machines
IV. Summary

Keeping things clean is a major responsibility of any person employed in the foodservice industry. Waitress or manager, cashier or busboy, dish machine operator or chef—all play an important role in cleanliness, which is not only a major part of the general environment of any foodservice establishment, but also directly affects the cleanliness of the eating service, cooking utensils, and food equipment. Dishes, silverware, glasses, and cups must be clean and shiny. Careful adherence to required dishwashing procedures is the main factor in this cleanliness.

Pond and Hathaway described an outbreak of 64 cases of acute gastroenteritis in a group of 322 graduate students. Very strong evidence indicated that transmission was through the recontamination of eating utensils by one of the students, a part-time worker in the cafeteria, whose job required the handling of eating utensils immediately after washing and sanitizing took place. An outbreak of infectious hepatitis, which resulted from contamination of washed and reused trays in a school cafeteria was described in chapter 8. Borrelia vincente (trench mouth) has been found on beverage glasses supposedly washed and sanitized.

The Food Service Sanitation Manual (U.S. Public Health Service, 1976) clearly states the regulations of the U.S. Public Health Service with regard to sanitation in the kitchen:

1. Tableware shall be washed, rinsed, and sanitized after each use.
2. To prevent cross-contamination, kitchenware and food-contact surfaces of equipment shall be washed, rinsed, and sanitized after each use and following any interruption of operations during which time contamination may have occurred.
3. Where equipment and utensils are used for the preparation of potentially hazardous foods on a continuous or production-line basis, utensils and the food-contact surfaces of equipment shall be washed, rinsed, and sanitized at intervals throughout the day. . . .

Every staff member must be concerned with the work of the dishroom crew. If tableware is recontaminated after proper washing and sanitizing, an expensive investment in dishwashing equipment and detergents will have been wasted and the sanitary procedures practiced by the dishroom crew will be useless. On the other hand, if a poor job is done by the dishwashing crew (Fig. 10-1), dirty tableware will result and customers will not return another time.

(a)

(b)

Figure 10-1 The result of ineffective dishwashing. (a) Cup with lipstick. (b) Fork with food remaining between tines.

LOCATION

To ensure a smoothly functioning, rapid dishwashing program, the operational area should be chosen carefully. Since most dishware is still carried by hand or transported on rolling carts, its location should be as close to the eating areas as possible. The problem of closeness is not a major one, however, if a conveyor system is used. In any case, dishware should be moved from the dining area to the dishwashing area as rapidly as possible in order to prevent bacterial growth on the uneaten food (soil) and possible contamination of the staff and premises.

Space should be provided for an area of temporary storage of dishware and utensils. Dishwashing machines and their accompanying scrapping and stacking racks are usually purchased to accomodate loads based on an average number of customers served in a given time period. During peak business hours, however, the number of items to be processed is frequently greater than the system can handle.

The pot-washing area should be close to the baking and main cooking areas since these departments use this type of utensil the most. It should also be close to the garbage area. Portable shelves are recommended to hold clean pots and pans and to facilitate transportation to those areas requiring them.

MANUAL DISHWASHING

The size and nature of some foodservice operations make washing dishes and silverware by hand the most feasible method. Satisfactory results can be achieved if the necessary procedures are followed to obtain safe and clean tableware. Even in a large establishment, manual dishwashing might be necessitated by equipment breakdown.

Regulations

In manual dishwashing, the first requisite is adequate space in which to operate. A three-compartment sink (Fig. 10-2) must be used. The first compartment is used for washing, the second for rinsing, and the third for sanitizing. These are three key processes that must be done properly in order to ensure a supply of sanitary dishware. Eating and drinking utensils that have not been properly washed, rinsed, and sanitized can become carriers of disease-causing microorganisms.

Chapter 5 of the Food Service Sanitation Manual states:

1. For manual washing, rinsing, and sanitizing of utensils and equipment, a sink with not fewer than three compartments shall be provided and used. Sink compartments shall be large enough to permit the accomodation of the equipment and utensils, and each compartment of the sink shall be supplied with hot and cold potable running water.
2. Drain boards or easily movable dish tables of adequate size shall be provided for proper handling of soiled utensils prior to washing and for cleaned utensils following sanitizing and shall be located so as not to interfere with the proper use of the dishwashing facilities.
3. Equipment and utensils shall be preflushed or prescraped and, when necessary, presoaked to remove gross food particles and soil.
4. Except for fixed equipment and utensils too large to be cleaned in sink compartments, manual washing, rinsing, and sanitizing shall be conducted in the following sequence:
   a. Sinks shall be cleaned prior to use.
   b. Equipment and utensils shall be thoroughly washed in the first compartment with a hot detergent solution that is kept clean.
   c. Equipment and utensils shall be rinsed free of detergent and abrasives with clean water in the second compartment.
   d. Equipment and utensils shall be sanitized in the third compartment . . . .

Dish tables and drainboards must be spacious enough to permit proper handling of soiled tableware before washing, and clean tableware after washing and sanitization. Construction of sinks, dish tables, and drainboards should be of galvanized metal or an easily

Figure 10-2 Three-compartment sink

cleanable, durable material. They should be sloped so as to be self-draining. All brushes and other tools, dish baskets, and so forth, designed for use in the dishwashing operation should be readily available, kept clean, and replaced whenever necessary.

Before washing, dishes should be sorted and stacked on the soiled dish table or drainboard. Stained dishes or those with dried-on or baked-on food soil will need presoaking. A large container holding a warm detergent solution may be placed on the floor under the drainboard and conveniently used for this purpose. Special brightening and stain removing detergents are available, if needed.

Silverware is also presoaked to loosen soil and tarnish. Usually a shallow aluminum pan or a stainless steel pan having the bottom lined with aluminum foil is placed on the soiled dish table. The silverware is soaked for about 15 minutes in a warm detergent solution. There are special silver dips for this job, if needed.

Washing Operation

The washing operations proceed as follows:

1. Scrape, prerinse, or both.
   Purpose: To remove loose soil from dishes, thus keeping wash water cleaner and more free of bacteria. The cleaner the wash water, the easier it is to maintain detergent strength at required levels.
   Method: Gross food soil should be scraped by hand or squeegeed through the table opening into the garbage container beneath the soiled dish table. Loose soil should be preflushed from dishes with the power spray.
2. Wash in sink number 1.
   Purpose: To remove all visible soil from dish surfaces, including the backs of dishes.
   Method: The sink should be cleaned and filled to the proper level with clean, hot water (120°–125°F.) [49–52°C]. To ensure correct water temperature, a permanent thermometer should be attached to the wall with the bulb immersed in the wash water behind a guard. The correct amount of detergent should always be used. The amount is based on the

quantity of water used in the sink. A sponge or a brush with bristles firm enough to provide the required friction should be used. The washed dishes should be placed in a dish basket so that wash water will drain from them and not weaken the rinse solution. This also prevents bacterial contamination of the rinse water.

3. Rinse in sink number 2.

   Purpose: To remove washing solution or other material from dish surfaces.

   Method: The dishes should be thoroughly rinsed in hot water (140°F.) [60°C] for several minutes, moving the long-handled dish basket up and down in the rinse water. For effective results, the rinse water must be kept clean.

4. Sanitize in sink number 3.

   Sanitizing and the approved methods of handling, storing, and reuse of sanitized dishes and utensils will be discussed in detail in chapter 12.

Glasses and Silverware

Glasses should be washed first, when the washing solution is the cleanest. A brush or sponge that will reach all inner surfaces should be used (Fig. 10-3). In addition to presoaking,

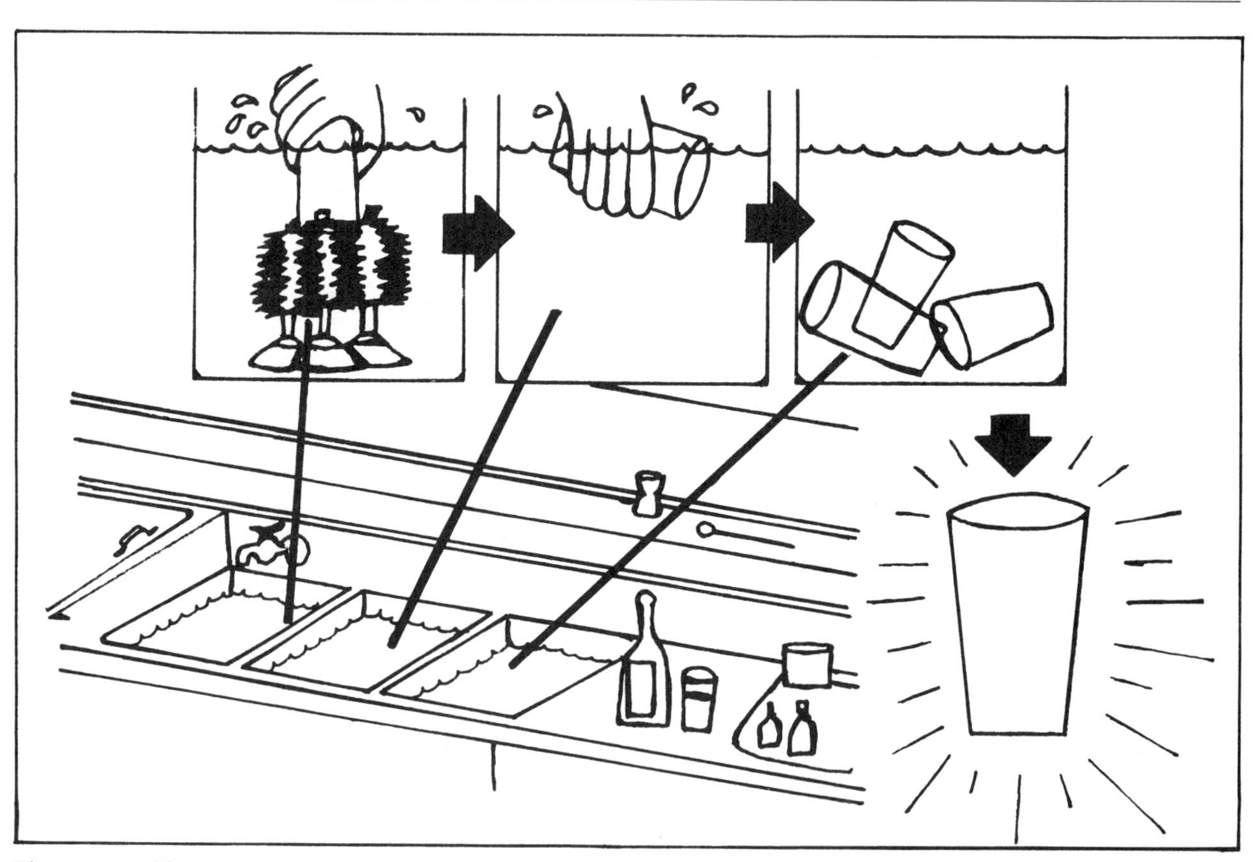

Figure 10-3 Three steps in washing of glasses. Brushes must be kept in good condition. No air should be trapped in glasses during washing and sanitizing.

Figure 10-4 Excellent pot-washing facilities. Note roomy drainboard on left for dirty pots. Next two compartments are for rinse and final rinse. Slanted draining shelf, open at left end, allows water to drain into washing sink.

silverware also requires a clean, hot detergent solution for washing and the application of friction to remove all soil. Careful inspection is necessary during the rinse operation to detect any soil that might remain on the tines of forks.

Pots and Pans

The same procedure of scrape, prerinse, wash, rinse, and sanitize that apply to the washing of dishware also apply to the washing of pots and pans (Fig. 10-4). After preliminary scrapping, pots and pans are often soaked in a warm detergent solution in sink 1. Even though this is an excellent practice, trouble develops if the same solution, grown cold and greasy, is then used for the washing operation. Pots and pans should be washed in a mild detergent solution, which will not damage metal, rinsed, and sanitized at 170°F. (77°C) for one minute or immersed in an approved sanitizing solution.

Types of Soil

The washing crew must contend with several types of soil. They vary in complexity, composition, and degree of adherence to surfaces. There is no single agent that will remove them all. Among the more common types of soil are:

1. Freshly deposited soil. This is the soil that remains on dishes and utensils after a meal or preparation of a meal. It is the easiest soil to remove because washing usually occurs while the materials are still moist.
2. A thin film remaining after the washing process is often difficult to detect, yet may be capable of sustaining bacterial life. Dishes coated with film lose their lustre, become stained, and require time-consuming soaking and special destaining and cleaning operations. Glasses no longer sparkle; pots and pans feel unclean to the touch and accumulate flavors of various foods. This type of soil usually results from ineffective cleaning methods, use of the wrong detergent, or flushing with plain water.
3. Built-up deposits. The employment of ineffective cleaning methods permits a day-by-day accumulation of soil. Frequently, minerals are incorporated with the various food materials and a hard, difficult to remove soil results.
4. Dried deposits. These soil accumulations result when dishes and utensils stand for long

periods of time or are exposed to high temperatures, allowing fresh, soft soil to dry, harden, and form a crusty deposit that is difficult to remove.

Water Temperatures

The maintenance of proper water temperatures during both manual and machine dishwashing is stressed by public health ordinances, sanitarians, and by dish machine and detergent manufacturers. These water temperatures play an important role in the breakdown and removal of soil from eating ware. One soil encountered in food establishments is fat. The various fats first must be softened or melted before the combination of wash solution and friction makes complete removal possible. If the recommended wash temperature of 120–125°F. (49–52°C) is maintained, these fats and oils will be removed easily from dishes, silverware, and utensils. See the following list of melting points.

Type of Fat or Oil	Melting point (Fahrenheit degrees and Centigrade equivalents)
butter fat	95°–107° (35–42°C)
cotton seed oil	41°–51° (5–11°C)
lard (outer fat)	86°–104° (30–40°C)
lard (outer leaf)	111°–118° (44–48°C)
tallow (beef)	104°–122° (40–50°C)
tallow (mutton)	111°–122° (44–50°C)

Hot water also increases the chemical activity of the detergent being used, permitting it to work at its peak efficiency. Handwashing compounds are formulated to act effectively at the wash temperatures specified for this type of cleaning. Water temperatures given for machine washing and rinsing can be much higher, since the hands are not involved with the wash solution. Different types of detergents, too, are formulated for use in the higher temperature water.

MECHANICAL DISHWASHING

The basic requirements of wash, rinse, and sanitize are common to both hand and machine dishwashing. However, since a wide range of dishwashing machine models are available to meet the needs of all types of foodservice installations, there are variations in the steps that accomplish these basic procedures. Properly operated and maintained, machine dishwashing is more reliable than manual dishwashing in removing soil and bacteria from tableware and utensils.

Types of Dishwashing Machines

There are several types of dishwashing machines. Built-in or under-counter types are the smallest of the commercial dishwashers. They are used in nursing homes, small lunch counters, snack shops, and for glass washing in taverns. They are similar to household dishwashers in that the door opens downward and the rack is pulled out onto the open door for loading and unloading. They are available as free-standing units or can be built into a counter or bar.

The wash and rinse waters are pumped through the same revolving wash arm. The tank is automatically drained after each wash or rinse cycle. The entire operation is controlled by an automatic timer. It can be equipped with a chemical sanitizing dispenser.

The roll-type machine is the smallest, above the table, spray-type model. It can be fitted into a small space, even in the food-serving area. The roll-type machine has a rounded top that opens for loading. A wash water tank holds the detergent solution. After the operator has placed a rack of dishes in the machine and closed the top, he pushes the start button or control handle. The wash solution is then circulated in the required volume through a single revolving wash arm in the bottom of the machine. It is forced through the wash sprays up the sides of the curved lid and falls back, by gravity, over the dishes.

Rinsing is accomplished through wing-type sprayers above and a revolving rinse arm below. This 180°F. (82°C) rinse, with its water flowing from a booster heater, is automatically timed.

The door-type machine is shaped like a box, with opposite doors sliding vertically upward for loading from either side (Fig. 10-5). It may also be equipped with adjacent doors for a corner operation. It, too, is classified as a single-tank, stationary-rack machine, but it has a greater capacity than the roll-type model.

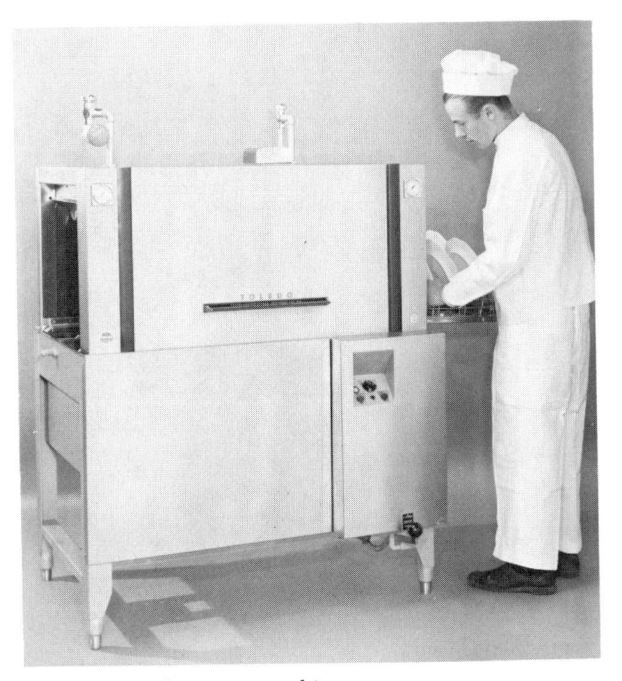

Figure 10-5 Door-type washing machine

The door-type model is used in many large operations as a secondary unit for glass washing. It is popular in restaurant operations that do not require the load capacity of the large conveyor-type machines. It meets the needs of many school lunch programs, inplant feeding installations, hospital isolation wards, and so forth.

The door-type machine has a single tank for the water and detergent, which are circulated in the required volumes and at the proper temperature by an electric motor-driven pump. The wash solution is pumped through the spray nozzles of two revolving spray arms, one above, the other below the rack of dishes. A final cycle of fresh, hot water rinses the dishes through a separate set of upper stationary and lower revolving rinse areas. All cycles have automatic time controls.

The door-type chemical sanitizing machine may be found in small nursing homes and similar operations, or it may be used for washing glasses. It is a single-tank, stationary-rack model. The washing operation is similar to other door-type machines, but all wash water containing detergent is discharged as waste following each wash cycle.

The final rinsing is accomplished with fresh water into which a sanitizing agent, either sodium hypochlorite or iodophor, is injected in sufficient quantity to provide the required sanitizing solution concentration. The operational cycle is automatically timed. Since this machine is designed to operate on building supply water temperature (120°–140°F.) [49–60°C] and to sanitize effectively at these temperatures, a booster heater is not required.

Conveyor-type machines, the largest models of dishwashing machines, use a conveyor to move dishware through the wash and rinse cycles. These machines are found in large hotels and restaurants, hospitals, or other types of heavy volume foodservice operations. There are two general types: the rack conveyor, into which the operator slides a loaded rack that the automatic conveyor carries through the wash and rinse cycle, and the rackless conveyor, on which about 70 percent of all dishes can be placed directly on the moving belt of the machine.

1. Rack conveyors. There are several types of rack conveyors. Single-tank rack conveyors are often used in school lunch programs, small institutions, inplant feeding operations, medium sized restaurants, and other similar institutions. A single tank contains the water and detergent at the required temperature. This wash solution is pumped through stationary or revolving upper and lower spray arms. Once the machine has been turned on, the wash solution flows constantly, whether a dish rack is in the machine or not.

After passing through the wash spray pattern, the dish rack is then conveyed to the "clean" end of the machine where it activates a fresh water (180°F.) [82°C] final rinse application through spray nozzles located above and below the dishes. The speed of the conveyor is automatically timed for the proper exposure of dishware to wash and rinse spray patterns.

Two-tank and multiple-tank rack conveyors are found in the dish rooms of larger foodservice operations (Fig. 10-6). In addition to a power wash and a final rinse, a two-tank machine has a power rinse that reuses the final rinse water, spraying it over the dishes to remove most of the detergent water and any remaining food soil. In the wash tank, the detergent solution flows through upper and lower stationary spray arms. Some models contain a power prewash, with its own motor, tank, and pump, and are referred to as multiple-tank conveyors (Fig. 10-7).

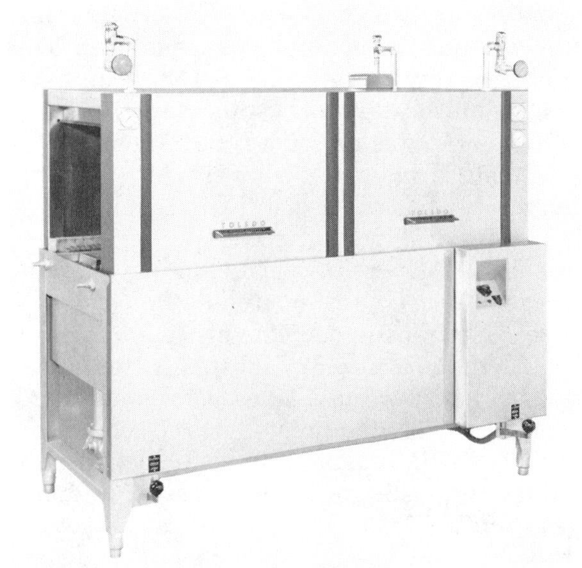

Figure 10-6 Two-tank, conveyor-type dishwasher

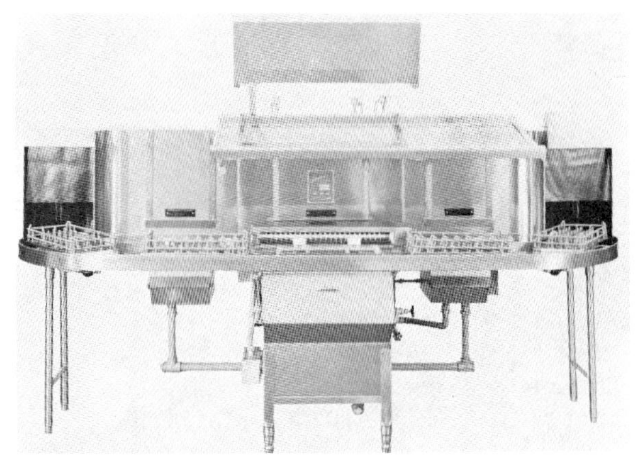

Figure 10-7 Multiple-tank, continuous conveyor dishwasher with prewash, wash, prerinse, and final rinse.

After the washing cycle, the rack is conveyed through a power rinse tank, where the dishes are rinsed free of most of the detergent solution by means of hot water pumped through stationary upper and lower spray arms, reusing water from the final rinse. The rack then passes through a final rinse, provided by hot water at a sanitizing temperature passing through upper and lower spray nozzles distributed evenly across the conveyor. An ample supply of hot water can be supplied through a booster heater. The conveyor is timed to carry the dishes through the series of wash and rinse cycles at a predetermined rate to ensure thorough washing and rinsing, and effective sanitizing.

2. Rackless conveyors. Rackless conveyors are available in single-tank and two-tank models and models with a power prewash (Fig. 10-8). Wash and rinse cycles closely approximate those of rack-type conveyors.

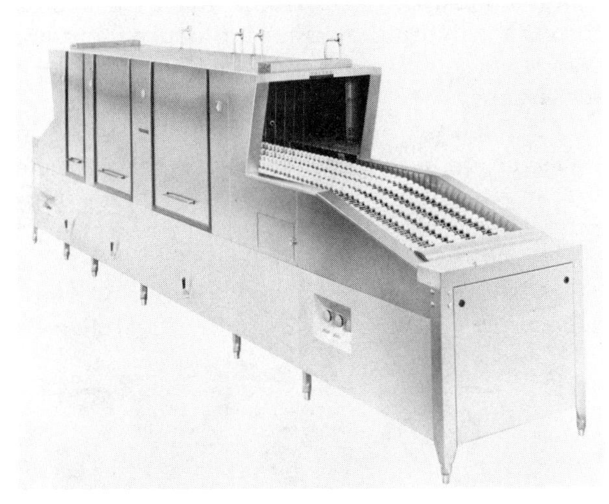

Figure 10-8 Rackless dishwasher, two-tank unit with prewash

Since trays are more easily handled when placed on the edge of the conveyor, rather than in a rack, many cafeterias prefer this type of dishwashing machine. A switch at the clean end of the machine is activated when a rack or dish strikes it, stopping the conveyor. Thus, dishes are prevented from being broken against the end of the machine if unloading is slow. If a blockage occurs in the machine, the machine automatically shuts off.

Regulations

Chapter 5 of the Food Service Sanitation Manual states these requirements of dishwashing machines:

1. Cleaning and sanitizing may be done by spray-type or immersion dishwashing machines or by any other type of machine or device if it is demonstrated that it thoroughly cleans and sanitizes equipment and utensils.
2. Machine or water line mounted numerically scaled indicating thermometers, accurate to 3°F, shall be provided to indicate the temperature of the water in each tank of the machine and the temperature of the final rinse water as it enters the manifold.
3. Rinse water tanks shall be protected by baffles, curtains, or other effective means to minimize the entry of wash water into the rinse water. Conveyors in dishwashing machines shall be accurately timed to assure proper exposure times in wash and rinse cycles in accordance with manufacturers specifications attached to the machines.
4. Drain boards shall be provided and be of adequate size for the proper handling of soiled utensils prior to washing.
5. Equipment and utensils shall be flushed or scraped and, when necessary, soaked to remove gross food particles and soil prior to being washed in a dishwashing machine unless a prewash cycle is a part of the dishwashing machine operation. Equipment and utensils shall be placed in racks, trays, or baskets, or on conveyors, in a way that food-contact surfaces are exposed to the unobstructed application of detergent wash and clean rinse waters and that permits free draining.

Hot Water Temperatures

The hot water temperatures required for mechanical dishwashing fall into four categories. There may be slight variations in these temperature ranges.

1. Prerinse water temperatures. This water is used to remove as much food soil as possible from dishes before they are washed. Suggested temperatures range from 110°–130°F. (43–54°C), depending on the method used to prewash.
2. Wash water temperatures. This water, which combines with the detergent to form the wash solution, must be hot enough to remove all food soil from the dishes, dissolve grease from dish surfaces, and activate the detergent to its most effective performance relative to both cleaning ability and economy.

   For commercial spray-type machines, the following temperatures are recommended:

single-tank conveyor 160°F. (71°C)
stationary rack machines 150°F. (66°C)
multiple-tank conveyor 150°F. (66°C)
door-type chemical sanitizing model 120°–140°F. (49–60°C)

3. Power rinse temperatures. The power rinse water temperature in multiple-tank machines should be 160°–170°F. (71–77°C).
4. Final rinse temperature. The final sanitizing rinse is specified as 180°F. (82°C) before it leaves the rinse nozzles. This should ensure a minimum sanitizing temperature of 170°F. (77°C) on the dishware.

Machine Dishwashing

An effective cleaning job should remove soil from dishware and utensils. These items should be washed as promptly as possible, thus making the washing easier and more effective.

The basic steps in machine dishwashing are:

1. Scrapping. Scrapping is the removal from dishware of all pieces of solid or semi-solid food refuse after service. It prevents the clogging of spray nozzles with food particles, thus misdirecting the spray. It also reduces the bacterial content of the wash water and keeps the wash solution cleaner, thus lessening the need for too frequent changes of water and additions of detergent.

Figure 10-9 Scrapping by use of a flexible spray shower

Food soil can be scraped by hand or squeegeed into the garbage container. Water scrapping, by use of a flexible shower spray (Fig. 10-9) or a prewash unit where the dishes pass through a hydraulic scrapping department are also effective. Prewash units can be built into a dishwasher.

2. Racking. Scrapped dishes are placed in racks or on a pegged conveyor for passage through the dishwasher. Correct racking permits the wash solution and rinse water to contact all surfaces being cleaned.

Dishes should be racked without overlapping (Fig. 10-10) since the backs must also be

Figure 10-10 Properly racked dishes

washed. Creamers or other miscellaneous pieces of dishware should not be stacked on top of properly racked dishes. Such jumbled arrangements result in ineffective washing and possible breakage. Cups, bowls, and any other deep dishes should be inverted so that wash sprays can clean inner surfaces. Silverware should be spread in a thin layer on a silver rack (Fig. 10-11) or sorted and placed, handles

Figure 10-11 Properly spread and racked silverware

down, in perforated cylinders or compartmented, rectangular carriers (Fig 10-12).

3. Washing. The wash cycle removes all food soil from dishes and dissolves grease on dishware. Grease acts as a binder to hold soil on dishware.

The detergent solution is circulated over and over again and keeps flowing through the upper and lower stationary or revolving spray arms at a controlled gallon per minute rate. The correct water temperature is maintained by the heat source in the tank. The dishes are conveyed through the wash pattern at an automatically controlled speed.

Figure 10-12 Silverware properly placed in carrier

4. Power Rinse. The power rinse completely removes most of the soil-laden detergent water from the dishware and builds up heat in the dishware, thus aiding the final sanitizing rinse and helping to ensure a quicker drying time.

After the dishes are passed from the washing area to the power rinse area, the hot pumped rinse water is sprayed over the dishes through stationary or revolving upper and lower spray arms in the correct volume and at sufficient pressure to rinse thoroughly all surfaces of the dishes.

5. Final rinse. By the time the dishes reach the final rinse, they should be clean. The high temperature final rinse water removes remaining detergent and kills pathogenic microorganisms remaining on the dishware. *All previous operational steps lead up to this final vital procedure.*

The final rinse comes from hot water at sanitizing temperature, flowing in the required volume and pressure through upper and lower spray nozzles distributed evenly across the conveyor. A drying agent is often injected into the final rinse to promote quicker, spot-free drying of dishes. The dishware should now be allowed to air-dry. Hand-drying should be avoided. The washed dishes should be placed in a dish basket so that the wash water will drain from them and not weaken the rinse solution. This also prevents bacterial contamination of the rinse water.

Pot- and Panwashing Machines

Developments in the construction and cleaning procedures of mechanical potwashing machines are aimed at acquiring thoroughly clean pots and pans in the shortest possible time. The preparation of pots and pans before washing is recommended with some models; in others, no soaking or preflushing is required.

The items to be washed are usually held on a suction cup or base and washed in 140°F.

(60°C) water. The nozzle pattern of the wash spray is structured so that the largest volume of detergent solution is directed with high pressure through a rotating spray head to contact the inside surfaces of the items positioned on the suction base. Secondary sprays are directed from the nozzles to contact outside surfaces of the utensils to be washed. A 180°F. (82°C) high pressure rinse follows the wash cycle.

It is important to remember that dishware that is not clean cannot be sanitized. Bacteria encompassed in food soil are not reached during the sanitizing operation. This soil can serve later as a growth medium for these bacteria and any others that may later contaminate the dishware. Foodborne illness can thus result from dishes that are washed, but not totally cleaned. It may take only one incidence of food poisoning, in which the sanitation of the establishment is at fault, to destroy a business. Proper washing and sanitizing procedures therefore should be followed carefully. Sanitizing procedures will be discussed in the next chapter.

SUMMARY

Keeping things clean is a major responsibility of any person employed in the foodservice industry. Food-poisoning illness has been transmitted through contaminated dishware. U.S. Public Health Service regulations state:

1. Tableware shall be washed, rinsed, and sanitized after each use.
2. To prevent cross-contamination, kitchenware and food-contact surfaces of equipment shall be washed, rinsed, and sanitized after each use and following any interruption of operations during which time contamination may have occurred.
3. Where equipment and utensils are used for the preparation of potentially hazardous foods on a continuous or production-line basis, utensils and the food-contact surfaces of equipment shall be washed, rinsed, and sanitized at intervals throughout the day . . . .

To ensure a smoothly functioning, rapid, dishwashing program, this area should be as close to the dining area as possible. Space should be provided for temporary storage of used dishware and utensils, which may accumulate during peak business hours.

In manual dishwashing, a three-compartment sink is used. The first compartment is for washing, the second for rinsing, and the third for sanitizing.

The Food Service Sanitation Manual states:

1. For manual washing, rinsing, and sanitizing of utensils and equipment, a sink with not fewer than three compartments shall be provided and used.
2. Drain boards or easily movable dish tables of adequate size shall be provided for proper handling of soiled utensils prior to washing and for cleaned utensils following sanitizing . . . .
3. Equipment and utensils shall be preflushed or prescraped and, when necessary, presoaked to remove gross food particles and soil.
4. Except for fixed equipment and utensils too large to be cleaned in sink compartments, manual washing, rinsing, and sanitizing shall be conducted in the following sequence:
   a. Sinks shall be cleaned prior to use.
   b. Equipment and utensils shall be thoroughly washed in the first compartment with a hot detergent solution that is kept clean.
   c. Equipment and utensils shall be rinsed free of detergent and abrasives with clean water in the second compartment.
   d. Equipment and utensils shall be sanitized in the third compartment . . . .

The washing operation proceeds as follows: (1) scrape, prerinse, or both, (2) wash in sink number 1 with clean, hot (120°–125°F.) [49–52°C] detergent solution, (3) rinse in hot (140°F.) [60°C] water in sink number 2, and (4) sanitize in sink number 3. Glasses should be washed first, when the washing solution is the cleanest.

There are several types of dishwashing machines: (1) built-in or under-counter types, (2) roll-type, (3) door-type, (4) door-type chemical sanitizing machine, and (5) conveyor-type machines.

The Food Service Sanitation Manual states:

1. Machine or water line mounted, numerically scaled indicating thermometers, accurate to 3°F., shall be provided to indicate the temperature of the water in each tank of the machine and the temperature of the final rinse water as it enters the manifold.
2. Drain boards shall be provided and be of adequate size for the proper handling of soiled utensils prior to washing.
3. Equipment and utensils shall be flushed or scraped, and when necessary, soaked to remove gross food particles and soil prior to being washed in a dishwashing machine unless a prewash cycle is a part of the dishwashing machine operation. Equipment and utensils shall be placed in racks, trays, or baskets, or on conveyors, in a way that food-contact surfaces are exposed to the unobstructed application of detergent wash and clean rinse waters and that permits free draining.

Review Questions

1. What does the Public Health Service Food Service Sanitation Manual state with regard to dishwashing sanitation in the kitchen?

2. What is the result of a poor job by the dishwashing crew?

3. a. Where should the dishwashing area be located? b. Where should the potwashing area be located?

4. How are the three sinks used in a 3-compartment sink?

5. What sequence is required by the U.S. Public Health Service in a manual dishwashing operation?

6. What dish cleaning procedures should be used prior to washing?

7. List the steps to be followed in manual dishwashing and tell what each step should accomplish.

8. What types of food soil must dishwashing procedures cope with? Describe these types of soil.

9. What is the importance of water temperature in the breakdown and removal of soil from eating ware?

10. List and describe briefly the types of dishwashing machines.

11. Compare water temperatures in manual and machine dishwashing operations.

References

Economics Laboratory. (1973). *Scientific Cleaning and Sanitation Procedures for the Food Service Industry.* Economics Laboratory, Inc. St. Paul, Minnesota.

Guthrie, R. K. (1972). *Food Sanitation.* Avi Publishing Co. Westport, Connecticut.

Longree, K. (1972). *Quantity Food Sanitation.* John Wiley and Sons, Inc. New York.

Lyons, D. C. (1936). "The Incidence and Significance of the Presence of Borrelia vincenti and other Spirochaetes on Beverage Glasses." *J. Bacteriol.* 31:523.

National Institute for the Foodservice Industry. (1974). *Applied Foodservice Sanitation.* National Institute for the Foodservice Industry. Chicago, Illinois.

National Sanitation Foundation. (1965). *Commercial Spray-Type Dishwashing Machines.* Standard 3. National Sanitation Foundation. Ann Arbor, Michigan.

Pond, M. A. and J. S. Hathaway. (1947). "An Epidemic of Mild Gastroenteritis of Unknown Etiology Presumably Spread by Contaminated Eating Utensils." *Amer. J. Pub. Health* 37:1402–1406.

U.S. Department of Health, Education and Welfare. (1976). *Food Service Sanitation Manual.* U.S. Government Printing Office. Washington, D.C.

Vester, K. G. (1967). *Food-Borne Illness.* Food Service Guides. Rocky Mount, North Carolina.

11 Dishwashing Compounds

Many types of dishwashing compounds are now sold. They include soaps and detergents, whose use is followed by some type of sanitizing procedure, and detergent sanitizers, which take care of washing and sanitizing in one step. The dishwashing compound chosen depends on several factors:

1. the type of establishment and variety of foods served;
2. the volume of customers;
3. the ability of the compound to accomplish, under existing conditions, a number of purposes essential to satisfactory cleaning of the dishware;
4. the formulation of the compound insofar as it effects the hands, eyes, or clothing of the persons using it, or any undesirable effects it might have on the objects to be washed; and
5. the quality of the compound in comparison to its cost.

The manufacture of detergents is a complex process. Detergent chemists formulate many different combinations of ingredients designed to cope successfully with a variety of food soils. Modern dishwashing compounds contain a number of ingredients blended and proportioned to perform their cleaning functions with a minimum of trouble, and a maximum of speed, effectiveness, and safety.

CHARACTERISTICS OF DETERGENTS

Water Hardness

Water in various parts of the country possesses different characteristics, but the characteristic most commonly involved with dishwashing is the degree of "hardness." Water containing soluble calcium (lime), magnesium, and iron salts is known as hard water. Hard water has traveled through or over rock strata long enough to dissolve into itself appreciable amounts of mineral salts.

Hardness in water is undesirable for two reasons. First the calcium, magnesium, and iron ions (charged particles) in it react with soaps, forming an insoluble, sticky, soapy curd that sticks to whatever objects it contacts, giving them a dingy appearance. Soap performance is therefore highly dependent on water hardness. Soap in excess of what reacts with the minerals must be added to obtain cleansing action. Second, hard water is responsible for the formation of scale in boilers, hot water pipes, and dishwashing machines. High temperatures cause some of the minerals present in the water to come out of solution and adhere to the sides of pipes, dishwashing machines, and so forth as scale.

Certain chemicals can remove the lime and magnesium salts from water by making them insoluble. When this happens, however, dishwashing efficiency is greatly impaired. There will be a gradual buildup of scale inside the machine, the dishware will be dull and filmed, and water spots will be present on dried dishes, glasses, and silver. Therefore, one of the most important functions of a dishwashing compound is to keep mineral salts dissolved in the water rather than allowing them to become insoluble and cause scale and filming problems.

An accurate analysis of water with regard to hardness can be obtained from the city health department or from water department records. Most dishwashing compounds are available in different grades, to control varying degrees of water hardness.

Softened Water

For dishwashing purposes, softened water is simply water that the dishwashing compound has conditioned (softened) by one or more of its components reacting with the mineral salts in the water. These reactions form compounds that remain dissolved and do not interfere with the washing process.

Water softening equipment is frequently used in food and drink establishments, particularly in areas where the water is extremely hard. Chemicals in the water softening units are exchanged with the hardening chemicals in the water.

The softening of hard water is often accomplished by a process called ion exchange. In the zeolite type of water softener, water containing calcium and magnesium ions filters slowly through thick layers of coarse granules of zeolite. As this occurs, sodium in the zeolite is replaced by calcium and magnesium from the hard water. These troublesome minerals are thus replaced with an ion (Na^+) that will remain dissolved and cause no problems.

The synthetic resin type of water softener removes all ions, both positive and negative, from the water, thereby completely demineralizing it. One of the resins in the water softener removes the positive ions from the hard water and replaces them with hydrogen ions. These ions then react with negative ions in the water, forming acids. The acids are removed by a second resin, thereby demineralizing the water.

Water softening can be accomplished on a small scale by adding washing soda (sodium carbonate), trisodium phosphate, or borax (sodium borate). These chemicals react with calcium and magnesium salts, freeing all of the soap for cleansing action. Polymetaphosphates such as $(NaPO_3)_x$, form soluble complexes with calcium and magnesium ions, preventing them from reacting with soap. Many cleansing formulations contain polymetaphosphates as a water softening agent.

Surface Tension and Wetting Agents

The molecules within the body of a liquid are attracted equally in all directions by neighboring molecules; the resultant force on any one molecule within the liquid is therefore zero. This is not true of the surface molecules, however. They are pulled only from the sides and from below. This tends to pull the surface tight and reduce its area. The attraction between the molecules then becomes harder to break. This contracting force is called surface tension and is the reason that a needle or razor blade can be made to float on water. The surface acts as if it were a stretched membrane. Some insects can also run about on the surface of water, supported by surface tension.

Surface tension prevents water from spreading completely and evenly over and into other surfaces. This condition must be corrected before the wash solution can work effectively on the removal of food soil. Grease, for instance, resists removal from a plate or utensil by plain water.

When some substances (wetting agents) are dissolved in a liquid, they lower the surface tension of the liquid, thereby allowing the water molecules to separate. They do this by lessening the attraction between the water molecules themselves, allowing them more freedom to adhere to dirt, grease, and particles of food soil.

A dishwashing compound must, therefore, contain ingredients that lower the surface tension of the water. Then the wash solution can contact every area of the surface to be cleaned; it will enter through small cracks or openings in the food soil, reach the surface of the dirty dish, and push on under the soil to loosen it from the dishware. This action permits all ingredients of the compound to work effectively on the soil, remove it from the dishware, and hold it in the wash solution until rinsed away. Wetting agents, along with water softeners, are incorporated into most modern cleansing agents.

Fats, Oils, and the Dishwashing Compound

Fats and oils are familiar to every member of the dishwashing crew, including pot and pan washers. These soils do not mix well with water. One of the necessary functions of ingredients in the dishwashing compound is to break the grease drops into tiny droplets, which can be distributed throughout the wash solution.

This process is called emulsification. It is aided by heat. The higher water temperature used in machine dishwashing causes solid fats to melt more easily into a liquid form than temperatures used in manual dishwashing. Thus, they can be more easily emulsified. In addition, dishwashing detergents carry emulsifying agents as a part of their makeup.

Formation of Soap During the Wash Cycle

The soap formed in the dishwashing machine's wash tank by the combination of fat, hot water, and the alkali chemical content of the dishwashing compound will result in a varying amount of suds. A small amount of suds causes no problem; too much foaming, however, can lessen the wash pump's ability to work properly and can act as a cushion between the wash water and the dishware.

The control of suds can be accomplished by including a chemical in the dishwashing compound that prevents suds formation. In food establishments where the machine and the wash solution must handle more than the usual soil load, these chemicals are especially valuable.

Pollution

Many of the modern cleaning compounds, so carefully formulated to meet today's cleaning needs, were, only a few short years ago, the object of great concern to the cleaning materials industry.

Thousands of communities take care of the removal of community waste and refuse by means of sewage systems. Wastes of all kinds are flushed or discharged into sewer pipes, passed along into sewage disposal plants, and treated to make them safe and unobjectionable. They then flow out to mingle with the waters of lakes, streams, or other bodies of water.

Communities that do not have sewage treatment plants dump waste water and refuse untreated into bodies of water. In some areas, manufacturing plants have been permitted to discharge their untreated industrial wastes into nearby bodies of water.

Many types of sewage and other wastes are decomposed by strains of bacteria that live in the soil and in bodies of water. Some compounds, however, such as DDT, are not decomposed and remain in the water. This was the case with many of the early detergents. They resisted the action of those bacteria that decompose soaps and many other chemicals.

Detergent wastes became a part of the water pollution problem. When large amounts of detergent wastes were discharged regularly into the waters, their presence was easily noticed by the foam on the water's surface. Present-day detergents are composed of molecules that can be decomposed by bacteria that exist in the water. Chemicals that can be decomposed in this manner are said to be biodegradable.

SYNTHETIC DETERGENTS

A detergent is any cleansing substance; water and soap are both detergents. Synthetic detergents are synthetic compounds that act like soap; they are synthetic cleansing agents. Unlike soaps, synthetic detergents do not react with the calcium and magnesium salts in hard water.

Characteristics

Synthetic detergents can dissolve in both aqueous (water) and nonaqueous materials. This occurs because these substances include two distinct groups in their molecular structure. One group is water-soluble and is called the hydrophilic (attracts or absorbs water) group. In a schematic representation, it can be called the "head" group (Fig. 11-1). The other group is not water soluble on its own and is called the hydrophobic (does not absorb, or repels water) group. In a schematic representation, it can be called the "tail" group (Fig. 11-1). The hydrophilic group enables the detergent to dissolve and the hydrophobic group removes soils from surfaces. Because of their high activity on the surfaces of objects that they contact, detergents are called surface-active agents, or surfactants.

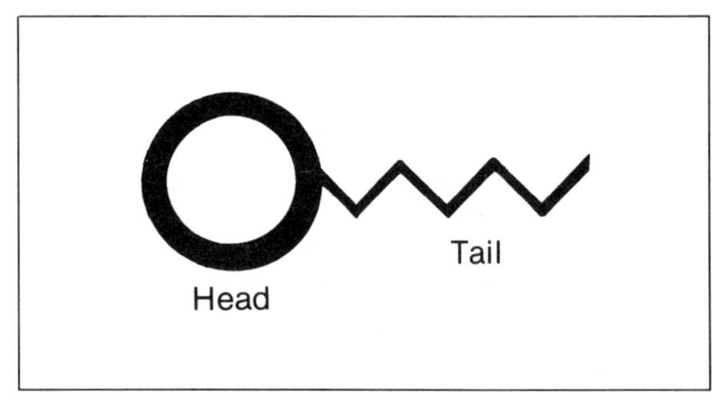

Figure 11-1 Schematic representation of a surface-active molecule showing "head" and "tail."

The hydrophilic group is usually added to a hydrophobic group through a chemical reaction or reactions, replacing a group that was originally present in the hydrophobic molecule. A group that will remove calcium or magnesium salts can also be attached, thus adding water-softening properties to the detergent molecule. Through the use of builders and additives, such as, phosphates, wetting agents, emulsifiers, foamers, optical brighteners, enzymes, bacteriostatic agents, ammonia, and a host of other chemicals, a cleaning product can be produced to meet virtually any cleaning requirement.

History

Synthetic detergents were developed by the Germans during World War I due to a shortage of fats in the soap industry. They were made by coupling certain alcohols with naphthalene and adding the sulphonate (HSO_3) group. They were good wetting agents but only fair detergents.

In the early 1930s, benzene replaced naphthalene in the detergent molecule and two types emerged: the alcohol sulphates and the alkyl aryl sulphonates. They were both sold as cleaning agents but did not appreciably affect soap sales. By the end of World War II, the alkyl aryl sulphonates had become the dominant detergent for general cleaning purposes and the alcohol sulphates were being used primarily by hair shampoo manufacturers.

By 1946, synthetic detergents had displaced soaps to a considerable extent in the field of fine laundering and dishwashing. In 1953, for the first time, the sales of synthetic detergents exceeded those of soaps. Detergent sales in the United States leaped from 4,500 tons in 1940 to 4,448,000 tons by 1972; soap sales fell from 1,410,000 tons to 587,000 tons during this same period. Since 1972, the sales of synthetic detergents have maintained a slow but steady rise, with liquid detergents accounting for much of the growth.

After the war, two major advancements occurred in detergent manufacturing that increased the efficiency and marketability of these products. The product then in use did not wash white cotton fabrics as white as possible. The dirt was being removed satisfactorily but the wash solution was not holding all of it in suspension and carrying it away. Some of it was resettling on the clothes and leaving them gray. The addition of a chemical (sodium carboxymethyl-cellulose or CMC) solved this problem by increasing the holding ability of the detergent.

The second advancement was the addition of such chemicals as washing soda, borax, sodium tripolyphosphate, and orthophosphates to the detergents. The success of synthetic detergents hinged, particularly, on the discovery and development of the complex phosphates, primarily sodium tripolyphosphate with its water-softening, soil-carrying, and emulsifying properties. These two advancements allowed synthetic detergents to be used for an ever greater variety of purposes.

From the late 1940s until the early 1960s, alkyl benzene sulphonate was the major ingredient used in dishwashing and laundry detergents. During the early 1960s, however, sewage treatment and water pollution problems became evident. The amount of foam on river waters was increasing noticeably, and foam began to appear in the tap water of homes and establishments near foamy rivers. It was obvious that current detergents possessed a major defect. Their molecules were not being decomposed (biologically degraded) by bacteria that exist in soil and water and that use many waste materials for nutrients. The branched-chain molecule of alkyl benzene sulphonate was highly resistant to attack by these bacteria. This problem was solved by the production of a straight-chain alkyl benzene sulphonate molecule, which is biodegradable.

In the late 1960s, another serious problem arose for the industry. Certain algae in many lakes began to reproduce at a tremendous rate. Their death clogged and fouled the water and killed other aquatic life. The blame was placed on the phosphate content of several types of manufactured products, among them detergents using sodium tripolyphosphate. The search for an adequate and safe replacement still continues.

The biggest single revolutionary trend in recent years in the detergent industry has been the use of enzyme additives for presoaking purposes. These enzymes act on proteins, starches, and fats, breaking them down and making it easier for the detergent to loosen and carry them away.

CLASSES OF SYNTHETIC DETERGENTS

The four main classes of synthetic detergents are: anionic, cationic, non-ionic, and amphoteric.

Anionic Detergents

Anionic detergents are cleansing compounds whose cleansing abilities are due to the anion or negatively charged component of its molecule. It is by far the largest and most commonly used class of detergents. Most of them are sulphated or sulphonated products.

The alkyl aryl sulphonates are far and away the most frequently used group of anionic detergents, since sodium alkyl benzene sulphonate, its main member, is the most widely used detergent at the present time. The hydrophilic part (the head) is sodium sulphonate, which is attached to a benzene ring. The hydrophobic part (the tail) is the alkyl group, which is also attached to the benzene ring. Benzene serves as a place of attachment for the two groups, since they do not combine easily with each other.

When this molecule dissolves in water it loses its sodium, which has a positive charge. The remainder of the molecule carries a negative charge. This negatively charged ion (charged particle) has the cleansing ability. Alkyl benzene sulphonate is therefore an ionic surfactant that possesses detergent properties.

If charged plates are placed in the detergent solution, the negatively charged alkyl benzene sulphonate ion will be attracted to the positive plate or anode. Because of this behavior, the alkyl benzene sulphonate ion and all other negative ions are called anions.

The hydrophilic part of this ion (the sulphonate group or head) lessens the attraction of water molecules for each other, allowing them to penetrate into and around the soil to a far higher degree than they would without the presence of the detergent. In other words, the detergent lowers the surface tension of the water; it wets it. The hydrophobic part of the molecule engages the soil and carries it away.

Other anionic detergents include:

1. Sulphated ethers. These compounds have high foaming properties. They are used in hair shampoo and in combination with other detergents in household dishwashing products to increase suds production.
2. Sulpho-succinates. These are compounds with excellent cleansing and high-foaming properties. They are used in toilet preparations.
3. Alkane sulphonates. Members of this group have excellent cleansing properties. They do not react with calcium and magnesium salts and, in addition, they are biodegradable.
4. Phosphate esters. These detergents possess good cleansing properties, especially on hard surfaces, are low in foaming action, and are biodegradable. They are used in metal cleaning and plating and as dry-cleaning agents.
5. Alkyl isethionates. The detergents in this group are gentle to human skin and are used in synthetic hand soap bars and hair shampoos.

Cationic Detergents

Cationic detergents are cleansing compounds whose cleansing abilities are due to the cation or positively charged component of its molecule. They command a very small share of the detergent market. Their cleansing properties are rather poor and their cost is high.

Their principal uses are as germicides, fabric softeners, and specialized emulsifiers. The most effective of the group are the quaternary ammonium salts.

Non-ionic Detergents

Non-ionic detergents contain no ionic components in their molecule. One type of non-ionic detergents, the fatty acid condensates, is commonly used as a component in household detergent powders. Another type, the alkyloamides, is used as an additive to other detergents for foam-boosting effects. Nonyl phenol is a very popular general detergent. The fatty amine oxides are, in themselves, good detergents, but their main uses are as foam boosters and skin-protecting agents in liquid detergents.

Amphoteric Detergents

These detergents have both acidic and basic groups in their molecule. They have characteristics of anionic detergents and cationic fabric softeners. Some of the amphoterics are excellent foamers but do not cause skin or eye irritation. These properties make them useful in hair shampoos and skin cleansers. Their foaming properties are also used in rug shampoos.

The usage of these four classes of detergents is approximately as follows:

Anionics	66 percent
Non-ionics	29 percent
Cationics	5 percent
Amphoterics	1 percent

DETERGENT ADDITIVES AND BUILDERS

If detergents that are purchased on the market contained only the active cleansing ingredients, they would be so concentrated that measuring them in the average working situation would be virtually impossible. Therefore, the original active ingredients are diluted so that they can be measured satisfactorily. In the case of liquid or paste detergents, water is used; for powders, sodium sulphates. In addition to water and sodium sulphates, detergents contain builders and additives to make them more effective.

Phosphates

Two classes of phosphates that are utilized in detergents are condensed phosphates and orthophosphates. The condensed phosphates are of far greater importance. They are able to hold substances like food soils in suspension and to emulsify oily or greasy substances. In addition, they have the ability to form soluble compounds, thus forming no scum or scale.

Many of the new machine dishwashing compounds, both solid and liquid, contain a mixture of trisodium phosphate, an orthophosphate, and sodium triphlyphosphate or tetrapotassium pyrophosphate, which are condensed phosphates. To date, there is no satisfactory replacement for phosphates in machine dishwashing detergents.

Silicates

Silicates, particularly sodium and potassium silicate, are added to detergents for a number of reasons. They soften water, suspend soil in solution, and have wetting and emulsifying properties, particularly on glass and glazed surfaces. These properties make them highly desirable for use in dishwashing powders. They also inhibit the corrosion of stainless steel and aluminum by synthetic detergents and complex phosphates, thus offering a protection to some types of equipment.

Antiredeposition Agents

Carboxymethyl-cellulose (CMC) has been previously mentioned in this chapter. Besides its use as an antiredeposition agent, it is often used in detergents as a thickening agent, mainly to increase the consistency of pastes and to prevent separation of the paste into phases. PVP is another redeposition agent that has become popular in recent years.

Optical Brighteners

These chemicals are now found in all washing powders. They are dyestuffs that are absorbed from the detergent solution and retained by textile fibers, enabling the fibers to reflect more light and thus to look brighter than they would without the dyestuff additive.

Chelating Agents

The purpose of adding a chelating agent to a detergent is to provide a chemical that can remove calcium, magnesium, and iron ions from hard water. The agent accomplishes this by reacting with the minerals and attaching the hardness-causing ions to its own molecule; it remains soluble and carries the minerals away in the waste water.

Enzymes

The most recent major development in the evolution of synthetic detergents has been the addition of enzymes to the product. Three types of enzymes are of interest to the foodservice industry: (1) proteases, which digest proteins, (2) amylases, which digest starches, and (3) lipases, which digest fats and oils. Enzymes are commonly incorporated into automatic washing machine powders.

Chlorine-Releasing Agents

Active or oxidizing chlorine is important in all dishwasher detergents. Its function is to react with soil traces and make them soluble. They help eliminate specks of soil that allow water to form droplets, resulting in spots on dishware.

Defoamers

These chemicals are designed to lessen the foam in dishwashing machines, thus allowing the spray to contact the surfaces of dishware with undiminished force. As soils are removed from the dishes, the pump and spray action of the dishware causes them to foam. This foam reduces the efficiency of the dishwashing machine. Many defoamers also help to remove soils from dishware surfaces, and provide emulsifying and carrying action.

DETERGENT DISPENSING SYSTEMS

Detergent dispensing is an extremely important operation in any foodservice establishment. The detergent content of the dishwashing machine is constantly diluted by rinse water draining back into that section of the machine. In order to maintain detergent concentration at the required strength, detergent must be added continually during the dishwashing cycle. Electronic controls maintain detergent concentrations by activating pumps or dispensers to feed detergent into the machine.

Time is an important factor in institutional-type dishwashing. In the wash section of an institutional dishwashing machine, the dishes are sprayed with detergent solutions only from 20 to 90 seconds, depending on the size of the machine. Because of the short time-period involved, institution-type detergents must be more highly concentrated than those designed for home use. These concentrations require safe handling. This safety is provided by remote dispensing and electronic dispensers.

A detergent dispenser is a device for adding detergents to wash water so that the desired strength of the detergent solution is maintained throughout the dishwashing machine's operating period. This is necessary because dishwashing compounds have been formulated to perform under certain conditions. The conditions determine the compound used. A stated amount of this compound must be present in a certain amount of wash water in order to create the proper strength wash solution for effective cleaning of dishes and utensils.

The first dispenser was undoubtedly the human hand, but this was certainly a most unsatisfactory device. The custom in the past was to add dishwashing powder to the wash solution "when needed." The problem was the vagueness of the phrase "when needed." A cupful or two might be used in the beginning, with another cupful supposedly to be added an hour later; or the compound might be added to every certain number of racks of dishes put through the machine.

If a cup was convenient, it would be used. But often a handful was scooped from the container or an estimated amount was poured from a box. As a result, many dishwashing machines operated for some part of the busiest periods with little or no detergent in the wash tank. As might be expected, the results were unsatisfactory. To correct these conditions, several of the leading detergent manufacturers developed the first hydraulic dispensers, hoping to keep the dishwashing compound at a more even level.

Hydraulic Dispenser

The term *hydraulic* refers to a liquid in motion or a device operated by a liquid, such as a hydraulic lift in a gas station. A hydraulic dispenser permits additional amounts of dishwashing compound to be fed into the wash tank each time the machine's hot water rinse system is operated.

An open stainless steel container, also termed a hopper or reservoir, is mounted on the machine. Tubing, connected to the pressure side of the final rinse line, leads to the detergent reservoir. Every time the rinse valve is opened, hot water flows through this tube into the dispenser reservoir, dissolving a certain amount of the detergent. An additional amount of fresh detergent solution is then fed through a second tube directly into the wash tank. In this way, more dishwashing compound is added at regular intervals to the wash solution without any human interference.

These hydraulic devices represented a great advance in detergent dispensing. Their development was based on the reasoning that: (1) the wash solution would be diluted somewhat by each release of hot rinse water and (2) each use of the rinse meant that one or more loads of dishware had been washed; therefore, the wash solution would be somewhat weakened by contacting and removing the food soil present on the dishes. It was thought that by adding a fresh charge of dishwashing compound to the wash solution with each use of the rinse, these conditions would be corrected.

Before very long, however, it was realized that the hydraulic dispenser did not provide all the answers. The main problems were:

1. The detergent hopper was not always kept filled and dishes sometimes went through the washing process without detergent.
2. The addition of small amounts of compound with each use of the final rinse did not guarantee the same, measured, detergent strength solution in the wash tank throughout the machine's entire operating period.
3. Multiple tank, conveyor-type machines were finding their place in the larger foodservice establishments.

Today, some form of electronic controlling device to maintain detergent concentration is found in most commercial dishrooms.

Electronic Dispenser

The early electronic dispenser consisted, in most cases, of a detergent reservoir situated on the machine. Jets of hot water, spraying periodically into this reservoir or hopper, would result in an amount of concentrated detergent being dissolved and released into the wash tank.

This type of dispenser measures the strength of the detergent solution in the wash tank. Whenever the concentration drops below the required level, detergent is added to the wash tank in the needed amount. When the proper concentration is reached, the feeding of detergent stops and the dispenser resumes its measurement of the wash solution strength, until time to add more dishwashing compound.

Electronic dispensers have three main parts:

1. A control head, which constantly measures the concentration of the detergent in the wash tank. When the detergent concentration falls below the level set in the control head, a message is sent to a device that controls the release of detergent.
2. An electrode, which is installed in the wash tank and connected by wires to the control head. The poles of the electrode measure the electrical conductivity of the wash solution. This varies with the concentration of the detergent in the solution. Through the electrodes, the control head monitors the detergent concentration.
3. An activating device, which is regulated by the control head. When the detergent concentration drops below the desired level, the control head transmits an electrical impulse to this instrument. This allows a flow of hot water to be released into the detergent reservoir. The hot water and detergent it dissolves form a concentrated solution that overflows through a tube into the wash tank. This flow continues until the concentration reaches the level set in the control head.

Detergent Concentration Meters

A detergent concentration meter is a device that shows the strength of the detergent solution in the wash tank. It can be used with or without a detergent dispenser.

The meter and electronic dispenser, if used, are first set at the desired concentration level. An electrode allows the meter to indicate continuously the strength, or concentration, of the detergent solution in the wash tank. An indicating dial is located on the side of the meter to show the strength of the wash solution. On the dial's face, a curved band, divided into sections, is plainly marked with such directives as "add compound," "OK," and "overcharged." While the machine is operating, the needle on the dial will point to the area indicating the strength of the detergent solution.

There are several valuable uses for detergent concentration meters. Where detergents are fed by hand into the wash tank, the meter serves as a guide as to when and how much compound to add. The needle should remain in the "OK" range all during the washing period.

When a hydraulic dispenser is used, the meter assists the operator in keeping the detergent solution at the proper level. The concentration meter can also be used as part of a dispensing system from the original shipping container to the wash tank.

Auger System Dry Powder Feed

This method also involves a detergent hopper, situated on the dishwashing machine, that contains a powdered detergent. When the dishwashing compound in the wash solution falls below the correct level, the electrode in the wash tank signals the control device. The controller signals the auger, located on top of the machine. The auger forces a predetermined amount of dry, concentrated detergent directly into the wash tank.

Dispensing from Original Container in Liquid Form

Whenever the electrode in the wash tank signals the electronic controller that the detergent concentration in the wash solution is low, the controller supplies power to the activator, allowing hot water to be released.

The liquid concentration in the shipping container is siphoned out, mixed with the water, and delivered to the wash tank, thus building the solution up to its required level. When the solution strength reaches this proper level, the activator is signalled to close and remains closed until the solution strength again falls below the predetermined level.

Dispensing Powdered Detergent from the Shipping Container

There are two ways in which a shipping container can be used in dispensing powdered detergents: (1) directly from the container to the dishwashing machine's wash tank, or (2) in the remote control operation, where the container is situated some distance from the dishwashing machine in another room, or even on another floor.

In the case of direct flow, when the electrode signals the electronic controller that the concentration of detergent in the wash solution is low, power is supplied to the activator, allowing it to open. Hot water is then admitted into the detergent container. The concentrated solution of hot water and detergent is fed into the wash tank, thus building up the solution strength. When the strength of the solution reaches the required level, the activator stops the flow.

In a remote control operation, when the strength of the detergent solution falls below the required level, the electronic controller energizes a relay that pumps concentrated detergent solution from a sump reservoir to the machine. In this instance, the shipping container is placed in an inverted position on a dispensing device. Brief spurts of water jet up into the container, producing a mixture of concentrated detergent and water, which is available for pumping to the dishwashing machine's wash tank.

Dispensing Liquid Detergent from a Remote Area to the Wash Tank

In this case, the electrode in the wash tank signals an electronic controller situated on a wall above the container of liquid detergent. When the electronic control receives the signal that the wash tank solution has fallen below the required detergent concentration, it activates a piston pump, which delivers the detergent from the shipping container to the dishwashing machine wash tank.

Dispensing from Container to Wash Sink in Manual Dishwashing and Pot- and Panwashing

This method of dispensing is becoming more and more widely used. Even though there are mechanical differences in these various control devices, they operate generally in this way. A special type of fitting, attached to the faucet used in filling the sink, is also connected by tubing to the drum or container of liquid detergent sitting on the floor.

When the water is turned on and flows through the fitting, the liquid detergent is drawn up through the tube, mixes with the water to form the desired wash solution, and flows into the sink, ready to use. A button on the fixture may be pushed to allow the detergent to begin its flow into the running water. When the water is turned off, the button pops out automatically, cutting off the flow of detergent. With this arrangement, clear water is available whenever the button is not pushed in by the dish or pot washer.

The amount of liquid detergent used is controlled or metered by various methods, but the end result is the correct proportioning of detergent to each gallon of water to form the required concentrations.

Rinse Injectors

A rinse injector is a mechanical device that automatically injects a small amount of drying agent into the dishwashing machine's final rinse, just before the rinse water sprays from the rinse nozzles over the tableware. This can be done in various ways (a pump, suction, or some other means).

Drying agents, or rinse additives, are liquid chemicals that are fed automatically, in very small amounts, into the final 180°F. (82°C) rinse water. They lower the surface tension of the water so that it runs off the tableware evenly, leaving no spots or streaks. When the hot rinse water spreads over the dishes in a thin sheet, the additional heat causes almost immediate drying of the ware.

SUMMARY

Many types of dishwashing compounds are presently available. Their effectiveness depends on their formulation and on the conditions under which they must function. One of these conditions is the hardness of the water being used.

Water containing dissolved calcium, magnesium, and iron salts is known as hard water. It is objectionable for two reasons: (1) the calcium, magnesium, and iron salts in it react with soaps forming a sticky, soapy scum, and (2) hard water is responsible for the formation of scale in boilers, hot water pipes, and dishwashing machines.

Synthetic detergents are synthetic compounds that act like soap. A synthetic detergent molecule contains two distinct groups in its molecule: a hydrophilic group, which is responsible for the dissolving ability, and a hydrophobic group, which engages the food soil. Since their activity occurs on eating ware surfaces, detergents are known as surface-active agents (surfactants). Detergents are successful, primarily, because they lower the surface tension of the water, giving it greater ability to penetrate into and under food soils that adhere to the surface of dishware, silverware, utensils, and so forth.

The four main classes of synthetic detergents are anionic, cationic, non-ionic, and amphoteric.

Anionic detergents are cleansing compounds whose cleansing abilities are due to the anion or negatively charged component of its molecule. This is the largest and most commonly used class of detergents. Most of them are sulphated or sulphonated products.

Cationic detergents are cleansing compounds whose cleansing abilities are due to the cation or positively charged component of its molecule. Their main uses are as germicides and fabric softeners.

Non-ionic detergents contain no ionic components in their molecule. Some are components of household detergent powders and others are used as foam boosters and skin-protecting agents.

Amphoteric detergents have both acidic and basic groups in their molecule. They have characteristics of anionic detergents and cationic fabric softeners. Some of the amphoterics are not only excellent foamers, but also cause no skin or eye irritation. They are used in skin cleansers, hair shampoos, and rug shampoos.

Among the builders and additives that make synthetic detergents more effective are:

1. phosphates for water softening, emulsifying, and soil holding
2. silicates for water softening, wetting, and soil holding
3. antiredeposition agents for soil holding
4. optical brighteners for brightening
5. chelating agents for water softening
6. enzymes for digesting fats, starches, and proteins
7. defoamers to lessen foam formation.

Detergent dispensers are devices for adding detergents to wash water so that the desired strength of the detergent is retained throughout the dishwashing machine's operating period. They are used for accuracy and safety in adding detergents to the wash water. Detergent dispensing is an important operation in any foodservice operation.

Two types of detergent dispensers in common use are the hydraulic dispenser and the electronic dispenser. The strength of the detergent can be shown visually through the use of a detergent concentration meter. These meters can be used with or without dispensers.

Review Questions

1. On what factors should the choice of a dishwashing compound depend?

2. a. What characteristic of water most affects its use as a dishwashing agent?

3. a. What mineral salts may be present in hard water? b. Why is hard water undesirable for use in washing dishware?

4. What is soft water?

5. a. What is meant by surface tension? b. How do wetting agents reduce surface tension?

6. Why should detergents be biodegradable?

7. a. What is a synthetic detergent? b. What are the roles of hydrophilic and hydrophobic groups in a detergent molecule?

8. How do the following aid the cleaning ability of detergent? a. CMC, b. enzyme additives, c. wetting agents, d. optical brighteners.

9. Describe the four main classes of detergents.

10. What is a detergent dispenser?

11. How does an electronic detergent dispenser work?

12. a. What is a detergent concentration meter? b. How does it function?

13. How is detergent dispensed from the shipping container?

14. How are faucets used to dispense detergent?

15. What is the value of using drying agents in the dishwashing process?

References

Cahn, A. (1974). "Basic Detergent Ingredients." In *Detergents in Depth. A symposium.* Soap and Detergent Association. New York.

Davidsohn, A. and B. M. Milwidsky. (1978). *Synthetic Detergents.* John Wiley and Sons, Inc. New York.

Longman, G. F. (1975). *The Analysis of Detergents and Detergent Products.* John Wiley and Sons, Inc. New York.

Lyng, A. L. (1974). "Detergents in Review." In *Detergents in Depth. A Symposium.* Soap and Detergent Association. New York.

Miller, S. J. (1963). "Sanitation and Dishes—Aspects Old and New: Part I." *J. Amer. Diet. Assoc.* 43:23–38.

Oberle, T. M. (1974). "Formulation Aspects of Machine Dishwashing Detergents." In *Detergents in Depth. A Symposium.* Soap and Detergent Association. New York.

Schwartz, A. M., J. W. Perry, and J. Berch. (1958). *Surface Active Agents and Detergents.* Interscience Publications, Inc. New York.

- I. Sanitizing
    - A. Terms
    - B. Activity of Antimicrobial Agents
        - 1. Coagulation of Cell Proteins
        - 2. Denaturation
        - 3. Injury to Cell Membranes
- II. Manual Sanitizing
    - A. Hot Water
    - B. Sanitizing with Approved Chemicals
    - C. Chemicals Used in Manual Sanitizing
        - 1. Chlorine Compounds
        - 2. Iodine and Its Compounds
        - 3. Quaternary Ammonia Compounds
- III. Regulations
- IV. Sanitizing in Dishwashing Machines
    - A. Hot water
    - B. Detergent Sanitizers
- V. Storage
- VI. Summary

SANITIZING

Sanitizing is the final step in the dishwashing operation, the last vital act of the tableware and equipment cleaning routine. The cleaning procedures leading up to sanitizing are carried out to provide the attractive looking tableware that management is proud to exhibit and customers are happy to use. Sanitizing enhances this attractive appearance by protecting the health of everyone who uses the tableware while eating a meal in the establishment.

Sanitization can be accomplished with heat or chemical processing, but it is not effective unless the articles or surfaces to be sanitized are first physically clean. Caked-on soils not removed by cleaning may shield bacteria from the sanitizing water or solution, or may later serve as a medium for bacteria growth if contamination occurs. Sanitizing is carried out to protect the health of everyone who eats in the establishment. For this reason, the performance of correct sanitizing routines is one of the greatest responsibilities of the dishwashing crew.

Sanitizing is a process or treatment that destroys microorganisms remaining on the surfaces of equipment and utensils after washing and rinsing. After removal of food soil that can be seen or felt, it is necessary to remove soil that cannot be seen or felt. Tableware and food-contact surfaces of equipment can be sanitized in several ways. The routine selected, however, must be approved by the local, state, or national public health service. With regard to sanitizing, the U.S Public Health Service Food Service Sanitation Manual States:

1. Tableware shall be washed, rinsed, and sanitized after each use.
2. To prevent cross-contamination, kitchenware and food-contact surfaces of equipment shall be washed, rinsed, and sanitized after each use and following any interruption of operations during which time contamination may have occurred.
3. Where equipment and utensils are used for the preparation of potentially hazardous foods on a continuous or production-line basis, utensils and the food-contact surfaces of equipment shall be washed, rinsed, and sanitized at intervals throughout the day. . . .
4. Moist cloths or sponges used for wiping food spills on kitchenware and food-contact surfaces of equipment shall be clean and rinsed frequently in one of the sanitizing solutions. . . .
5. Moist cloths or sponges used for cleaning non-food-contact surfaces of equipment such as counters, dining table tops and shelves . . . shall be stored in the sanitizing solution between uses.

Terms

In public health work, a number of different terms are used in connection with the methods of materials employed to destroy pathogens. Some of the terms, though not exactly the same in meaning, are often used interchangeably. The following are terms that foodservice people should know.

1. *Germicide.* A germicide is any agent that kills microorganisms. The term refers in particular to chemical agents that destroy disease germs. Some germicides are formulated for use on inanimate objects and are called disinfectants; others can be applied to living tissues and are called antiseptics.

2. *Disinfectant.* A disinfectant is a germicide, usually a chemical, which is used to kill disease-producing organisms. An agent may be a disinfectant for one type of organism but not another. The term is usually used to describe a chemical, radiation, or other process that destroys bacteria and is normally used on inanimate objects (dishes, cooking utensils, equipment, floors, etc.). Disinfectants tend to cause irritation or damage to living tissues. They represent the oldest group of antimicrobial agents. Their main requirement is that they should act on all microbes in high dilutions, preferably within a short period of time. Bi-

chloride of mercury, calcium and sodium hypochlorite, phenols, quaternary ammonium compounds, and iodophors are disinfectants.

3. *Antiseptic.* An antiseptic is a germicide that is applied to living tissues, such as open sores, cuts, and wounds in order to destroy microorganisms or stop their growth on tissues. Familiar antiseptics (tincture of iodine, merthiolate, and ethanol [ethyl alcohol]) can be found in most home medicine cabinets.

4. *Bactericide.* A bactericide is an agent that injures microorganisms so badly that they lose their ability to multiply and eventually die. The injury is irreversible. The term is applied especially to chemical agents that kill bacteria. Bactericidal treatment is the application of a bactericide, in accordance with public health regulations, to such objects as foodservice equipment, dishes, or utensils. Hence, when effective bactericidal treatment is applied to such objects, they are sanitized. A bactericide is a type of disinfectant designed to kill bacteria.

Some chemical agents have a bacteriostatic effect. This refers to a condition in which microbes are temporarily unable to multiply but are not immediately killed. The injury is reversible; when the bacteriostatic agent is removed, the microbes recover and resume growth. Low temperatures, such as in refrigerators, and weak concentrations of many disinfectants have a bacteriostatic effect.

5. *Sanitizer.* A sanitizer is an agent, usually a chemical, that is applied in a separate operation after the objects to be sanitized have first been cleaned. The regulations covering food and beverage establishments do not require sanitizing agents to destroy all of the microorganisms present on eating and drinking utensils and equipment. They must, however, reduce the number of these organisms to a level acceptable to health departments. Sanitizers destroy both disease-causing bacteria and those regarded as harmless. Sanitizing some types of foods extends their time of usage by reducing the number of spoilage bacteria on them.

Sanitizing agents may be combined with detergents to form products called detergent sanitizers, detergent germicides, detergent disinfectants or some other equally descriptive term, depending on the formulation of the product.

6. *Sterilize.* This is a process, physical or chemical, that destroys all forms of life, including viruses and bacterial spores. Sanitization is not sterilization. Sanitizing reduces the bacterial population to a safe level; sterilization kills or inactivates all forms of life.

Activity of Antimicrobial Agents

Bacterial life depends on a complex of enzyme actions, and enzymes have the structures and properties of proteins. Enzymes regulate the activities of the cell and the inactivation of one or more essential enzymes can result in the death of the cell. There are a variety of enzyme systems in the different kinds of microbial life, and so even the best of disinfecting agents is not successful against all forms of microbes. Antimicrobial agents can effect injury or death of a microbe in several ways.

Coagulation of Cell Proteins. Proteins are vital chemicals in all cells and are a part of many cell activities. They are normally suspended in the cell water. A solution with proteins suspended in this manner is called a *sol.* This is the state of a normal, healthy cell. If the proteins lose their ability to remain suspended, they coalesce and form a network of protein molecules. This state is called a *gel*—a semisolid, jellylike substance. Upon heating, the gel changes to a sol again. For example, the protein gelatin suspended in water is in the sol state when warm, but coalesces into a gel when cool.

High temperatures and many chemical disinfectants, especially at high concentrations, cause cell proteins to coagulate into a solid mass. When its cellular proteins are brought together into a solid gel, a microorganism can no longer function and it dies. Among the chemical agents having a strong coagulating action are formalin, phenols, and alcohols.

Denaturation. When proteins are heated or treated with strong acid or alkali, they undergo a major structural change. A changed or denatured protein molecule is formed and, having lost its original structure, cannot return to its former composition or configuration. The denatured molecule loses the properties that characterize the proteins of living cells.

Many active chemicals combine with proteins and enzymes, denaturing them in the process. This disrupts vital cell activities, causing death. Some of these chemicals, such as chlorine and iodine, are used in sanitizing agents. Cresol, phenols, and alcohols denature proteins and enzymes.

Injury to Cell Membranes. Any agent that disrupts the normal functioning of the cell membrane is injurious or lethal to a microorganism. Some chemicals combine with the cell membrane, thereby denaturing or dissolving some of the membrane compounds. Phenols, alcohols, quaternary ammonium compounds, soaps, and synthetic detergents act in this manner.

MANUAL SANITIZING

Hot Water

At the conclusion of the manual dishwashing discussion, the dishes had been properly washed and thoroughly rinsed. The dishes, clean to sight and touch, were left ready for sanitizing in sink number 3.

When hot water is the sanitizing agent, a water temperature of at least 170°F. (77°C) must be maintained in the sink in order to destroy bacteria. Eating and drinking utensils and equipment items must be immersed totally in this water for at least 30 seconds. The water must be kept clean.

These sanitizing requirements necessitate the use of the following items:

1. A sufficient amount of clean, hot water at 170°F. (77°C) temperature to keep the sanitizing water in sink number 3 at its proper heat level during the entire dishwashing period. Any drop in water temperature during the sanitizing treatment lessens its effectiveness, and compliance with public health codes is lost.
2. A wall thermometer, mounted on the number 3 sink, so that water temperature can be monitored frequently. This thermometer must be accurate to within ±3°F. (±2°C). Use of a wall-type thermometer will reduce the costs of breakage or instrument loss.

   Frequent reference to the thermometer is necessary. The temperature of the dishes and utensils placed in the hot water is less than 170°F. (77°C) and this lowers the temperature of the sanitizing water. Therefore, water flowing into the sink should be at least 180°F. (82°C) to compensate for the continual drop in temperature (Fig. 12-1).

Figure 12-1 Water flowing into the sink should be at least 180°F. (82°C) for sanitizing purposes.

3. Dish baskets with long handles, designed so that the dishes, utensils, and small pieces of equipment placed in the baskets are completely immersed in the sanitizing water. The long handles on the baskets allow the dish-

washing crew to carry on the sanitizing treatment without touching the hot water or the dishware.

4. Tongs or hooks for handling large or heavy equipment that can be sanitized in the third sink but cannot fit into a long handled basket.
5. A clock with a second hand located within convenient view of the dishwashing crew. It takes time to kill microorganisms. Dishes and utensils can remain in the sanitizing water longer than 30 seconds, but the time cannot be shortened. The 30-second time allowance is based on the rate of bacterial kill at 170°F. (77°C).

Equipment, even when disassembled, which is too large to be sanitized in the sink, can be treated satisfactorily:

1. by rinsing carefully with boiling water or
2. with live steam from a hose, if the equipment can confine the steam.

After being immersed in 170°F. (77°C) water for at least 30 seconds, the loaded dish baskets should be placed on a clean, slanting drainboard while their contents dry. Under no circumstances should a towel be used to dry dishes or utensils. It is impossible to keep towels clean or free from hand contamination. Hands are a major source of *Staphylococcus aureus*. The use of towels invalidates the sanitizing treatment.

The hotter the sanitizing water, the faster the drying and the less chance of water spotting. The drying process can be hastened by the addition of a small amount of rinse additive, containing a drying agent, to the sanitizing water.

Sanitized dishes, utensils, and equipment items should be removed carefully from the long handled baskets in order to avoid contamination. Food-contact surfaces should not be touched (Fig. 12-2). Sanitized spoons, knives, and forks should be picked up or touched only by their handles. Sanitized cups, glasses, and bowls should be handled so that fingers do not contact inside surfaces or lip-contact surfaces

Figure 12-2 Food contact surfaces of dishes should not be touched after they have been sanitized.

(Fig. 12-3 and 12-4). Plates should be handled by their rims.

Sanitizing With Approved Chemicals

Sanitizing with approved chemicals is also carried on in many manual operations. Hot water and chemical sanitization are equally effective if properly applied.

Except where specified in local food sanitation regulations, the three-compartment sink is a standard requirement for a routine wash, rinse, and chemical sanitizing operation. The clean, running, and constantly overflowing rinse water in sink number 2 will remove any traces of loosened food soil and detergent solution remaining on the dishware after washing. The chemical solution in sink number 3 can then do an effective job of sanitizing.

The fact that chemical sanitizers do not work satisfactorily on unclean surfaces is important to remember at all times. Thorough washing and rinsing of the dishware must take place prior to sanitizing. In addition, the wash, rinse, and sanitizing sinks must be kept clean, as well as the tools and other equipment (brushes, long handled baskets, etc.) that are used in the wash-rinse-sanitize routine.

To make up the correct sanitizing solution, the exact amount of water that goes into the sanitizing sink must be known. Health department ordinances regulating food and drink establishments generally state the required amount of sanitizer in the solution as so many parts per million (ppm). Manufacturers' use instructions, however, usually translate this requirement into specified numbers of tablespoons or ounces to stated amounts of water.

The effectiveness of the sanitizer gradually weakens as it contacts more and more dish and equipment surfaces. Hence, the solution should be changed whenever it falls below the required strength. For solution-testing purposes, inexpensive, easy-to-use test kits can be purchased or obtained from the manufacturers. It is advisable to test the solution strength after about 400 articles have been sanitized.

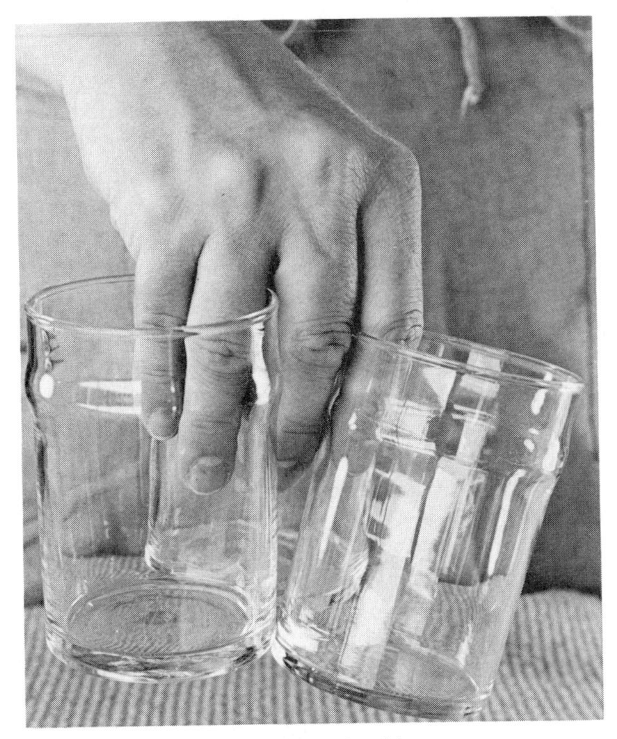

Figure 12-3 Sanitized glasses should not be handled in such a way that fingers contact inside surfaces.

Figure 12-4 After cups have been sanitized, fingers should not contact inside surfaces. (a) Improper and (b) proper manner of handling sanitized cups.

The strength of the sanitizing solution and the immersion time must meet the requirements stated in the U.S. Public Health Service Food Service Sanitation Manual. In their examples, the immersion times and solution strengths are the minimum allowed for effective chemical sanitization. Various state and local ordinances may expand these requirements.

Chemicals Used in Manual Sanitizing

Although the scientific application of disinfectants dates back only about 150 years, practices that were more or less effective go back to ancient times. The Bible contains many references and regulations concerning cleanliness. Alexander the Great required his armies to boil their drinking water and to bury wastes.

In 1825, Labarraque used chlorinated soda solution in treating infected wounds and recommended it for general disinfection. In 1835, the American physician, Oliver Wendell Holmes (father of the famous jurist of the same name), reported that the transmission of puerperal fever (childbed fever) by nurses to other patients was eliminated by washing hands in a chloride of lime solution following each visit to an infected patient.

Phenol or carbolic acid, like chlorine, was first used as a deodorant to prevent the foul odors of sewage and garbage, and then to prevent wound infection. Although it was discovered in 1834, it was 1860 before it was used as a wound disinfectant. The later work of Lister in disinfecting wounds, bandages, surgical thread, and operating rooms gave carbolic acid great popularity. The use of carbolic acid on human tissues eventually ceased as other, less irritating antiseptics were found.

Tincture of iodine, a 5 percent solution of iodine in diluted alcohol, was in use by 1830, but it was not used as a surgical antiseptic until 1839. It was used with notable success in the United States Civil War to treat battle wounds. Its value against a wide variety of microorganisms was soon recognized. By 1900, the principal disinfecting agents were heat, hypochlorites, carbolic acid, iodine, and alcohol. There are, currently, over 10,000 antimicrobial agents on the market.

There are many different kinds of microbial pathogens. Representatives are found among the fungi, yeasts, rickettsia, protozoa, viruses, and bacteria. Some bacteria form spores that are very difficult to destroy. Many of these pathogens vary in their resistance to antimicrobial agents.

There are also many different kinds of antimicrobial chemical agents, affecting biological systems in different ways. Methods of action are not always clear. However, the effectiveness of chemical agents does seem to depend on the following factors:

a. the chemical formulation of the agent
b. the concentration in which it is used
c. the amount of time the microorganisms are exposed to it
d. the temperature at which it is used
e. the pH of the environment
f. the presence of organic matter, such as food soil, which may combine with the agent, thereby reducing its effectiveness
g. the type of microorganism present
h. the number of microorganisms present.

Some highly effective chemical agents are unusable in certain situations because of their toxicity to man or animals and because of their corrosive action on some surfaces. The following chemical agents are in common use in foodservice establishments.

Chlorine Compounds. Chlorine, in gas or liquid form, or incorporated into unstable compounds is a very effective germicidal agent. Free chlorine is an extremely active substance, combining readily with proteins and enzymes in bacterial cells, causing injury and death. Chlorine gas and liquid chlorine are widely used to purify drinking water, swimming pool water, and sewage.

Calcium and sodium hypochlorite are commonly used sanitizing agents in the foodservice industry. They are sold in solid form and in solution. Of the chlorine compounds, hypochlorites are the most reactive and rapid in their germicidal action. Hypochlorous acid, released by the hypochlorite, seems to be the reactive agent. It penetrates the bacterial membrane and attacks essential enzymes in the cell.

Organic matter, even in small amounts, reduces the effectiveness of hypochlorites. Surfaces must be clean, wettable, and free of detergent residue. Hypochlorites work best in an acid environment and at temperatures of 75°F. (24°C) or better. They may be used for water purification, the treatment of milk and food utensils, drinking glass disinfection, and for treating toilets and floors.

Chlorine also exists in compounds as organic chlorine complexes. Among these are trichloromelamine, chlorinated isocyanurate, and chlorinated hydantoin. These products are generally available in powder form. They are formulated for use at a pH close to 7 or lower. The chlorinated isocyanurates and chloromelamines form hypochlorous acid in solution.

Hypochlorites are the oldest and most widely used of active chlorine compounds in the field of chemical disinfection. They are (1) proven and powerful germicides, controlling a wide spectrum of microorganisms, (2) deodorizers, (3) nontoxic to man at normal use concentrations, (4) colorless and nonstaining, (5) easy to handle, and (6) economical.

Iodine and Its Compounds. Iodine has long been a well-known germicide. It is chemically related to chlorine and many of their reactions are similar. Like chlorine, iodine combines with cell proteins, causing injury and death. Alcoholic solutions (tinctures) of iodine are widely used as antiseptics and disinfectants.

Iodine has been used in various forms as an antiseptic for skin and for wounds, for sterilization of the air, for disinfection of drinking water and swimming pool water, and for sanitization of eating and drinking utensils.

Iodine functions best in acidic solutions at temperatures of 75°F. (24°C) or better. It kills more quickly than either chlorine or the quaternary ammonia compounds and is more effective against viruses, fungi, and spores. The presence of organic matter, such as food soils, lessens its effectiveness, however.

Iodophors are compounds resulting from a complex of elemental iodine with one or more wetting agents to lower the surface tension of water. The wetting agents act as carriers and, in solution, release the iodine. The antimicrobial action is due to elemental iodine.

Iodophors exhibit their greatest biological effectiveness in acidic solutions, and for this reason phosphoric acid is frequently a constituent of their formulation. The presence of the acid increases the iodine stability in the presence of organic soils. Food soils do inhibit the effectiveness of iodophors, but the influence is less than on hypochlorites or quaternaries.

Temperature differences between 35° and 125°F. (2–52°C) seem to have less influence on the germicidal effectiveness of iodophors than on hypochlorites. They exhibit greater antimicrobial activity at the high end of the temperature range, however.

Iodophors are relatively nonselective and are effective against a wide range of bacteria. They are also effective against small viruses, many fungi, and the spores of several pathogenic bacteria.

Iodophors are less irritating and staining than are solutions or tinctures of iodine. They will, however, tarnish silver, silver plate, and copper.

Quaternary Ammonia Compounds. The quaternary ammonia compounds have been known since 1935. They are ammonia salts, many of which are effective as disinfectants. Quaternaries are surface-active agents that tend to reduce the surface tension of fluids to which they are added, thus wetting surfaces that they contact. Those that wet the fatty compounds in a cell membrane are likely to dissolve them, thereby allowing cell contents to leak out. They may also penetrate into the cell and unite with cell enzymes, thereby injuring the cell proper. Either mode of action can cause death. Their chemical behavior seems to depend on their concentration.

Quaternaries are more effective in alkaline solutions than in acidic ones. They are heat-stable and can be used at temperatures as high as 212°F. (100°C). They are relatively slow and ineffective in cold solutions. Their effectiveness is increased with increasing temperatures. Organic matter reduces their effectiveness.

Quaternaries are approved for use as sanitizing agents in foodservice establishments. They have no color or odor, are highly stable, and are nontoxic when used in recommended concentrations. They also serve well as deodorants and can be used in sanitizing washrooms and other areas where odor is a problem.

REGULATIONS

With regard to manual sanitizing, the Food Service Sanitation Manual states:

1. Equipment and utensils shall be sanitized in the third compartment . . . .
2. The food-contact surfaces of all equipment and utensils shall be sanitized by:
   a. Immersion for at least one-half minute in clean, hot water of at least 170°F; or
   b. Immersion for at least 1 minute in a clean solution containing at least 50 parts per million of available chlorine as a hypochlorite and at a temperature of at least 75°F; or
   c. Immersion for at least 1 minute in a clean solution containing at least 12.5 parts per million of available iodine and having a pH not higher than 5.0 and at a temperature of at least 75°F; or
   d. Immersion in a clean solution containing any other sanitizing agent allowed . . . that will provide the equivalent bactericidal effect of a solution containing at least 50 parts per million of available chlorine as a hypochlorite at a temperature of at least 75°F for 1 minute; or
   e. Treatment with steam free from materials or additives other than those specified . . . in the case of equipment too large to be sanitized by immersion, but in which steam can be contained; or
   f. Rinsing, spraying, or swabbing with a chemical sanitizing solution of at least twice the strength required for that particular sanitizing solution . . . in the case of equipment too large to sanitize by immersion.
3. When hot water is used for sanitizing, the following facilities shall be provided:
   a. An integral heating device or fixture installed in, on, or under the sanitizing compartment of the sink capable of maintaining the water at a temperature of at least 170°F; and
   b. A numerically scaled indicating thermometer, accurate to ±3°F convenient to the sink for frequent checks of water temperature; and
   c. Dish baskets of such size and design to permit complete immersion of the tableware, kitchenware, and equipment in the hot water.

These regulations are minimum standards for health and safety imposed by the United States Public Health Service. State health departments may obtain their guidance from this source or their requirements may be more stringent. For example, the Wisconsin Division of Health requires a 2-minute chlorine immersion of at least 100 ppm, a 2-minute iodine immersion, or a 2-minute quaternary ammonium immersion of at least 200 ppm.

SANITIZING IN DISHWASHING MACHINES

Hot Water

Hot water is the sanitizing agent in the majority of dishwashing machines. The hot water may be the fresh water rinse, as in the door-type machines, or the final rinse, as in the conveyor-type machines.

The temperature of the final sanitizing rinse, sprayed at required volume and pressure, should be at least 180°F. (82°C) as it leaves the rinse nozzles (Fig. 12-5). If all other factors in the dishwashing operation are being performed correctly, the 180°F. (82°C) hot water will provide effective sanitization of the items being rinsed.

It is important that the final rinse be at least 180°F. (82°C), but it should not exceed 195°–200°F. (90–93°C). Water at these higher temperatures, when released from pressure in the

Figure 12-5 The temperature of the sanitizing rinse water should be at least 180°F. (82°C) as it leaves the rinse nozzles.

rinse line, will atomize and become vapor, with the result that the necessary heat impact of a liquid rinse is lost before it contacts dish and utensil surfaces.

An easily readable, enclosed, sealed type of thermometer, accurate to within ±3°F. (2°C) should be installed on the dishwashing machine to show the temperature of the sanitizing rinse water as it enters the spray arm or manifold of the final rinse.

The Food Service Sanitation Manual lists these regulations with regard to dishwashing machine sanitizing:

1. Cleaning and sanitizing may be done by spray-type or immersion dishwashing machines or by any other type of machine or device if it is demonstrated that it thoroughly cleans and sanitizes equipment and utensils.
2. Machines (single-tank, stationary-rack, door-type machines and spray-type glass washers) using chemicals for sanitization may be used: provided that,
   a. Chemicals added for sanitization purposes shall be automatically dispensed.
   b. Utensils and equipment shall be exposed to the final chemical sanitizing rinse in accordance with manufacturers' use specifications for time and concentration.
   c. The chemical sanitizing rinse water temperature shall be not less than 75°F. nor less than the temperature specified by the machines' manufacturer.
   d. A test kit or other device that accurately measures the parts per million concentration of the solution shall be available and used.
3. After sanitization, all equipment and utensils shall be air dried.

Detergent Sanitizers

A detergent sanitizer is a balanced chemical cleaning compound that has a sanitizing agent added to its formulation. Products in this group are called germicidal detergents, detergent disinfectants, or other descriptive names, depending to a large extent on the type and special qualities possessed by the sanitizing agents in the product.

The term *sanitize* is used to imply a combination of cleaning and disinfecting. In restaurant dishwashing, it is necessary to remove soil and also to kill microorganisms. This is an important public health measure.

These operations are usually carried out in two steps, the first consisting of a cleaning operation and the second of a disinfecting operation. Cleaning and disinfecting can be carried out in one step, however, provided a composition is available that cleans and simultaneously disinfects in a satisfactory manner. Such compositions are referred to as detergent sanitizers. Several types of detergent sanitizers are available for use on eating ware, utensils, and equipment.

To be effective, a sanitizer should have rapid action. Therefore, it must be combined with cleaning ingredients that will not lessen or slow down the killing ability of the sanitizing chemical. Thus, since certain sanitizers work best in acidic solutions whereas others are more effective in alkaline solutions, a correct solution pH becomes an important factor.

Chlorinated detergent sanitizers have been used successfully for stain control on dishware and glasses, and for the lowering of bacterial populations of wash solutions, as well as for other cleaning routines.

Iodine has also been combined with suitable detergents. The detergent acts as a dissolving medium for the iodine and greatly reduces its irritating action on the skin. The mixture can therefore be used as a topical antiseptic for personal use, as well as for cleaning and sanitizing utensils.

Quaternary ammonium compounds are also combined with other chemicals to create detergent sanitizers. They are highly effective against a wide variety of microorganisms. The quaternaries act more effectively in alkaline solutions than in acidic types.

STORAGE

All sanitized dishware, utensils, and equipment must be protected from recontamination. Once cleaned and disinfected, most health codes require that dishware and utensils be stored so that they remain sanitary. This requirement means minimal handling, and clean, dry, dust-free storage that is protected from insects, rodents, and droplet contamination from customers or employees (Fig. 12-6). Unless such storage facilities are available, kept disinfected, and properly used, the efforts expended in sanitizing the utensils will be wasted.

Knives, forks, spoons, and all other utensils should be picked up by their handles and stored in a clean, dry place, protected from dust. Small equipment and utensils should be stored on shelves or racks off the floor and protected from dust. Pots, pans, and other containers should be hung on pegs or inverted.

The Food Service Sanitation Manual states:

1. Handling. Cleaned and sanitized equipment and utensils shall be handled in a way that protects them from contamination. Spoons, knives, and forks shall be touched only by their handles. Cups, glasses, bowls, plates, and similar items shall be handled without contact with inside surfaces or surfaces that contact the user's mouth.
2. Storage.
   a. Cleaned and sanitized utensils and equipment shall be stored at least 6 inches above the floor in a clean, dry location in a way that protects them from contamination by splash, dust, and other means. The food-contact surfaces of fixed equipment shall also be protected from contamination. Equipment and utensils shall not be placed under exposed sewer lines or water lines, except for automatic fire protection sprinkler heads that may be required by law.
   b. Utensils shall be air dried before being stored or shall be stored in a self-draining position.
   c. Glasses and cups shall be stored inverted. Other stored utensils shall be covered or inverted, whenever practical. Facilities for the storage of knives, forks, and spoons shall be designed and used to present the handle to the employee or consumer. Unless tableware is prewrapped, holders for knives, forks, and spoons at self-service locations shall protect these articles from contamina-

Figure 12-6 Sanitized silverware and trays protected in the proper manner.

tion and present the handle of the utensil to the consumer.

SUMMARY

Sanitizing is a process or treatment carried out to destroy microorganisms remaining on the surfaces of equipment and utensils after washing and rinsing. The Food Service Sanitation Manual states: (1) tableware shall be washed, rinsed, and sanitized after each use, and (2) to prevent cross-contamination, kitchenware and food-contact surfaces of equipment shall be washed, rinsed, and sanitized after each use.

A sanitizer is an agent, usually a chemical germicide, applied in a separate operation after the objects to be sanitized have first been cleaned. Sanitizing agents are not required to destroy all of the microorganisms present on eating and drinking utensils and equipment. They must, however, reduce the number of microorganisms to a safe level acceptable to health departments.

Antimicrobial agents can effect injury or death to a microbe in several ways: (1) coagulation of cell proteins, (2) denaturation, and (3) injury to the cell membrane.

Either hot water at 170°F. (77°C) or acceptable chemical disinfectants can be used to sanitize eating ware, utensils, and equipment. These agents can be used in either manual dishwashing or machine dishwashing. Their effectiveness depends on: (1) chemical formulation, (2) concentration, (3) exposure time, (4) temperature used, (5) amount of food soil present, (6) type of microorganism present, and (7) number of microorganisms present.

The three types of sanitizers most commonly used in food and beverage establishments are: (1) chlorine compounds, (2) iodine compounds, and (3) quaternary ammonium compounds.

Regulations concerning immersion time of eating ware and concentration of sanitizers can be found in state health department sanitation codes, and should be followed.

Review Questions

1. What does sanitizing accomplish?

2. What kitchen objects or ware are required by the U.S. Public Health Service to be sanitized?

3. Define: a. germicide, b. disinfectant, c. antiseptic, d. bactericide, e. sanitizer.

4. How can antimicrobial agents effect injury or death of microorganisms?

5. a. If hot water is used as a sanitizing agent, what temperature must it reach? b. How long must objects be immersed in water of this temperature? c. What equipment should be available in order to use hot water as a sanitizing agent?

6. How can equipment be treated satisfactorily if it is too large to be sanitized in a sink?

7. a. How should sanitized cups, glasses, and bowls be handled? b. How should sanitized plates be handled?

8. Why must dishware and utensils be washed and rinsed thoroughly before sanitizing?

9. On what factors does the effectiveness of antimicrobial agents depend?

10. What chemical agents are in common use in foodservice establishments?

11. How does temperature affect the activity of most antimicrobial agents?

12. What is a detergent sanitizer?

References

Brock, T. D. (1970). *Biology of Microorganisms.* Prentice-Hall, Inc. Englewood Cliffs, New Jersey.

Burden, K. L. and R. P. Williams. (1964). *Microbiology.* Macmillan Publishing Co., Inc. New York.

Chaplin, C. E. (1951). "Observations on Quaternary Ammonium Compounds." *Can. J. Bot.* 29:373–382.

Frazier, W. C. (1967). *Food Microbiology.* McGraw-Hill Book Co. New York.

Gershenfeld, L. and B. Witlin. (1955). "Iodine Sanitizing Solutions." *Soap and Chem. Specialties* 31:189–190, 195–196, 217–218, 223.

Graubard, M. (1958). *Foundations of Life Science.* D. Van Nostrand Co. Princeton, New Jersey.

Hugo, W. B. (1971). *Inhibition and Destruction of the Microbial Cell.* Academic Press, Inc. New York.

Lawrence, C. A. and S. S. Block. (1968). *Disinfection, Sterilization and Preservation.* Lea and Febiger. Philadelphia, Pennsylvania.

Miller, S. J. (1963). "Sanitation and Dishes—Aspects Old and New: Part I." *J. Amer. Diet. Assoc.* 43:23–28.

Miller, S. J. (1963). "Sanitation and Dishes—Aspects Old and New: Part II." *J. Amer. Diet. Assoc.* 43:29–32.

Sartwell, P. E. (1965). *Preventative Medicine and Public Health.* Appleton-Century-Crofts. New York.

Schwartz, A. M., J. W. Perry, and J. Berch. (1958). *Surface Active Agents and Detergents,* vol. 2 Interscience Publications, Inc. New York.

State of Wisconsin. *Food Sanitation Manual.* Division of Health. Wisconsin Department of Health and Social Services. Madison, Wisconsin.

Sykes, G. (1958). *Disinfection and Sterilization.* D. Van Nostrand Co. Princeton, New Jersey.

U.S. Department of Health, Education and Welfare. (1976). *Food Service Sanitation Manual.* U.S. Government Printing Office. Washington, D.C.

- I. Methods of Preservation
  - A. Pasteurization
  - B. Canning
  - C. Dehydration
  - D. Freezing
  - E. Refrigeration
- II. Types of Food
  - A. Dairy Products
  - B. Meat and Meat Products
  - C. Raw Foods
  - D. Growth Medium
  - E. Time-Temperature Relationships
- III. Food Preparation
  - A. Clean Hands
  - B. Clean Equipment
  - C. Personal Hygiene
  - D. Proper Cooking
- IV. Food Protection
  - A. Food Purchasing
  - B. Food Storage
  - C. Display and Holding
- V. Summary

The possible sources of contamination of food and drink are numerous. Food of animal or plant origin may be contaminated when it enters the processing plant, retail store, or restaurant; animals such as rodents or insects may carry pathogenic microorganisms onto the premises; and man, himself, may transfer food-poisoning organisms to food or equipment if he is a carrier or is actually ill. Both food processor and retailer must therefore be constantly on the alert for any carelessness or breakdown in sanitary procedures.

Many of the public health problems with which food processors and retailers must cope are similar: cleanliness, maintenance of high temperatures when necessary, adequate refrigeration, and others. However, the food processor is more concerned with techniques of preservation involving long storage or shelf life, whereas the retailer, including the restaurateur, is more concerned with short-term preservation. Both are concerned with placing safe food before the public.

The U.S. Public Health Service Food Service Sanitation Manual states:

> At all times, including while being stored, prepared, displayed, served, or transported, food shall be protected from potential contamination, including dust, insects, rodents, unclean equipment and utensils, unnecessary handling, coughs and sneezes, flooding, drainage, and overhead leakage or overhead drippage from condensation. The temperature of potentially hazardous food shall be 45°F. or below or 140°F. or above at all times. . . .

METHODS OF PRESERVATION

Pasteurization

Pasteurization is the heating of food, especially milk and milk products, to a temperature that destroys all harmful bacteria. Milk can be pasteurized at either a relatively low temperature (145°F. [63°C] or higher) for at least 30 minutes or at higher temperatures (161°F. [72°C] or higher) for at least 15 seconds. These temperatures (145°F. and 161°F.) do not sterilize food but do kill 90 to 99 percent of the microorganisms present, including most bacteria, yeasts, and molds. The few survivors include small numbers of thermophilic and spore-forming bacteria. After pasteurizing, the products are quickly cooled to 45°F. (7°C) or lower.

Some foods, such as cream used for butter, ice cream, and coffee cream, are subjected to high temperatures because of their protective effect on microorganisms. On the other hand, since acidic foods are generally capable of destroying bacteria at a lower temperature than nonacidic foods, fruit juices, pickles, sauerkraut, and so forth are often pasteurized rather than subjected to the usual higher processing temperatures.

The U.S. Public Health Service Food Service Sanitation Manual states:

> Fluid milk and fluid milk products used or served shall be pasteurized and shall meet the Grade A quality standards as established by law. Dry milk and dry milk products shall be made from pasteurized milk and milk products.

Canning

The invention of canning, like many inventions and discoveries, came as the result of war. Napoleon had noticed that more of his soldiers died from inadequate diet or from foodborne illness than died in battle. As a result, a prize of 12,000 francs was offered by the French government for the discovery of an adequate method of preserving food.

The winner of the prize was a French candymaker, Nicolas Appert. His experiments led him to believe that air was involved in the spoilage of food. This was an astute observation since it preceded the experiments of Pasteur by a good 50 years.

Appert first cooked the food and packed it in tightly corked bottles. He then placed the bottles in boiling water and cooked the food again. Most foods remained edible for months after this operation. He was awarded his prize on January 10, 1810, after 15 years of work.

About the same time, an Englishman, Peter Durand, was developing metal containers called cannisters. These proved more satisfactory for handling and transporting than bottles and soon achieved widespread use. These containers became known as cans.

The purpose of canning is to destroy the microorganisms that cause food spoilage and food poisoning. Canning combines both cooking and sterilizing, with the foods to be preserved sealed in cans or jars. Pressure cooking is essential for the canning of these foods that may be contaminated with *Clostridium botulinum*, since the spores of this organism can withstand high temperatures and are potentially dangerous. Pressure cooking destroys all heat-resistant bacteria, provided the food is cooked for a sufficient length of time. High temperatures cause coagulation of cell proteins and the melting of fatty compounds, thereby killing the microorganism.

Dehydration

Dry heat is sometimes used to remove moisture from foods in order to preserve them. The temperatures used in this process are not high enough to kill large numbers of bacteria. The heat merely removes one of the factors (water) necessary for bacterial growth.

The dehydration of foods to preserve them probably dates to the early days of man. The first American colonists preserved cooked corn and salted dry fish in this manner. Settlers and Indians, alike, dried thin strips of buffalo meat and beef (jerky) by hanging them in the sun.

Dehydration of vegetables was practiced during the Civil War when soldiers' rations contained small quantities of dried vegetables. During World War I, 60 percent of the vegetables consumed by the German army consisted of dehydrated products. The United States did not enter this field seriously until World War II.

The first egg-drying plant in the United States was started in about 1878. Production in later years was erratic and finally ceased around 1915 because of the importation of cheaper Chinese egg products. When these imports decreased between 1927 and 1931 because of civil war in China and high tariff duties, American production resumed, finally reaching 4 million pounds a year during World War II.

The drying of milk dates back to the thirteenth century. Milk tablets were commercially produced in France in 1810. American production of dried milk started in 1856. Development of the industry has been associated with that of the condensed milk industry; the difference between the two is only in the amount of water present. The industry, with its production of instant dry nonfat milk, is now well established in the United States, although there were some early problems with staphylococcal contamination (Anderson and Stone, 1955; Armijo et al., 1957).

The salting and drying of meat and fish is an old procedure. Dried fish were prepared and sold by the Puritans in their early American settlement. Meat and fish are also preserved by smoking.

Freezing

Although freezing food as a means of preservation is not new, it was not until the development of the quick-freezing process that frozen fruits and vegetables could be processed on a commercial scale. Since 1930, quick-freezing has been used extensively as a method of vegetable, fruit, and fruit juice preparation. The frozen food industry has developed so rapidly that it is now possible to buy virtually any type of food in quick-frozen form. It should be remembered, however, that although freezing may reduce bacterial populations, it rarely kills all forms of microorganisms present. Therefore, frozen foods should not be allowed to thaw at room temperatures for long periods of time. Extensive bacterial growth may result.

The Public Health Service Food Service Sanitation Manual states:

1. Frozen food shall be kept frozen and should be stored at a temperature of 0°F. or below.
2. Potentially hazardous foods shall be thawed:
   a. In refrigerated units at a temperature not to exceed 45°F; or
   b. Under potable running water of a temperature of 70°F. or below, with sufficient water velocity to agitate and float off loose food particles into the overflow; or
   c. In a microwave oven only when the food will be immediately transferred to conventional cooking facilities as a part of a continuous cooking process or when the entire, uninterrupted cooking process takes place in the microwave oven; or
   d. As part of the conventional cooking process.

Most frozen foods can be kept satisfactorily for one year. At a temperature of −20°F. (−29°C), frozen meat, fruits, and other products have been held for two years without any noticeable change of quality. Frozen foods are not sterile, however. Most of the bacteria are killed by very low temperatures but some of them form spores and continue to live. During the thawing process, bacterial activity is resumed. It is imperative, therefore, that this process occur as rapidly as possible.

Refrigeration

Although refrigerators keep perishable foods from spoiling for varying lengths of time, food cannot be kept indefinitely in the refrigerator without showing some signs of mold or bacterial activity; milk sours, meat becomes slimy, butter becomes rancid, and many foods become moldy. Refrigeration temperatures only check the growth of bacteria, they do not destroy them. Some species, however, such as *Clostridium perfringens*, are more sensitive to cold than other species. *Clostridium botulinum* type E, on the other hand, seems to be more resistant to refrigeration temperatures than most food-poisoning species.

Fresh fruits and vegetables need to be refrigerated to maintain their storage life and taste, but potentially hazardous foods, such as, meat or meat products, eggs, milk, or other high protein foods, must be maintained at a temperature of 45°F. (7°C) or below or food poisoning may result.

Foods should be refrigerated at all times except when they are actually being prepared. Hot food that is to be saved for future use should be refrigerated immediately in as small a container as practical. In this way, the food is in the danger zone for only a short period of time.

Many factors affect the rate at which food cools. Among these are the amount of food in the container, the shape of the container, the nature of the food, and the use of stirring.

Small amounts of food cool more quickly than do large quantities. Food placed in shallow pans or small containers will cool much more rapidly than food placed in deep or large containers. Food cooled in large stock pots may be in the danger zone for four hours or longer as it cools. This is long enough to allow extensive bacterial growth and the production of considerable enterotoxin.

Cooling food in containers is hastened considerably if it is stirred periodically. This brings warmer food to the surface, allowing it to lose heat that would otherwise be retained within the mass of the food, thereby keeping it within the danger zone for an undesirable period of time. When the food has been cooled adequately, it should be refrigerated immediately.

The U.S. Public Health Service Food Service Sanitation Manual states:

Since any temperature between 45°F. and 140°F. presents a hazard to public health in terms of microbial growth, food must remain in the critical temperature zone as little time as possible.

TYPES OF FOOD

Foodborne illness will not occur unless three major requirements are present: first, food must be contaminated with food-poisoning organisms; second, the food must be a good growth medium for the type of microorganism present; and third, the food must remain within the temperature range of the organism for a long enough period of time, usually several hours, to allow large scale multiplication (Fig. 13-1). A major responsibility of persons in the foodservice industry is to prevent these requirements from being fulfilled.

Dairy Products

Dairy products tend to be highly perishable. They are an excellent medium for *Staphylococcus aureus* and for many other food-poisoning bacteria. Few foodborne illnesses in the United States are currently associated with fluid milk and cream, however. Milk herds and barns are kept in clean, healthy conditions, pasteurization is extremely effective, and refrigeration is practiced from milking to consumer.

When milk is taken from a healthy cow, it contains few bacteria. If the udder is infected, however, pathogenic bacteria such as *S. aureus*

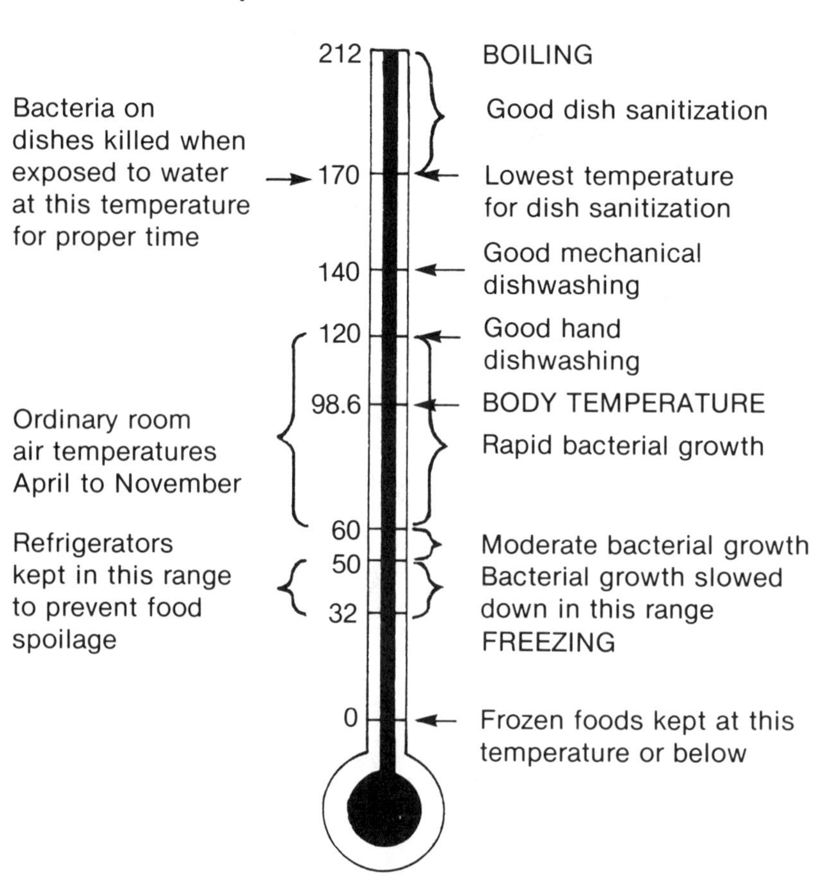

Figure 13-1 Some temperatures affecting microorganisms and their control.

can enter the milk. Bacteria can also be transferred through milking machines from an infected cow to a healthy one. In the case of hand-milking, the hands can serve as an agent of contamination. If milk is not properly refrigerated, food-poisoning bacteria can multiply and large numbers result, some possibly reaching the public. This rarely happens in the United States, however, because of regulation of dairy herds and effective pasteurization.

Several types of milk products can be prepared from whole milk. Among these are evaporated milk, condensed milk, and dried milk. Evaporated milk is made from pasteurized, whole milk that is partially evaporated, homogenized, and canned. Condensed milk is evaporated milk to which sugar is added before canning. Because of its sugar content, condensed milk is highly susceptible to invasion by such microorganisms as yeasts, molds, and bacteria.

Storage time is brief for condensed milk and evaporated milk once they have been opened. These products should be kept refrigerated at all times once opened. Leaky or bulging cans should be discarded because of the high probability of contamination.

Several dried milk products are now on the market, including whole milk, skim milk, cream, buttermilk, whey, malted milk, and ice cream mixes. Factories producing these foods are closely regulated, but their products can contain bacteria. When water is added to these foods they become highly perishable. They should be refrigerated immediately and used as soon as possible.

Cheese made from unpasteurized milk has caused large-scale food-poisoning outbreaks in the past. *Staphylococcus aureus* and *Escherichia coli* have been problem contaminants. *S. aureus* has also been isolated from cheese made from pasteurized milk, which indicates that either recontamination sometimes occurs during manufacturing or that pasteurization was not totally effective. Cheese should be kept refrigerated.

Meat and Meat Products

Livestock on the farm tend to be healthy animals. As these animals proceed to the slaughterhouse, they are subject to more crowded conditions and the incidence of carriers of pathogens, particularly salmonellae, increases. Although diseased animals are rejected at the stockyards, carriers may pass through undetected.

The main contamination of meat occurs, however, after the carcass has been opened and eviscerated. Washing of the cavity after evisceration commonly spreads intestinal bacteria, such as salmonellae and clostridia, to other parts of the carcass. Efforts to reduce the bacterial content of meat at the retail level have concentrated on more sanitary handling of livestock and carcasses at the packing plant.

Many surveys have been conducted to determine bacterial counts on market meats (see chapters 3–6). Jay examined chicken, pork liver, pork chops, veal steak, and lamb chops. A total of 32 percent of the samples contained staphylococci. Duitschaever et al. investigated the incidence of *S. aureus* in fresh ground beef and in hamburger patties. A total of 108 samples of fresh ground beef and 99 samples of commercial frozen hamburger patties were examined. *S. aureus* was present in 46 percent of the fresh ground beef and 94 percent of the frozen hamburger patties.

Galton et al. examined 217 samples of fresh pork sausage and 127 samples of smoked sausage. A total of 23 percent of the fresh sausage and 12.5 percent of the smoked sausage samples were positive for salmonellae. Felsenfeld et al. found salmonellae in 18 percent of the hamburger samples they examined.

A salmonella outbreak, which occurred in 1970 (Bailey et al., 1972), clearly shows the hazards that exist during food production and the difficulty and expense that sometimes occur. Many incidences of contamination cannot be traced to the ultimate source.

In April, 1970, a village grocer became ill and *Salmonella panama* was isolated from her stool. She blamed her illness on cooked ham that had been produced at a nearby bacon factory. Shortly thereafter, her parents, husband, and two children became symptomless excreters (temporary carriers). The organism was also isolated from a road worker in the village. His wife and three children also became symptomless excreters. The next victims were two children in a nearby village who developed gastroenteritis. *S. panama* was also isolated from them. The common factor among all these persons was the consumption of ham produced by the same bacon factory. This factory produced a full range of raw and cooked pork.

Stool samples were obtained from each employee in the factory. One man, who worked in the baked ham area, was found to be a symptomless excreter of *S. panama*. He was taken off the job until three successive stool samples were negative. The problem appeared solved.

During the summer, however, new cases began to appear. The bacon factory was again implicated as the source. Swabs from the factory revealed the presence of *S. panama* behind a baking oven, and on a meat-packing table, meat-packing machine, and conveyor belt. After examination of hams, produced during the week, revealed that some contained *S. panama*, the entire stock of factory hams was recalled from retail outlets and destroyed.

About this time, two of the factory workers became ill with gastroenteritis, and *S. panama* was isolated from their stools. Examination of the workers at the plant now showed 82 carriers and one additional case of intestinal illness. As a result, the factory was closed and new equipment was purchased. Two weeks later, workers with three negative stool examinations were allowed to return to work. It was many more weeks before full production could be resumed.

This case is a good example of why all food-handling businesses, from producer to retailer, must constantly make sure that all techniques associated with proper food sanitation are followed—no exceptions should be allowed (Fig. 13-2). In this instance, the ultimate source of the infection was never found. Whatever the source of the contamination, however, it was able to spread throughout sections of the plant. If sanitizing procedures had been scrupulous, salmonellae would not have been isolated from the meat-packing table, meat-packing machine, and conveyor belt. It would not have been necessary to replace expensive equipment, an expensive and damaging recall would not have occurred, workers would not have lost paychecks, and many cases of salmonellosis, a potentially dangerous dis-

Figure 13-2 A sometimes overlooked facet of sanitation. Walk-in meat storage coolers should not be so overcrowded that inadequate circulation between meat cuts results. If meat cuts touch, the temperature of those areas may remain above 45°F. (70°C), resulting in bacterial growth.

ease, would not have spread throughout the area.

Raw Foods

Since the origin of raw fruits and vegetables that enter the kitchen is rarely known to the staff, these foods must be cleaned thoroughly before use. Insecticide residues and bacteria may still be present on their surface. These types of food have usually been handled by many people, from harvest to serving, and the possible sources of contamination are many.

Growth Medium

The nature of the food is the factor that determines the type and success of the microorganisms that will grow on it. The food must supply:

1. Nutritional requirements, including carbohydrates, fats, proteins, vitamins, minerals, and so forth. All bacteria have certain nutritional requirements and foods that supply these requirements to food-poisoning bacteria are potentially hazardous foods.
2. Moisture content. Most cells consist mainly of water, and the vegetative cells of bacteria are no exception. Moist foods are highly susceptible to heavy bacterial growth, providing other requirements are present.
3. Proper pH. The acidity or alkalinity of a food affects the degree to which bacteria are able to utilize it successfully. Food-poisoning bacteria cannot live in highly acidic foods. Growth is most rapid in foods whose pH is near normal.
4. Growth temperatures. The temperature of the food regulates the rate of bacterial growth. Foods that remain within the temperature range of 45° to 140°F. (7 to 60°C) for long periods of time are most susceptible to growth of food-poisoning bacteria.

Foods vary, therefore, in their ability to support bacteria. Many bacteria have special nutritional requirements, particularly amino acids, which must be supplied by their food medium. Such foods as meat, poultry, fish, seafood, and milk, which supply these requirements, must be prepared with extreme care to avoid contamination and subsequent growth of food-poisoning bacteria. The great majority of food-poisoning outbreaks caused by *S. aureus*, *C. perfringens*, *B. cereus*, and the *Salmonella*, result from improper handling of these foods.

Time-Temperature Relationships

Most foods are subject to attack by bacteria. Unless they are handled properly from harvest to serving, foods can become dangerous products, causing illness or death. All persons associated with the handling and care of foods should recognize the dangers involved and should use appropriate measures.

Records of foodborne illnesses usually show that the food that transmitted the bacteria or toxin was heavily infected because it was allowed to remain for long periods at temperatures that allowed massive bacterial growth. It should be realized, therefore, that the factors of time and temperature are closely related and that these factors must be regulated carefully to ensure safe food for human consumption.

Food-poisoning bacteria grow over a rather wide range of temperatures. The temperature range of the most common food-poisoning bacteria extends from about 45°F. (7°C) to about 140°F. (60°C). This temperature range is called the danger zone or critical zone for bacterial growth. Foods should be allowed to remain within this temperature zone only for brief holding periods.

The United States Public Health Service Food Service Sanitation Manual states:

1. Potentially hazardous food requiring refrigeration after preparation shall be rapidly cooled to an internal temperature of 45°F or below.
2. The internal temperature of potentially hazardous foods requiring hot storage shall be 140°F. or above except during necessary periods of preparation. Potentially hazardous food to be transported shall be held at a temperature of 140°F. or above. . . .

3. Since any temperature between 45°F. and 140°F. presents a hazard to public health in terms of microbial growth, food must remain in the critical temperature zone for as little time as possible.

Temperatures below 45°F. (7°C) either inhibit or considerably retard the growth of all food-poisoning bacteria except *C. botulinum* type E. This organism can grow and produce toxin at temperatures as low as 38°F. (3°C). A temperature of 45°F. (7°C) will prevent growth of *S. aureus* in most foods and prevent toxin formation by the organism in all foods. Some molds can grow at refrigerator temperatures, however.

Although freezing does not destroy all of the microorganisms present in a food, it does prevent the multiplication of the remaining bacteria. Frozen foods should be thawed in accordance with the regulations set forth in the United States Public Health Service Food Service Sanitation Manual. Thawing at room temperature will give any bacteria present an opportunity to multiply, particularly those in the outer layers of the food.

A temperature of 140°F. (60°C) will inhibit the growth of all food-poisoning bacteria. This is not a lethal temperature, however. It will reduce the number of bacteria, but if the temperature falls to within the growth range the bacteria will resume multiplication. In addition, spores are unharmed by a temperature of 140°F. (60°C).

Since spore-forming bacteria such as *Bacillus cereus* and *Clostridium perfringens* survive the temperatures used in cooking many foods, rapid cooling of foods to prevent the growth of remaining bacteria is essential. The rate at which a food will heat or cool is also important in time-temperature relationships. Such foods as turkey, meat loaf, and large hams must reach an internal temperature high enough to kill all pathogens that might be present.

Turkey is frequently incriminated in food-borne illness. It is an excellent medium for *S. aureus*, *C. perfringens*, and the *Salmonella*. Raw turkeys are often contaminated with these bacteria when they arrive in the kitchen. When cooked, an internal temperature of 165°F. (74°C) should be reached.

The United States Public Health Service Food Service Sanitation Manual states:

1. Potentially hazardous foods requiring cooking shall be cooked to heat all parts of the food to a temperature of at least 140°F., except that:
   a. Poultry, poultry stuffings, stuffed meats and stuffings containing meat shall be cooked to heat all parts of the food to at least 165°F. with no interruption of the cooking process.
   b. Pork and any food containing pork shall be cooked to heat all parts of the food to at least 150°F.
   c. Rare roast beef shall be cooked to an internal temperature of at least 130°F., and rare beef steak shall be cooked to a temperature of 130°F. unless otherwise ordered by the immediate consumer.

FOOD PREPARATION

Given the universal occurrence of bacteria and the large number of foods on which they can grow, it is virtually impossible to produce bacteria-free food except under certain sterile hospital conditions. However, all foods can be rendered safe from a public health viewpoint, that is, the bacterial count is low enough so that the food can be consumed safely. To accomplish this requires rigorous and unceasing sanitary practices.

Clean Hands

Hands must be kept clean while on the job. They should be washed thoroughly and often with soap and warm water, particularly when changing from one task to another. Microorganisms are easily transferred from one food to another through contaminated hands, counters, utensils, or equipment such as grinders and slicers. This type of contamination is known as cross-contamination and has been responsible for many food-poisoning outbreaks.

Unnecessary hand contact with food should be scrupulously avoided. Food should be handled, if at all possible, with clean and sanitized utensils (tongs, scoops, or tableware) (Fig. 13-3). Plastic gloves can also be worn but

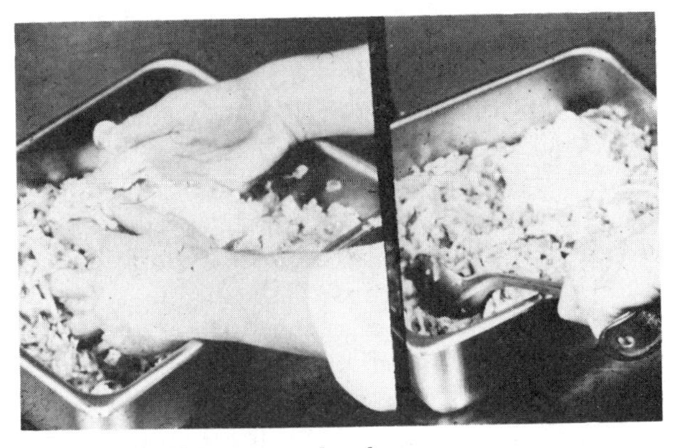

Figure 13-3 Unnecessary hand contact with food should be avoided. Clean and sanitized utensils should be used or plastic gloves worn.

even these become contaminated through food contact and should be changed when switching from one job to another.

Clean Equipment

The importance of clean equipment has been demonstrated during many outbreaks, several of which were mentioned in previous chapters. Two of these outbreaks involved slicing machines. In one outbreak, 507 cases of typhoid fever occurred over several days from a slicer that became contaminated from canned corn beef. The slicer then transferred the bacteria to other cold cuts that were sliced on it. The other outbreak involved a delicatessen-restaurant. A contaminated slicer was implicated in this outbreak, also. More than 600 cases of salmonellosis occurred as a result of the unsanitary practices in this establishment.

A knife was the agent of cross-contamination in an outbreak involving watermelon. A total of 17 cases of salmonellosis occurred in five families in Massachusetts. Contamination of the knife resulted from slicing a melon that was infected with S. *miami*. The knife then spread the bacteria to other melons. Contaminated cafeteria trays were the agents responsible for contaminating food items in a hepatitis outbreak in a Texas school system. A total of 27 cases resulted.

Surfaces of tables and equipment used to prepare all raw foods, especially meat, seafoods, and poultry, should be cleaned and sanitized before being used to hold cooked foods (Fig. 13-4). These foods frequently contain microorganisms when they enter the food-service establishment. Once a counter or piece of equipment is contaminated, other foods contacting it may become contaminated. The food-contact surfaces of food equipment should be cleaned frequently during the day (Fig. 13-5).

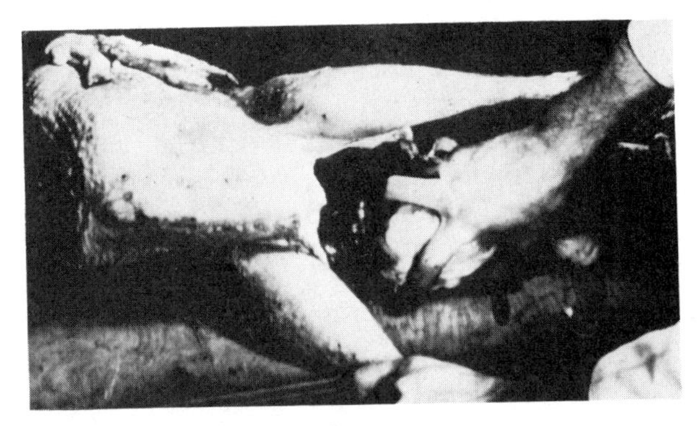

Figure 13-4 All surfaces that come into contact with food should be kept clean and sanitized.

Figure 13-5 All equipment, such as this roll warmer, should be kept clean and closed.

Personal Hygiene

This subject will be discussed in more detail in chapter 14. Personal cleanliness is of the utmost importance in any establishment in which food is handled.

Bacteria are common on the skin, hair, clothing, and in the nostrils of most people. They are easily transferred from these areas to food and equipment. During working hours, hand-hair and hand-nostril contact should be avoided and, of course, hands should contact food as infrequently as possible. Smoking should be avoided in food preparation areas (Fig. 13-6).

Proper Cooking

Proper cooking not only makes food more palatable and digestible, it also destroys bacteria that cause foodborne illness. The destruction of these bacteria, however, requires cooking temperatures that will destroy both vegetative cells and spores.

Some foods, particularly fowl and large cuts of meat, heat slowly. Thus, they tend to spend rather long periods of time with their temperature within the danger zone of 45° to 140°F. (7 to 60°C). Foods also vary in the degree to which they exert a protective effect on bacteria. It is vitally important, therefore, that meat and fowl be cooked until their recommended internal temperatures have been reached. Temperatures should be taken with a thermometer designed for internal recording.

Pork is the source of *Trichinella spiralis,* a very dangerous worm that in its larval stage, attacks the muscles and occasionally the brain of man. It can cause blindness and death. The larval worm is microscopic and cannot be detected during normal methods of processing pork. A temperature of 150°F. (66°C) is required to kill it. This temperature will also kill *Taenia solium,* the pork tapeworm. A higher temperature, however, is required to destroy the spores of the food-poisoning bacterium, *Bacillus cereus.* A thermometer inserted into the center of the meat should be used to ensure that the meat has reached a safe temperature.

Poultry are frequently contaminated with salmonellae at the time they leave the processing plant. Undercooked fowl are a major cause

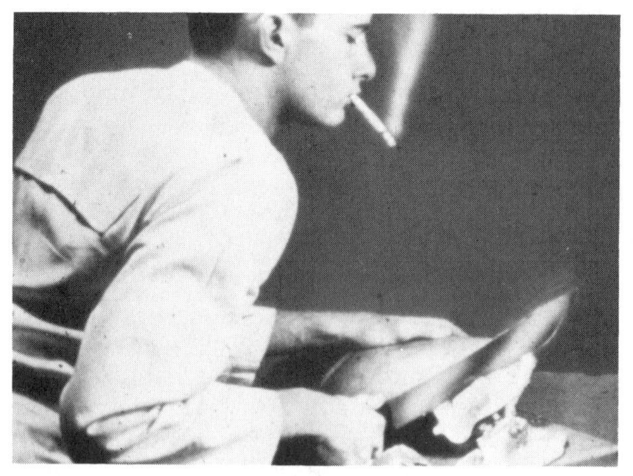

Figure 13-6 Ashes can fall onto food, tables, utensils, and equipment making them unsightly and offensive to customers.

of salmonellosis. An internal thermometer should be used to make sure that a temperature of 165°F. (74°C) has been reached. If the fowl is stuffed, the dressing should also reach this temperature.

FOOD PROTECTION

Food Purchasing

Although the operator of a foodservice establishment rarely knows the origin of the food he purchases and has no control over the conditions under which it was raised, processed, or transported, he does have direct control over his source. It is the responsibility, therefore, of the owner or manager of the business to purchase all food from sources approved or recognized as clean and sanitary by the proper federal, state, or local agencies. Among these agencies are the United States Public Health Service, United States Department of Agriculture, and state and local health and agricultural departments. All food should be clean and wholesome, free from spoilage, adulteration, and misbranding, and should otherwise be safe for human consumption.

The United States Public Health Service Food Service Sanitation Manual states:

> Food shall be in sound condition, free from spoilage, filth, or other contamination and shall be safe for human consumption. Food shall be obtained from sources that comply with all laws relating to food and food labeling. The use of food in hermetically sealed containers that was not prepared in a food processing establishment is prohibited.
>
> To control foodborne illness and prevent food spoilage, which may result from improperly processed, handled, or transported food, food service establishments must be concerned with the sources of the food they use. The sound conditions, proper labeling, and safety of food are basic requirements for the protection of the public health.
>
> The use of hermetically sealed, noncommercially packaged food is prohibited because of the history of such food in causing foodborne illness. Additional specific requirements for food supplies, such as the pasteurization of milk and milk products or the use of only clean, whole-shell eggs, are included because these products are exceptionally good media for the growth of pathogens. Labeling requirements, particularly for shellfish, provide assurance that the source of any such food is under the control of a regulatory authority, thus providing for the protection of the public health.

Food Storage

Proper storage of foods is a definite factor in contributing to their safety. Since many foods are not served immediately after arrival on the premises of the foodservice establishment, they must be stored in a manner that will ensure their wholesomeness when they are used.

Food placed in dry storage should be arranged so that the food stored first is removed first. This lessens the possibility of contamination and bacterial growth during the storage period (Fig. 13-7).

The United States Public Health Service Food Service Sanitation Manual states:

1. Food, whether raw or prepared, if removed from the container or package in which it was obtained, shall be stored in a clean covered container except during necessary periods of preparation or service. Solid cuts of meat shall be protected by being covered in storage, except that quarters or sides of meat may be hung uncovered on clean sanitized hooks if no food product is stored beneath the meat.
2. Containers of food shall be stored a minimum of 6 inches above the floor in a manner that protects the food from splash and other contamination, and that permits easy cleaning of the storage area . . . .

Display and Holding

Food held on a serving line must be retained so that it is not contaminated by coughing or sneezing. It must also be protected from unnecessary hand contact by customers. This can be accomplished by protective guards that allow the customer to reach only the closest foods in the display counter.

Figure 13-7 Stored foods should be kept off the floor and arranged so that foods stored first will be used first.

Cold foods on a serving line must be kept chilled to a temperature of 45°F. (7°C) or less. This prevents the growth of food-poisoning bacteria in any foods that might have become contaminated.

Hot foods on a serving line should be maintained at a temperature of 140°F. (60°C) or above. This includes display counters, holding devices, barbecue racks, or any piece of equipment that holds hot food for future serving. The temperature should be measured with a thermometer and should be the actual temperature of the food, not the temperature of the heating unit. A temperature of 140°F. (60°C) or above will prevent the growth of any food-poisoning bacteria that might be present.

The United States Public Health Service Food Service Sanitation Manual states:

> Potentially hazardous food shall be kept at an internal temperature of 45°F. or below or at an internal temperature of 140°F. or above during display and service, except that rare roast beef shall be held for service at a temperature of at least 130°F.

The following preventive measures should be practiced to ensure that safe food reaches the public:

1. Keep food clean.
2. Keep utensils and equipment clean and sanitized.
3. Keep cold food at a temperature of 45°F. (7°C) or below.
4. Keep hot food at a temperature of 140°F. (60°C) or above.

As you can see, most of these practices are related to temperature control. This is the most potent weapon the restaurateur has in his arsenal for defense against foodborne illness. Proper time-temperature control accompanied by sound purchasing and storage practices, proper maintenance and cleaning procedures, and good personal hygiene among the employees should ensure safe food for the public.

The United States Public Health Service Food Service Sanitation Manual states:

Proper food protection measures should include:

1. Application of good sanitation practices in the handling of food.
2. Strict observation of personal hygiene by all foodservice employees.
3. Keeping potentially hazardous food refrigerated or heated to temperatures that minimize the growth of pathogenic microorganisms.
4. Inspecting food products as to their sanitary condition prior to acceptance at the establishment.
5. Provision of adequate equipment and facilities for the conduct of a sanitary operation.

SUMMARY

Pasteurization is the heating of food to a temperature that destroys all harmful bacteria. Pasteurization does not sterilize food but it does kill 90 to 99 percent of the microorganisms present.

The purpose of canning is to destroy the microorganisms that cause food spoilage and food poisoning. Canning combines both cooking and sterilizing, with the foods to be preserved sealed in cans or jars.

Dehydration is used to remove moisture from foods in order to preserve them. The temperatures used in this process are not high enough to kill large numbers of bacteria. The heat merely removes water, which is necessary for bacterial growth.

Freezing is an effective way of preserving foods. Temperatures should be maintained at 0°F. (−18°C) or below. Although freezing may reduce bacterial populations, it rarely results in killing all forms of microorganisms present.

Refrigeration keeps perishable foods from spoiling for varying lengths of time. Refrigeration temperatures only check the growth of bacteria; they do not destroy them. Refrigerators should be maintained at a temperature of 45°F. (7°C) or below. The amount of food in a container, the shape of the container, the nature of the food, and the use of stirring affect the rate at which a food cools.

Foodborne illness will not occur unless three major requirements are present: a food must be contaminated, it must be a good growth medium, and it must remain within the growth range of the bacteria long enough to allow large-scale multiplication.

Dairy products tend to be highly perishable. They are excellent media for many food-poisoning bacteria. Meat and poultry are also excellent growth media. Contamination frequently occurs at the packing plant. Foods to be eaten raw should be thoroughly washed to remove bacteria and traces of insecticide.

The nature of the food determines the type and success of the microorganism that will grow on it. Among the requirements the food must supply are: nutritional requirements, water, proper pH, and temperature.

The temperature range of 45° to 140°F. (7–60°C) is known as the danger zone. Foods should remain within this zone for as short a time as possible. Temperatures of 45°F. (7°C) or below, or 140°F. (60°C) or above, inhibit the growth of food-poisoning bacteria.

Clean hands, clean equipment, and good personal hygiene are vital in the preparation of safe food. Proper cooking is also necessary. Proper cooking destroys bacteria that cause foodborne illness.

Food should be purchased only from sources approved by the proper federal, state, or local agencies. This food should be stored properly in a place safe from contaminating agents.

Review Questions

1. a. What is pasteurization? b. How is it accomplished?

2. a. Describe Nicolas Appert's method of canning food. b. What is the purpose of canning? c. How is this accomplished?

3. How does dehydration preserve food?

4. a. At what temperature should frozen foods be kept? b. If frozen foods are thawed in a refrigerator, what temperature should be used? c. Why should frozen foods not be thawed at room temperature except by specified procedure?

5. How are some bacteria able to survive freezing conditions?

6. How does refrigeration prolong the life of foods?

7. What factors affect the rate at which food cools?

8. What three major requirements must be present for foodborne illness to occur?

9. Why are dairy products highly perishable?

10. When is contamination of meat most likely to occur?

11. Why must raw foods be washed thoroughly?

12. What bacterial requirements must a food provide to be a good growth medium?

13. a. What is the temperature range of most food-poisoning bacteria? b. Why is this temperature range called the danger zone?

14. How do temperatures below 45°F. (7°C) and above 140°F. (60°C) affect food-poisoning bacteria?

15. a. To what temperature must poultry and stuffings be cooked to ensure their safety for consumption? b. To what temperature must pork be cooked? c. To what temperature must

rare cuts of beef be cooked? d. To what temperature must all other hazardous foods be cooked?

16. Why must the hands of a food handler be kept clean at all times?

17. What is the importance of clean equipment in food preparation?

18. What is the importance of proper cooking in ensuring the safety of food?

19. What is the function of protective guards on a serving line?

References

Anderson, P. H. R. and D. M. Stone. (1955). "Staphylococcal Food Poisoning Associated with Spray-Dried Milk." *J. Hyg.* 53:387.

Armijo, R., D. A. Henderson, R. Timothee, and H. B. Robinson. (1957). "Food Poisoning Outbreaks Associated with Spray-Dried Milk—An Epidemiologic Study." *Amer. J. Pub. Health* 47:1093–1100.

Castellani, A. G., R. R. Clark, M. I. Gibson, and D. F. Meisner. (1953). "Roasting Time and Temperature Required to Kill Food Poisoning Microorganisms Introduced Experimentally in Stuffing in Turkey." *Food Res.* 18:131–138.

Crisley, F. D. (1963). "Factors Affecting the Quality of Processed Foods." *J. Environ. Health* 26:181–186.

Duitschaever, C. L., D. H. Bullock, and D. R. Arnott. (1977). "Bacteriological Evaluation of Retail Ground Beef, Frozen Beef Patties, and Cooked Hamburger." *J. Food Protect.* 40:378–381.

Felsenfeld, O., V. M. Young, and T. Yokimura. (1950). "A survey of Salmonella Organisms in Market Meat, Eggs, and Poultry." *J. Amer. Vet. Med. Assoc.* 116:17–21.

Galton, M. M., W. D. Lowry, and A. V. Hardy. (1954). "Salmonella in Fresh and Smoked Pork Sausage." *J. Infect. Dis.* 95:232–235.

Jay, J. M. (1962). "Further Studies on Staphylococci in Meats. III. Occurrence and Characteristics of Coagulase-Positive Strains from a Variety of Nonfrozen Market Cuts." *Appl. Microbiol.* 10:247–251.

Rogers, R. E. and C. S. McClesky. (1957). "Bacteriological Quality of Ground Beef in Retail Markets." *Food Technol.* 11:318–320.

U.S. Department of Health, Education and Welfare. (1976). *Food Service Sanitation Manual.* U.S. Government Printing Office. Washington, D.C.

I. Personal Hygiene
   A. Importance
   B. Protection of Hands

II. Sanitary Facilities
   A. Water Supply
   B. Sewage
   C. Plumbing
   D. Toilet Facilities
   E. Lavatory Facilities
   F. Garbage and Refuse

III. Construction and Maintenance of Physical Facilities
   A. Floors
   B. Walls and Ceilings
   C. Lighting
   D. Dressing Rooms and Locker Areas
   E. Storage of Poisonous or Toxic Materials

IV. Summary

In previous chapters, we discussed the characteristics of many of the microorganisms that cause foodborne illness. We studied the major foodborne illnesses and how they can be contracted; we examined the characteristics of many foods and learned which types are potentially hazardous. The role of insects and rodents as agents of contamination was studied; proper techniques of dishwashing and sanitizing were investigated; and, finally, we examined methods of protecting food from storage to serving.

This chapter deals with the environment of all these happenings. This environment must not only be clean and sanitary, but it must also be attractive to customers. Cleanliness and attractiveness are closely related and success in achieving a clean establishment frequently leads to success in attracting customers.

PERSONAL HYGIENE

Importance

Personal hygiene is a subject of great importance to persons who work in a foodservice establishment. First, it affects the general health of all persons working there. Good health means more alert, thinking employees who are more likely to perform their tasks effectively and less likely to make the kind of mistakes that can lead to contamination of food. Second, good personal hygiene lessens the possibility of the spread of communicable diseases within the establishment and leads to a healthier environment.

Personal hygiene refers to a group of working habits that should be practiced daily until they are automatic. They include not only the obvious practices of washing hands thoroughly after using the restroom and after preparing raw foods, but also such diverse practices as getting the proper amount of sleep, remaining home when sick, and wearing clean clothes.

Adequate refrigeration, proper cooking routines, correct washing and sanitizing of tableware and equipment, control of insects and rodents, can all be done and it would still be possible to have disease transference in foodservice operations if the foodservice worker forgets or neglects to follow the simple and common-sense rules of personal hygiene.

It follows, then, that the foodservice worker must be competent. The competent worker makes a definite contribution to the place of work. The incompetent worker can be responsible for any number of unsanitary conditions in the establishment—conditions that can lead to the illness of some customers or fellow workers. Such incompetence can eventually mar the favorable image of any establishment and substantially lower the customer count.

Competency involves two things. First, the employee must know his job. He must be aware of his duties, and he must perform them efficiently so that the foodservice operation progresses smoothly. Second, his efficiency must include the carrying out of sanitation rules as they pertain both to the job and to the employee.

Foodservice is exacting work, impossible to perform efficiently unless the worker is in good physical condition. In addition, the sanitation trained worker knows that the state of his health directly influences the health of fellow workers or those who use the operation's facilities. Therefore, a few basic rules should be followed.

1. Have a thorough physical checkup once a year. This checkup will not guarantee day-by-day freedom from infections that may cause foodborne diseases, but it will provide the employee with necessary information on his general physical condition.
2. Have a dental checkup twice a year. Infections of the teeth and gums can lead to serious trouble elsewhere in the body. No one can do an efficient job with a nagging toothache. Preoc-

cupation with any physical ailment can lead to carelessness in the performance of duties, some of which require sanitary procedures.

3. Eat proper amounts of the right foods. A person whose diet includes the necessary amount and type of protein, fat, and carbohydrate, along with needed vitamins and minerals, will be alert, efficient, and capable of carrying out all the duties of his job in a satisfactory manner. A well-nourished person is usually cheerful and attentive. Cheerful people attract customers and have a positive effect on the morale of the staff. Attentive people know and follow proper procedures and can serve as models for other staff members.
4. Get an adequate amount of sleep. A tired person cannot work at a high level of efficiency. Thinking processes are slowed by fatigue and careless mistakes are sometimes made. A tired person will commonly have hand-face contact, thereby contaminating the hands with bacteria. A tired person will frequently do things the quick or easy way, bypassing sanitary safeguards.
5. Stay home when ill. Germs are easily spread. Many food-poisoning outbreaks, which resulted from ill people handling food, would not have occurred if these persons had remained home for the proper length of time. Any person who exhibits any symptoms of a communicable disease should not be in a food preparation or serving area.

The United States Public Health Service Food Service Sanitation Manual states:

No person, while infected with a disease in a communicable form that can be transmitted by foods or who is a carrier of organisms that cause such a disease or while afflicted with a boil, an infected wound, or an acute respiratory infection, shall work in a foodservice establishment in any capacity in which there is a likelihood of such person contaminating food or food-contact surfaces with pathogenic organisms or transmitting disease to other persons.

Even though these general rules could apply to anyone in any line of work, the foodservice worker, above all others, needs to observe those requirements of personal hygiene and cleanliness that result in a healthy, well-groomed worker. Other requirements include:

1. A daily bath. It is essential that any person involved with food be clean.
2. Frequent shampoos. Unclean and oily hair can result in bacterial buildup, dandruff, and infections that can be transmitted to food.
3. Clean uniforms. Stained and dirty uniforms can harbor bacteria. Uniform appearance can also have an important effect on customers.
4. Proper hair covering. Female workers should wear hair nets. Male chefs, kitchen workers, food runners, dishwashing personnel, and others should wear caps. Hairs contain surface bacteria. If a loose hair falls from the head and settles in or on a food item, bacterial growth can result in the food.

The United States Public Health Service Food Service Sanitation Manual states:

Employees shall use effective hair restraints to prevent the contamination of food or food-contact surfaces.

5. Do not wear jewelry on the job. Pieces of jewelry can break or come loose, fall into food, and end up in a customer's mouth or throat. A costly lawsuit and damaging publicity can result. Jewelry is also in contact with skin or clothing during the day. Bacteria collect on their surfaces. Any contact with food can therefore initiate bacterial growth in the food.

Protection of Hands

Proper hygiene of the hands has been mentioned several times in previous chapters. Here is a summary of the practices that should be followed faithfully.

1. Wash the hands thoroughly after restroom visits, after smoking, or after handling any unclean object before resuming work with food or food utensils. This practice must become automatic.

The United States Public Health Service Food Service Sanitation Manual states:

Employees shall thoroughly wash their hands and the exposed portions of their arms with soap and warm water before starting work, during work as often as necessary to keep them clean, and after smoking, eating, drinking, or using the toilet. Employees shall keep their fingernails clean and trimmed.

2. Keep the hands away from the mouth, nose, ears, eyes, and hair while on the job. If any of these areas are handled, wash the hands before resuming duties.
3. Cover coughs and sneezes, and wash the hands if spray from coughs or sneezes contaminates them. If plastic gloves are being worn when coughs or sneezes are covered with the hands, discard the gloves for a fresh pair.
4. Immediately clean, disinfect, and bandage any cuts or sores on fingers, hands, or arms. Bacteria from unbandaged and infected cuts and sores may be transferred to food and cause illness. An incident of this sort caused an outbreak of staphylococcus food poisoning in Peoria and Springfield, Illinois, in 1978. A worker in the butter department of a dairy plant in Kentucky transmitted the staphylococci to the butter from an infected cut on a finger. The worker was directly involved in the production of whipped butter. The butter was eventually distributed to several foodservice establishments in Peoria and Springfield and caused the food-poisoning outbreak.

Wearing plastic gloves provides an additional safeguard against contamination of fresh food, clean dishes, and utensils by infective material that could seep through a bandage. Punctured or torn gloves should be discarded immediately. Bacterial infection can pass through the tear to contaminate food, utensils, or equipment.

5. Wash the hands after removing soiled dishes from tables, after wiping dirty tables, or after cleaning restrooms. These soiled surfaces have probably been in contact with the mouths of other people or contaminated in various other ways. Such contamination must not be transferred from the foodservice worker to food, utensils, and equipment.

The United States Public Health Service Food Service Sanitation Manual states:

Employees shall handle soiled tableware in a way that minimizes contamination of their hands.

6. Wash the hands before preparing food that is to be eaten without cooking. Any bacteria on the hands or under the nails can easily be transferred to the food. Therefore, it is essential that the hands be clean before such food preparation is begun and that no contamination of the hands occurs during the working period. If it should, the hands must be washed before resuming work.
7. Do not dip fingers into food in order to taste it. A spoon should be used. This spoon should not be placed in the food again unless it is washed and sanitized first. Bare hands should not be used for dipping ice. Plastic gloves should be worn and then removed or tongs should be used.

SANITARY FACILITIES

The existence of clean, well-kept, sanitary facilities affects not only the quality of the food that is served, but also the morale and attitude of the staff and the attractiveness of the establishment. These facilities require effective cleaning. Effective cleaning requires that walls, ceilings, and floor coverings be made of specified materials that are easily cleaned. The water supply must be sanitary and sewage disposal effective. This requires proper plumbing.

All of these topics are important enough in maintaining a sanitary establishment to be regulated by local and state codes and by the United States Public Health Service. Many of these regulations will be mentioned in the appropriate places and should be studied carefully.

Water Supply

It is vital to the sanitary and safe operation of any foodservice establishment that all water used on the premises be of high quality. Water is used for so many purposes (dishwashing, cooking, cleaning, and drinking) that it must be safe. Diseases such as typhoid fever, hepatitis, and shigellosis are easily transmitted through contaminated water. Some waterborne disease outbreaks have involved tens of thousands of persons.

Most of the federal and state regulations affecting water supplies relate to providing a safely constructed and operated system and to specifying safety regulations involving sources of water.

The United States Public Health Service Food Service Sanitation Manual states:

1. All potable water not provided directly by pipe to the foodservice establishment from the source shall be transported in a bulk water transport system and shall be delivered to a closed-water system. Both of these systems shall be constructed and operated according to law.
2. Bottled and packaged potable water shall be obtained from a source that complies with all laws and shall be handled and stored in a way that protects it from contamination. Bottled and packaged potable water shall be dispersed from the original container.

Sewage

Proper disposal of sewage is necessary to prevent contamination of ground and water supplies, such as reservoirs and wells, that obtain their water from runoff or trickling through the soil. Improper disposal of sewage can contaminate these water supplies and result in the subsequent contamination of food, utensils, and equipment.

The United States Public Health Service Food Service Sanitation Manual states:

All sewage, including liquid wastes, shall be disposed of by a public sewerage or by a sewage disposal system constructed and operated according to law.

Plumbing

Proper plumbing contributes to the safety of customers and to the environment of the foodservice establishment. Improper plumbing or maintenance can result in potential health hazards such as cross-contamination, back-siphonage, or leakage. A case discussed in chapter 8 noted an improper connection made to the water intake pipe of a soft drink machine. It resulted in 11 cases of severe gastroenteritis caused by an additive to the radiator system of the golf club in which the machine had been placed.

Improper plumbing can result in the contamination of food, utensils, and equipment; it can be responsible for noxious odors in the establishment; and, it can result in an insufficient flow of water to equipment, such as dishwashing machines and garbage grinders, thereby adversely affecting their performance.

The United States Public Health Service Food Service Sanitation Manual states:

1. Plumbing shall be sized, installed, and maintained according to law. There shall be no cross-connection between the potable water supply and any nonpotable or questionable water supply nor any source of pollution through which the potable water supply might become contaminated.
2. The potable water system shall be installed to preclude the possibility of backflow.
3. Except for properly trapped open sinks, there shall be no direct connection between the sewerage system and any drains originating from equipment in which food, portable equipment, or utensils are placed.

Toilet Facilities

Well-designed, clean, and sanitary restrooms contribute to the sanitary environment of the establishment and the attitude of the employees toward sanitation; they also contribute much to customer confidence in the cleanliness of the entire establishment. Customers who have this confidence usually return. A dirty, untidy restroom, with stains in the sinks and toilets, fixtures that do not flow or flush properly, and no towels or toilet tissue, can indicate a lack of concern for sanitation on the part of management. This can lose customers and cause employees to be disinterested, themselves, in cleanliness and sanitation.

Adequate, sanitary toilet facilities are necessary for the proper disposal of human waste.

If the toilets are not kept clean and in good working order, excrement may not be disposed of properly and immediately. Human excrement frequently carries pathogenic microorganisms. These organisms can be spread to food, utensils, food-contact surfaces, and clothing in lockers by insects such as flies and cockroaches, which are attracted to dirty, unsanitary environments. Hands are also easily contaminated in dirty restrooms and can serve as vectors of transmission to other areas such as the kitchen.

The United States Public Health Service Food Service Sanitation Manual states:

1. Toilet facilities shall be installed according to law, shall be the number required by law, shall be conveniently located, and shall be accessible to employees at all times.
2. Toilets and urinals shall be designed to be easily cleaned.
3. Toilet rooms shall be completely enclosed and shall have tight-fitting, self-closing, solid doors, which shall be closed except during cleaning or maintenance, except as provided by law.
4. Toilet fixtures shall be kept clean and in good repair. A supply of toilet tissue shall be provided at each toilet at all times. Easily cleanable receptacles shall be provided for waste materials. Toilet rooms used by women shall have at least one covered waste receptacle.

Lavatory Facilities

The human hand is probably the most common agent responsible for the contamination of food, utensils, and food-contact surfaces. Hands become contaminated many times during the day because they come into contact with so many different objects. They should be washed frequently and thoroughly. Clean lavatories should be provided for this purpose. Clean lavatories, like clean toilet facilities, make a definite contribution to the sanitary environment of the establishment and to the attitude of the staff toward the practice of proper sanitation.

Clean lavatories should be available to any employee whenever he needs to wash his hands. Each lavatory should have hot and cold running water and should provide an adequate flow for thorough hand washing. Soap or detergent should be provided, along with sanitary towels or a hand-drying machine.

The United States Public Health Service Food Service Sanitation Manual states:

1. Lavatories shall be at least the number required by law, shall be installed according to law, and shall be located to permit convenient use by all employees in food preparation areas and utensil-washing areas.
2. Lavatories shall be accessible to employees at all times.
3. Lavatories shall also be located in or immediately adjacent to toilet rooms or vestibules. Sinks used for food preparation or for washing equipment or utensils shall not be used for handwashing.
4. Each lavatory shall be provided with hot or cold water tempered by means of a mixing valve or combination faucet.
5. A supply of hand-cleaning soap or detergent shall be available at each lavatory. A supply of sanitary towels or a hand-drying device providing heated air shall be conveniently located near each lavatory. Common towels are prohibited. If disposable towels are used, easily cleanable waste receptacles shall be conveniently located near the handwashing facilities.
6. Lavatories, soap dispensers, hand-drying devices, and all related fixtures shall be kept clean and in good repair.

Garbage and Refuse

Proper storage and disposal of garbage and refuse are important measures in the constant battle against insects and rodents. They remove or make inaccessible food and harborage. Carelessness in these areas of sanitation can result in an increase in the fly, cockroach, rat, and mouse population, with resulting problems in the foodservice establishment. These animals can transmit disease, soil food preparation areas with their droppings, and cause food losses in storage areas. Improper storage and disposal of refuse can also cause bad odors that can seep into the establishment and cause customers not to return.

Storage areas for garbage and refuse containers must be constructed so that they can be cleaned easily. Hot water or steam should be available and good drainage should be provided. All garbage containers should be in good condition and provided with tight-fitting lids.

When in use, garbage cans should be placed on a platform of sturdy construction that is elevated from the ground six inches. The platform should be constructed so that it can be cleaned underneath. This will aid in insect and rodent control. The entire area of refuse and garbage collection should be kept clear of boxes, barrels, or other litter. Almost any kind of litter can serve as a home for rats and mice.

The United States Public Health Service Food Service Sanitation Manual states:

1. Garbage and refuse shall be kept in durable, easily cleanable, insect-proof and rodent-proof containers that do not leak and do not absorb liquids.
2. Containers used in food preparation and utensil-washing areas shall be kept covered after they are filled.
3. Containers stored outside the establishment, and dumpsters, compactors, and compactor systems shall be easily cleanable, shall be provided with tight-fitting lids, doors, or covers, and shall be kept covered when not in actual use.
4. Soiled containers shall be cleaned at a frequency to prevent insect and rodent attraction. Suitable facilities, including hot water and detergent or steam, shall be provided and used for washing containers.
5. Garbage and refuse on the premises shall be stored in a manner to make them inaccessible to insects and rodents.
6. Garbage or refuse storage rooms, if used, shall be constructed of easily cleanable, non-absorbent, washable materials, shall be kept clean, shall be insect-proof and rodent proof. . . .
7. Garbage and refuse containers, dumpsters, and compactor systems located outside shall be stored on or above a smooth surface of nonabsorbent material such as concrete or machine-laid asphalt that is kept clean and in good repair.
8. Garbage and refuse shall be disposed of often enough to prevent the development of odor and the attraction of insects and rodents.

CONSTRUCTION AND MAINTENANCE OF PHYSICAL FACILITIES

It is important when designing or decorating a foodservice establishment to remember that it must be cleaned periodically. Floors must be cleaned more frequently than walls or ceilings, but all must be cleaned at some time. They must be constructed, therefore, of easily cleanable materials.

Floors

Floors should be properly constructed, smooth, nonabsorbent, and maintained in good condition. Cracks, breaks, and worn spots can hold dirt, food soil, and bacteria. Floors should be made of materials that can be easily cleaned without leaving a residue of water and detergent. For this reason, floors should be constructed of smooth rather than rough materials. Concrete, terrazzo ceramic tile, and high quality linoleum are satisfactory materials.

Carpeting should be of a tightly woven material so that it can be easily cleaned. It should not be used in food preparation areas because it tends to be absorbent and will retain liquids, thereby providing food for microorganisms. Carpeting should be used only in areas in which it can be thoroughly cleaned every day.

The United States Public Health Service Food Service Sanitation Manual states:

1. Floors and floor coverings of all food preparation, food storage, and utensil-washing areas, and the floors of all walk-in refrigerating units, dressing rooms, locker rooms, toilet

rooms, and vestibules shall be constructed of smooth durable material such as sealed concrete, terrazzo, ceramic tile, durable grades of linoleum or plastic, or tight wood impregnated with plastic, and shall be maintained in good repair.

2. Carpeting, if used as a floor covering, shall be of closely woven construction, properly installed, easily cleanable, and maintained in good repair. Carpeting is prohibited in food-preparation, equipment-washing and utensil-washing areas where it would be exposed to large amounts of grease and water, in food storage areas, and toilet room areas where urinals or toilet fixtures are located.
3. Properly installed trapped floor drains shall be provided in floors that are water-flushed for cleaning or that receive discharges of water or other fluid waste from equipment, or in areas where pressure spray methods for cleaning equipment are used.

Walls and Ceilings

These structures also play a role in the sanitary environment of the foodservice establishment. Walls and ceilings must be periodically cleaned and should therefore be constructed of suitable materials. Walls and ceilings of food preparation, utensil-washing, equipment-washing, and toilet areas should be light in color and made of smooth, nonabsorbent materials. They can then be cleaned easily and thoroughly. Light colored walls are an aid in locating dirt and facilitating cleaning.

The United States Public Health Service Food Service Sanitation Manual states:

1. Walls and ceilings, including doors, windows, skylights, and similar closures, shall be maintained in good repair.
2. The walls, including nonsupporting partitions, wall coverings, and ceilings of walk-in refrigerating units, food preparation areas, equipment-washing and utensil-washing areas, toilet rooms, and vestibules shall be light colored, smooth, nonabsorbent, and easily cleanable.
3. Light fixtures, vent covers, wall-mounted fans, decorative materials, and similar equipment attached to walls and ceilings shall be easily cleanable and shall be maintained in good repair.
4. Cleaning of floors and walls, except emergency cleaning of floors, shall be done during periods when the least amount of food is exposed, such as after closing or between meals. Floors, mats, duckboards, walls, ceilings, and attached equipment and decorative materials shall be kept clean. Only dustless methods of cleaning floors and walls shall be used, such as vacuum cleaning, wet cleaning, or the use of dust-arresting sweeping compounds with brooms.

Lighting

Proper lighting is necesary to locate areas that need to be cleaned and to provide the needed candlepower to perform the cleaning task adequately. Poor light can result in poor cleaning and a subsequent buildup of dirt and food soil. This can be unsightly and can provide a medium for pathogenic microorganisms.

Good lighting is also necessary for the proper preparation and handling of food, and for the successful cleaning and sanitizing of equipment and utensils. Adequate light is certainly needed for reading and identifying labels and colors and for recognizing the condition of stored food.

The United States Public Health Service Food Service Sanitation Manual states:

1. Permanently fixed artificial light sources shall be installed to provide at least 20 foot candles of light on all food preparation surfaces and at equipment or utensil-washing work levels.
2. Shielding to protect against broken glass falling into food shall be provided for all artificial lighting fixtures located over, by, or within food storage, preparation, service, and display facilities, and facilities where utensils are cleaned and stored.

Dressing Rooms and Locker Areas

Street clothes and personal belongings can contaminate food, equipment, and food preparation surfaces. Therefore, an easily cleanable room, separate from all food preparation areas, should be designated for changing and storing clothes and for storing personal effects. Toilet and lavatory facilities should be available in this room so that employees can be reminded to wash their hands prior to starting their first task. Lockers should be provided for the storage of personal belongings (purses, coats, shoes, jewelry).

The United States Public Health Service Food Service Sanitation Manual states:

1. If employees routinely change clothes within the establishment, rooms or areas shall be designated and used for that purpose. These designated rooms or areas shall not be used for food preparation, storage, or service, or for utensil washing or storage.
2. Enough lockers or other suitable facilities shall be provided and used for the orderly storage of employee clothing and other belongings. Lockers or other suitable facilities may be located only in the designated dressing rooms or in food storage rooms or areas containing only completely packaged food or packaged single-service articles.

Storage of Poisonous or Toxic Materials

All materials of this nature should be completely labeled and stored in cabinets that are used for no other purpose. Different types of toxic materials, such as, insecticides, sanitizers, and polishes, should be stored separately from each other. The storage cabinets should be frequently inspected for signs of leakage.

In order to reduce the possibilities of accidents and contamination, only those poisons and toxic materials necessary for the operation of the establishment and having a designated use should be purchased and stored. Proper labeling, use, storage, and handling of these materials are essential to prevent the accidental contamination of food, equipment, and utensils.

The United States Public Health Service Food Service Sanitation Manual states:

1. There shall be present in foodservice establishments only those poisonous or toxic materials necessary for maintaining the establishment, cleaning and sanitizing equipment and utensils, and controlling insects and rodents.
2. Containers of poisonous or toxic materials shall be prominently and distinctly labeled according to law for easy identification of contents.
3. Poisonous or toxic materials consist of the following categories:
   a. Insecticides and rodenticides.
   b. Detergents, sanitizers, and related cleaning or drying agents.
   c. Caustics, acids, polishes, and other chemicals.

   Each of the three categories . . . shall be stored and physically located separate from each other. All poisonous or toxic materials shall be stored in cabinets or in a similar physically separate place used for no other purpose. To preclude contamination, poisonous or toxic materials shall not be stored above food, food equipment, utensils or single-service articles. . . .
4. Bactericides, cleaning compounds, or other compounds intended for use on food-contact surfaces shall not be used in a way that leaves a toxic residue on such surfaces or that constitutes a hazard to employees or other persons.
5. Poisonous or toxic materials shall not be used in a way that contaminates food, equipment, or utensils, nor in a way that constitutes a hazard to employees or other persons, nor in a way other than in full compliance with the manufacturer's labeling.

SUMMARY

Good personal hygiene is of great importance to persons who work in foodservice establishments. It can affect the general health of the personnel and lessen the spread of communicable diseases within the establishment.

Personal hygiene refers to a group of working habits that should be practiced every day. These habits include: (1) a thorough checkup once a year, (2) a dental checkup twice a year, (3) proper amounts of the right foods, (4) adequate sleep, and (5) remaining home when feeling ill.

There are several rules for protecting the hands. (1) Wash the hands thoroughly after restroom visits, after smoking, or after handling any unclean object before again working with food or food utensils. (2) Keep the hands away from the mouth, nose, ears, eyes, and hair while on the job. (3) Cover coughs and sneezes, and wash the hands if any spray from coughs or sneezes contaminates them. (4) Immediately clean, disinfect, and bandage any cuts or sores on fingers, hands, or arms. (5) Wash the hands before preparing food that is to be eaten without cooking.

The existence of clean, well-kept, sanitary facilities affects the quality of the food that is served. It is vital to the sanitary and safe operation of any foodservice establishment that all water used on the premises be of high quality. Improper plumbing or maintenance can result in potential health hazards such as cross-contamination, back-siphonage, or leakage.

Restrooms should be well designed, clean, and sanitary. A dirty, untidy restroom can indicate a lack of concern for sanitation on the part of management. Adequate, sanitary toilet facilities are necessary for the proper disposal of human waste. Human excrement frequently carries pathogenic microorganisms.

The human hand is probably the most common agent responsible for the contamination of food, utensils, and food-contact surfaces. Clean lavatories should be provided for washing the hands. They should be available to all employees when needed.

Proper storage and disposal of garbage and refuse are important measures in the battle against insects and rodents. They remove or make inaccessible food and harborage.

Floors should be properly constructed, smooth, nonabsorbent, and maintained in good condition. Concrete, terrazzo, ceramic tile, and durable linoleum are among the acceptable materials. Walls and ceilings of food preparation, utensil-washing, equipment-washing, and toilet areas should be light in color and made of smooth, nonabsorbent materials.

Proper lighting is necessary to see areas that need to be cleaned and to provide the needed candlepower to perform the cleaning task adequately. At least 20 footcandles of light on all food preparation surfaces and at equipment-, or utensil-washing work levels are required.

An easily cleanable room, separate from all food preparation areas, should be available for changing and storing clothes. Toilet and lavatory facilities should be available in this room. Lockers should be provided for the storage of personal belongings.

All materials of a poisonous or toxic nature should be completely labeled and stored in cabinets that are used for no other purpose.

Review Questions

1. Why is a knowledge of personal hygiene important to the foodservice worker?

2. What is meant by personal hygiene?

3. Why should a foodservice worker: a. have a thorough physical checkup once a year? b. have a dental checkup twice a year? c. eat proper amounts of the right food? d. obtain adequate sleep? e. stay home when feeling ill?

4. List seven hygenic rules involving the hands and explain the reasons for these rules.

5. What is the importance of clean, well-kept, sanitary facilities?

6. Why must the water supply of a foodservice establishment be of the highest quality?

7. What is the importance of proper sewage disposal in a foodservice establishment?

8. How does proper plumbing contribute to the healthful environment of the foodservice establishment?

9. What is the importance of clean, sanitary restrooms in the foodservice establishment?

10. Why must clean lavatories be available for all employees?

11. a. Why is garbage and refuse disposal important in maintaining a sanitary foodservice establishment? b. What requirements must garbage and refuse containers meet?

12. What requirements must the following meet? a. floor and floor coverings, b. carpeting, c. walls.

13. Why is good lighting important in a foodservice establishment?

14. What are the reasons for having dressing rooms in a foodservice establishment?

References

Anonymous. (1971). *Tips on Cleaning*. July/Aug. Soap and Detergent Association. New York.

Anonymous. (1973). *Scientific Cleaning and Sanitation Procedures for the Food Service*

Industry. Economics Laboratory, Inc. St. Paul, Minnesota.

Crisley, F. D. and M. J. Foter. (1965). "The Use of Antimicrobial Soaps and Detergents for Hand Washing in Food Service Establishments." *J. Food Milk Technol.* 28:278–284.

Horwood, M. P. and V. A. Minch. (1951). "The Numbers and Types of Bacteria Found on the Hands of Food Handlers." *J. Food Sci.* 16:133–136.

Kotchevar, L. H. and M. E. Terrell. (1961). *Food Service Layout and Equipment.* John Wiley and Sons, Inc. New York.

Longree, K. (1972). *Quantity Food Sanitation.* John Wiley and Sons, Inc. New York.

Sartwell, P. E. (1965). *Preventive Medicine and Public Health*, 9th ed. Appleton-Century-Crofts. New York.

Sassano, J. M. (1963). "Today's Floor-Cleaning Units." *Food Engineering* 35:81–82.

Smith, G. A. (1963). "Planning: Key to Successful CIP." *Food Engineering* 35:72–74.

Stuart, S. L. (1962). "Federal Regulation of Bactericidal Chemicals Used in Building, Industrial and Institutional Sanitation Programs." *J. Milk Food Technol.* 25:308–312.

U.S. Department of Health, Education and Welfare. (1976). *Food Service Sanitation Manual.* U.S. Government Printing Office. Washington, D.C.

- A. As a Sale
    1. Inns and Victual Houses
    2. Early English Decisions
    3. Early American Decisions
    4. Later American Decisions
    5. The Uniform Commercial Code
- B. Implied Warranty
    1. Early English Decisions
    2. Early American Decisions
    3. Later American Decisions
    4. The Uniform Commercial Code
- C. Summary

15 Liability: History I

This chapter deals with the liability of manufacturers, wholesalers, and retailers, including restaurateurs, for illness caused by their products. This area of law is called products liability and is a relatively new area. The basis of products liability is the responsibility placed on one who sends goods out into the channels of trade for use by others. One court has stated that the product liability of a manufacturer, and the corresponding right of the consumer, is simply the liability that the law imposes on a manufacturer in favor of a consumer for loss suffered by reason of a defective product attributable to that manufacturer.[1] Food is considered to be a product and, as such, must be free of defects.

Most state codes define food in about the same terms as does the Illinois Pure Food Law.[2] This law defines food as (1) articles used for food or drink for man, (2) chewing gum, and (3) articles used for components of any such article.

As a Sale

The liability involved in the dispensing of food by hotels, restaurants, cafeterias, and so forth has an interesting legal history. All states consider the dispensing of food by these establishments to be a sale. The customer gains title to the food and may remove all or part of it from the premises. The customer chooses from a menu and money is exchanged after the meal. Since it is a sale, the food must be of merchantable quality and must fit its intended purpose as defined by the pure food laws and sales codes of the various states. Judge Barrow of the court of Civil Appeals of Texas has stated the current American philosophy clearly: "It is well-settled law in Texas that parties who sell food or drink intended for human consumption may be held liable, upon an implied warranty (a warranty or guarantee inferred by law irrespective of any intention of the seller to create it), for damage caused to a person who eats or drinks the same, if such food or drink is unfit for human consumption."[3]

This has not always been the law by any means. It has been a long and sometimes confusing trail to this point. In early times, the public serving of food was a function of the innkeeper and victualer (one who sells food or drink prepared for consumption on the premises). These two types of establishments were the forerunners of our present-day hotels and restaurants.

Inns and Victual Houses

Inns were a product of old English traveling conditions. Villages were far apart. The roads were poor and passed through many forested areas. These forested areas were unsafe for travel at night due to robbers. Because of these conditions, it was usually necessary for the traveler to find a house in which to spend the night. Certain houses became known for accepting travelers. This led to a type of established business—the inn. Thus, some private houses became devoted to the business of furnishing food, drink, and overnight accommodations to travelers.

Professor Beale, in his book on inns, had this to say with regard to food:

> The innkeeper must provide as much food as he can reasonably foresee to be necessary. He must provide and keep on hand enough food for such guests as may be reasonably expected. He is not an insurer of the quality of his food, but he would be liable for knowingly or negligently furnishing bad or deleterious food. As an innkeeper does not lease his rooms, so he does not sell the food he supplies to his guest. It is his duty to supply such food as the guest needs, and the corresponding right of the guest is to consume the food he needs and to take no more. Having finished his meal, he has no right to take food from the table, even the uneaten portion of the food supplied to him; nor can he claim a certain portion of the food as his own, to be handed over to another in case he chooses not to consume it himself. The title to food never passes as a result of an ordinary transaction of supplying food to a guest.[4]

The eating house or early restaurant was evolving at about the same time and was also an outgrowth of the private home. The English victualer furnished food and drink only; no lodging was provided. With regard to food served by the victualer, Professor Beale continues "The restaurant keeper is not an insurer of the quality of the food he furnishes, but he is liable for knowingly or negligently furnishing bad or deleterious food."

Due to maintaining business in their homes, the innkeeper and victualer occupied an unusual niche in English commercial law. They were not considered traders or dealers, and were therefore free from some of the responsibilities placed on merchants. The innkeeper and victualer were supposed to exert all possible care in the preparation of their food. They did not have to guarantee the wholesomeness or purity of their food, however. In case of illness, the only legal recourse for the unfortunate victim was to prove negligence or deceit on the part of the innkeeper or victualer. This situation occurred because the English courts placed inns and eating houses outside the mainstream of normal commercial business because these establishments were conducting their business in private homes.

Early English Decisions

In 1635, in a bankruptcy case filed against an innkeeper for indebtedness, it was ruled that ". . . an innkeeper doth not get his living by buying and selling; for although he buy provision to be spent in his house, he doth not properly sell it, but utters it at such rates as he thinks reasonable gains, and the guests do not take it at a certain price, but they must have it or refuse it if they will."[5] Since the innkeeper was not a merchant, he was not subject to bankruptcy proceedings. Since he was not a merchant, he did not have to guarantee his food. This, then, was early English law.

In 1690, in another bankruptcy suit, it was ruled that

> . . . an innkeeper could not be bankrupt, for he is not like a trader; . . . he is not taken notice of by the law, as a trader, but as a host, hospitator; and is paid not merely for his provisions, but also for his care, pain, protection and security; and he buys meat and drink, not for sale and trading, but for accommodation. And an innkeeper cannot make a contract ad libitum (at will), nor does he buy or sell at large, but to guest only."[6]

And again, in 1701, it was ruled that ". . . an innkeeper, as such, cannot be a bankrupt, because he does not sell but utters his provisions."[7]

We can see that the inn, developing as it did from rather obscure beginnings, had not yet been brought under the same regulations that affected other merchants and dealers of those times. The innkeeping business was considered to be a service type of business and so was relatively free of restrictions.

In 1767, Lord Mansfield, the presiding judge, granted the victualer the same legal status as the innkeeper when he stated during another bankruptcy case:

> The analogy between the two cases of an innkeeper and victualer is so strong that it cannot be got over. He does not deal upon contracts. . . . He buys, only to spend in his house; and when he utters it again it is attended with many circumstances additional to the mere selling price.[8]

And thus the pattern was set for another one hundred years. The innkeeper and victualer were not merchants or dealers, did not sell their products, and were not, therefore, regulated by sales codes.

These English decisions formed the basis for early American court decisions, which uniformly followed their English counterparts. The dispensing of food by the victualers and innkeepers was not a sale and therefore was not regulated under the sales acts that were beginning to be passed by some of the states. This is not to say that the victim of the unwholesome food had no legal recourse. He did. However, the unfortunate plaintiff was restricted to recovery suits in negligence and deceit. These were very difficult to prove, in as much as the plaintiff rarely had access to the food preparation area for inspection purposes.

Early American Decisions

The English philosophy was continued and firmly imbedded as American legal philosophy by an Illinois case, decided in 1896. The incident involved the Boston Oyster House, a public restaurant in Chicago. Two ladies lunched at this establishment, each dining on oyster stew. One of the two ladies became violently ill and continued so for about three weeks. Her health remained bad and she sued the restaurant. As part of the suit, the plaintiff contended that since innkeepers are liable for losses that happen to the goods of their guests, they should be liable for injury resulting from unwholesome food furnished by them. The Illinois Supreme Court, in rejecting this contention stated:

> The law is well settled that the keepers of public inns are required to safely keep the property of their guests. . . . As respects the goods of a guest, which he takes with him when he stops at the inn, the innkeeper is practically an insurer. . . . But as to food served at a restaurant, such as oysters, ice cream, and the like, we are not aware that a similar rule establishing liability ever existed. There is no similarity between the two cases, and the principle that governs one does not apply to the other. If this rule is adopted, the plaintiff would be relieved from proving the most important element of her declaration, the negligence of defendants, which is really the foundation of the action. This would, in effect, make the restaurant keeper an insurer. Such a rule is not correct in principle, nor has it been sustained, so far as we are advised by any respectable authority."[9]

Later American Decisions

In 1879, a brick was loosened from the protective wall around the public eating houses. A charge was brought by the state of Massachusetts against a man who was dispensing food and intoxicating beverages in his dwelling house for a single price. He did not possess a license for selling alcoholic beverages. The defendant maintained that occasionally parties were held in his dwelling house and that he supplied meals of victuals only; that as part of these meals he furnished wine, lager beer, and other liquors; and that no bar was maintained on the premises for the sale of liquor. The state charged that payment for the meals included payment for the liquors and constituted an illegal sale as the defendant did not possess a tavern license. The Massachusetts Supreme Court ruled:

> The purchase of a meal includes all the articles that go to make up the meal. It is wholly immaterial that no specific price is attached to those articles separately. If the meal included intoxicating liquors, the purchase of the meal would be a purchase of the liquors. It would be immaterial that other articles were included in the purchase, and all were charged in one collective price."[10]

This decision meant that henceforth, in Massachusetts at least, the dispensing of food in public eating houses was to be considered a sale. It would pave the way for the imposing of a guarantee of wholesomeness on food in public eating places.

In 1890, a case occurred that somewhat expanded the decision in the previous case. At that time, the sale of oleomargarine in Pennsylvania was prohibited by the Pennsylvania pure food law. A restaurant in Pittsburgh was serving oleomargarine as a part of its meals. The state filed a complaint. In 1890, the dispensing of food for immediate consumption was considered to be a service, as it was in every other state except Massachusetts. However, the Supreme Court of Pennsylvania followed the Massachusetts lead and ruled:

> That the food furnished to McRay and Spence, or so much of it as they saw fit to appropriate, was sold to them cannot be reasonably questioned.

When it was set before them, it was theirs to all intents and purposes, to eat all, or a part, as they chose, subject only to the restaurateur's right to receive the price, which it is admitted was promptly paid. They might not eat all of the article set before them, but they had an undoubted right to do so; and, even assuming that the meal is the portion of food taken, in the sense stated, the transaction must be regarded as a sale. . . ."[11]

This decision drastically overthrew, in the commonwealth of Pennsylvania, the old law that the patron did not gain title to the food that was set before him; under this decision the patron gained title. This meant that a sale had taken place. The new viewpoint now rested on firm ground.

In 1900, Vermont joined the list of states classifying the dispensing of food for immediate consumption as a sale when, in a case similar to the Massachusetts case previously mentioned, the Vermont Supreme Court ruled:

The question was whether the transaction was a sale, not a gift, and the case was submitted upon the ground that the ale and wine were furnished for pay, as a part of the meals. If a person in his private dwelling furnishes a man his dinner, and with it, and as a part of it, intoxicating liquors, and receives pay for it, such transaction is a sale of the liquors so furnished.[12]

In 1914, in Kentucky, a hotel was serving quail on its menu. At the time, quail was out of season in Kentucky. Even though the quail had been purchased from a wholesaler in Illinois, the conviction on the grounds of violation of the Kentucky game laws was upheld by the Kentucky Court of Appeals. The court ruled that a sale had taken place.

The guest at the hotel or restaurant who was served with quail for compensation as certainly purchases it, and the proprietor of the hotel or restaurant as certainly exposes it for sale and sells it, as if it were purchased for compensation from a dealer who had it for sale. . . .[13]

A similar case in the state of New York, in 1917, involved a restaurant serving partridge. The partridge was a protected species under the New York conservation law and its sale was prohibited. The court ruled: "Clearly, if in a hotel where meals are served a la carte a partridge is ordered, prepared and served as food and paid for as such, it would constitute a sale. . . ."[14]

The previous five cases illustrate the second stage in the evolution of the legal responsibility of foodservice establishments. In the beginning, the dispensing of food for immediate consumption was not considered a sale, as this type of business was not considered to be a merchant type of business. Recovery for damages due to unwholesome food was allowed only if the "guest" could prove negligence or deceit on the part of his "host." This was usually very difficult.

As we have seen, English food law was carried to America and continued here. However, abuses began to occur. You may have noticed that in the preceding five cases, all of the complaints were filed by the states involved. Many food establishments were dispensing alcoholic beverages without a license, contending that they were not selling but dispensing a service. Illegal foods were being served by some restaurants. Laws were being broken. This was occurring because the dispensing of food for immediate consumption on the premises was not a sale and, therefore, not affected by the sales statutes. The sales laws were written to regulate merchants and dealers. Since the title to the food never passed from the innkeeper or victualer to the patron, no sale was considered to have taken place. It was an attempt to curb these abuses that led some states to take courtroom action in order to classify the dispensing of food in public eating houses as a sale.

It might be noted that many states resisted the change in status of the restaurants. These states held firmly to the old concept. In 1914, another noted decision occurred, this time in Connecticut. The plaintiff contended that the food she ate was sold to her and that there should be an implied warranty that the food was wholesome. The Connecticut Supreme Court rejected both contentions in no uncertain terms, stating:

> The customer does not become the owner of the food set before him, or of that portion which is served for his use, or of that which finds a place upon his plate, or in the dishes set about it. No designated portion becomes his. He is priveleged to eat, and that is all. The uneaten food is not his. He cannot do what he pleases with it. That which is set before him or placed at his command is provided to enable him to satisfy his immediate wants, and for no other purpose. He may satisfy those wants; but there he must stop. He may not turn over unconsumed portions to others at his pleasure, or carry away such portions. The true essence of the transaction is service in the satisfaction of a human need or desire—ministry to a bodily want. A necessary incident of this service or ministry is the consumption of the food required. This consumption involves destruction, and nothing remains of what is consumed to which the right of property can be said to attach. Before consumption title does not pass; after consumption there remains nothing to become the subject title. What the customer pays for is the right to satisfy his appetite by the process of destruction. What he thus pays for includes more than the price of the food as such. It includes all that enters into the conception of service, and with it no small factor of direct personal service. The transaction between the plaintiff and the defendants did not involve a sale of goods.[15]

Figure 15-1 The customer was unable to prove negligence when she received a mouse with her food and so lost her case. *Kenny* v. *Wong Len.* 128 A 343. 1925. Based on original drawing by Cindy Paddack.

The above case, *Merrill* v. *Hobson,* is quoted at length because it formed one of the cornerstones of what became known as the Connecticut-New Jersey rule. This rule definitely rejected the concept of a sale occurring when food was dispensed at a public eating place. The rule retained the old English view of the innkeeper and victualer *uttering* his provisions in response to the needs of his guests. Many courts continued to adhere to this doctrine. The Connecticut-New Jersey rule was frequently cited by these courts.

In a 1925 case, a customer in a New Hampshire restaurant encountered a dead mouse in her food (Fig. 15-1). The case reached the New Hampshire Supreme Court. This court also rejected the sale concept, ruling:

> Without considering the technical point whether one served with food at a restaurant obtains the right to destroy it rather than acquires title to it, it is a well settled rule that, when service is the predominant, and transfer of title to personal property the incidental, feature of a transaction, the transaction is not a sale of goods within the application of statutes relating to sales.[16]

The customer was not able to prove negligence in this case and so she and the mouse were both losers.

Finally, a 1927 case involved a woman who had eaten oysters in a New Jersey restaurant and sustained food poisoning. The case reached the New Jersey Court of Errors and Appeals. In its decision, which formed the other half of the Connecticut-New Jersey rule, the court stated: "We think enough has been said to indicate that the service of food at eating houses has never been and cannot be regarded as a sale at common law. . . ."[17]

The Uniform Commercial Code

In 1942, the National Conference of Commissioners on Uniform State Laws and the American Law Institute began the preparation of a uniform commercial code, which it hoped would be adopted by all states in the United States. The object of the Uniform Code was to achieve uniformity in all the states and territories of the United States. This would keep legal decisions in the various states in harmony with each other and eliminate the disagreement that existed in food cases where the courts of some states ruled that the dispensing of food for immediate consumption was a sale, while other state courts ruled that it was a service.

The Uniform Commercial Code was completed and endorsed by the National Conference of Commissioners on Uniform State Laws, the American Law Institute, and the American Bar Association in 1952. It was revised and updated in 1962. Article 1 deals with sales and states: "The serving for value of food or drink to be consumed either on the premises or elsewhere is a sale."[18]

Thus ended one long and confused chapter in the development of the legal responsibility placed on that segment of the food industry that serves food for immediate consumption on the premises. The following portion of this chapter traces the evolution of the implied warranty, probably the single most important legal recourse the consumer has at his disposal and whose development was linked to the development of the sale concept.

IMPLIED WARRANTY

A warranty is a guarantee that the article that has been sold is fit for its intended purpose; that it contains no defects not mentioned during the sale. When a seller issues a warranty, he then becomes liable for all damages resulting from any defects not mentioned.

There are two types of warranty—express and implied. An express warranty is a written or oral expression issued at the time of the sale. Most people are familiar with this type of warranty in the form of a written guarantee that accompanies the sale of automobiles, refrigerators, tires, watches, and so forth. The "money back if you are not satisfied" advertisements also constitute an express warranty, as does a promise by a salesperson that "this should work" or "this will solve your problem." The latter are oral express warranties.

An implied warranty needs no oral or written expression. What is necessary, however, is that the seller know the purpose for which his goods are to be used and that the buyer has relied on the skill and judgment of the seller when making the purchase. To illustrate such a case, assume that a customer, shopping at the local supermarket, stops at the meat counter and requests three frying chickens. The butcher selects three chickens, wraps them, and hands them to the customer. The butcher is presumed to know the purpose for which these chickens have been purchased. This would be for use as food. There would be an implied warranty of merchantability accompanying the chickens. No words about this need to be spoken. The butcher is the agent of the supermarket. Therefore, if any of the chickens are unwholesome when purchased and cause food poisoning to any person who eats the chickens, the owner of the supermarket is liable for damages.

The same implied warranty accompanies food served in a restaurant, cafeteria, or lunch counter. When a person orders a meal it is presumed that the server or vendor knows the purpose for which the food is intended; it is for human consumption. Therefore, an implied warranty of merchantability or fitness accompanies the food.

The Uniform Commercial Code states: "Unless excluded or modified a warranty that the goods shall be merchantable is implied in a contract for their sale if the seller is a merchant with respect to goods of that kind."[19] You can see from the wording of the statute that its application to food sold in restaurants depends on the food that is dispensed there being classified as a sale and, from this, a restaurateur's being classified as a merchant or dealer.

The Code also states that:

> Where the seller at the time of contracting has any reason to know any particular purpose for which the goods are required and that the buyer is relying on the seller's skill or judgment to select or furnish suitable goods, there is unless excluded or modified under the next section an implied warranty that the goods shall be fit for such purpose.[20]

Early English Law

In order to follow the history of the implied warranty as it pertains to food, particularly food served by public eating establishments, we must go back to the year 1266. At that time, a statute was passed that ordained that no person should sell corrupt victuals. The statute applied to vintners (wine merchants), brewers, butchers, and cooks. Legal action was restricted, however, to suits showing negligence or deceit.

The idea of an implied warranty attached to food first appeared in an English case in 1593. This case involved a man who had falsely replaced a vicar (the minister of an English parish) who had died: "to which vicarage the defendant, on the ninth of June, 35 Elizabeth affirmed that he was the lawful incumbent, and had a right to the tithes [a tax one-tenth of the yearly production of land, animals and personal wages paid for the support of the church] from the death of Thomas Vaughan the incumbent."[21] Among the tithes was food, the result of butchering livestock to meet the tithe obligation. The defendant sold part of this food. Some of the cuts were spoiled and the purchaser filed a complaint. In ruling for the defendant, on the basis that one could not be held liable for selling an article that was not legally his, the court stated, however: "If a man sells victuals which is corrupt, without warranty, an action lies. . . ."

In early 1815, a decision was reached in an English court that affected some later American court decisions in food cases. A London merchant, on the basis of samples shown to his agent, purchased 12 bags of waste silk. Upon arrival, the silk was found to be much inferior to the samples. At the time of purchase, a sale note was written that merely mentioned that 12 bags of waste silk were sold without referring to the samples or specifying particularly to the quality of the silk. No warranty was involved. The merchant filed a suit for recovery against the dealer. In allowing the recovery, the court stated:

> The sample was not produced as a warranty that the bulk corresponded with it, but to enable the purchaser to form a reasonable judgment of the commodity. I am of the opinion, however, that under such circumstances, the purchaser has a right to expect a saleable article answering the description in the contract. Without any particular warranty, this is an implied term in every such contract. Where there is no opportunity to inspect the commodity, the maxim of caveat emptor [the doctrine that the purchaser buys at his own risk] does not apply. He cannot without a warranty insist that it shall be of any particular quality or fineness, but the intention of both parties must be taken to be, that it shall be saleable in the market under the denomination mentioned in the contract between them. The purchaser cannot be supposed to buy goods to lay them on a dunghill. The question then

is, whether the commodity purchased by the plaintiff be of such quality as can be reasonably brought into the market to be sold as waste silk? The witnesses describe it as unfit for the purpose of waste silk, and of such quality that it cannot be sold under that denomination. Verdict for the plaintiff.[22]

We see in this case two statements of importance, both of which were to be developed further in subsequent decisions. One statement was: "Without any particular warranty, this is an implied term in every such contract." This came to be the law in any sale of personal property in which a written contract existed. It did not, as yet, apply to sales of food or provisions, but that would soon be remedied. The second critical pronouncement was: "When there is no opportunity to inspect the commodity, the maxim of caveat emptor does not apply." Previously, in the absence of an express warranty, the maxim of caveat emptor was the governing rule. Now, under certain circumstances, it was removed and the dealer must furnish satisfactory goods or he is liable. This case laid the groundwork for future extensions of this ruling.

In late 1815, in New York, the philosophy set forth in *Gardiner* v. *Gray* (quoted previously) was extended to one segment of food sales. The defendant had slaughtered a sick cow (Fig. 15-2). From this cow he sold a quarter beef to the plaintiff. The plaintiff and his family became ill from the beef. The resulting legal action was based on deceit but the court remarked during its discussion that: "In the sale of provisions for domestic use, the vendor is bound to know that they are sound, at his peril. This is a principle, not only salutary, but necessary to the preservation of health and life."[23]

What the court said was that a dealer in foods is responsible for the product that he sells. He is to be held liable for the wholesomeness of his food. The principle was extended to include only dealers (retailers). Public eating

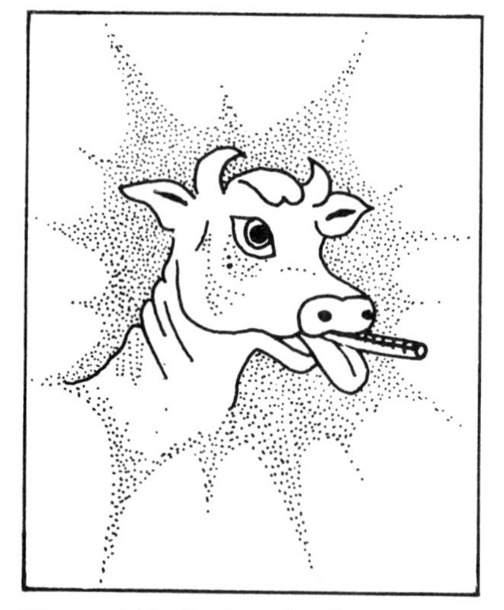

Figure 15-2 In the sale of provisions for domestic use, the vendor is bound to know that they are sound, at his peril. *Van Bracklin* v. *Fonda*. 12 Johns 467. 1815. Based on original drawings by Cindy Paddack.

places were many years away from inclusion under this rule.

In 1847, a decision in an English case confirmed the new rule to be applied to the food dealer. Farmer A bought, in the public market of a country town, the carcass of a pig. He left the pig hanging from the butcher's stall while conducting other business. Farmer B attempted to buy the pig from the butcher and was referred by him to farmer A. Farmer A sold the carcass, at a profit, to farmer B. The pig turned out to be unfit for human consumption. Farmer B sued farmer A. The court ruled for farmer A, the defendant, on the basis that he was not a dealer but during the course of its discussion stated:

> . . . that there is no other difference between the sale of victuals for food, and other articles than this, that victualers, butchers, and other common dealers in victuals, are not merely in the same situation that common dealers in other commodities are, so that if an order be sent to them to be executed, they are presumed to undertake to supply a good and merchantable article.[24]

This ruling definitely placed the sale of foods by dealers within the same legal framework as the sale of any other article of goods. This meant that there were conditions under which a food dealer could be held liable on an implied warranty if the food was not fit. The law at this time, however, was not as yet clearly formulated. Courts were still groping for a clear-cut philosophy with regard to foods.

In 1862, in England, a court complaint resulted in a firmly worded decision on the subject of implied warranties and food sold by a dealer. A dealer contracted to supply provisions to the East India Company. The provisions were to be shipped with troops to India. Upon inspection, they were deemed unwholesome. East India Company contended that there was an implied warranty that the provisions should have been fit for voyage. The court in its ruling stated:

> "First, whether upon a contract to supply provisions there is an implied warranty that they shall be reasonably fit for the purpose for which they are intended. Upon that point not a doubt can be entertained. When a buyer buys a specific article, the maxim caveat emptor applies, but when the buyer orders goods to be supplied, and trusts to the judgment of the seller to select goods which shall be applicable to the purpose for which they are ordered, there is an implied warranty that they shall be reasonably fit for that purpose.[25]

In *Gardiner* v. *Gray* (see note 22) it was ruled that where there is no opportunity for the buyer to inspect the goods and, as a result, the buyer relies on the judgment of the seller to select the goods, an implied warranty exists. In *Gardiner* v. *Gray*, the product involved was silk. We now see that the concept set forth in that case has been extended to the buying and selling of food.

To sum up our cases, as matters stood in 1862, there was no difference between a sale of provisions and the sale of other articles in respect to there being an implied warranty that they are fit. We also find that the rule making a dealer liable for selling unsound provisions is the rule that is applied where any chattel (any article of movable property) is ordered of a manufacturer or dealer for a particular purpose. In such a case, there is an implied warranty that the article furnished will be fit for the particular purpose.

Early American Decisions

The case of *Scheffer* v. *Willoughby*, discussed in the preceding section, took place in 1896 in Illinois. In this case, the plaintiff attempted to have the dispensing of food by a restaurant classified as a sale. This contention was rejected by the court. The court also stated in this case: "But where an action is brought to recover damages the burden is upon the person bringing the action to establish carelessness or negligence."

This ruling upheld previous cases discussed in that only manufacturers and dealers were held liable for their goods on an implied warranty. At this time, restaurateurs were not yet considered dealers and so were liable only for negligence and deceit. However, by 1896, some states were classifying the dispensing of food by restaurants as a sale. The days of the "innkeeper and victualer" philosophy were drawing to a close.

In 1912, an attempt was made to force an implied warranty on eating establishments. A food-poisoning complaint was filed in Maine against a railroad by a traveler who became sick after eating in the dining car of a train. In rejecting the complaint, the Supreme Court of Maine stated:

> We know of no rule of law which will imply a warranty of that of which it is impossible for a defendant to know by the exercise of any skill, knowledge, or investigation, however great. If . . . the only possible conclusion is that the parties understood the matter precisely alike, and that the defendant sold, and the plaintiff bought, exactly what she ordered. She therefore assumed the risk of its imperfections. . . . It is the opinion of the court that in the absence of an express warranty the defendant is not liable.[26]

In 1914, another complaint was filed against a railroad, this time for food poisoning sustained while eating in a buffet car. The action was filed in a federal court, alleging an implied warranty on the food. This court, with Judge Augustus Hand writing the decision, also rejected the contention, stating:

> My own feeling is that protection to the public lies not so much in extending the absolute liability of individuals, as in regulating lines of business in which the public has a particular interest in such a way as reasonably to insure its safety. In other words, pure food laws, and rigorous inspection of meats, canning factories, and other sources of food supply, would seem to me a much more effective way of protecting the public than by the imposition of the liability of an insurer upon those who furnish food. In my opinion there is no well-considered authority and no public policy which afford any justification for imposing upon the defendant the absolute liability of an insurer of its food, and I deem that the only obligation of the defendant, or any keeper of a restaurant or inn, is to exercise the reasonable care of a prudent man in furnishing and serving food.[27]

Later American Decisions

In 1913, a lower New York court finally opened the door and ushered the implied warranty into a public eating establishment. In an historic decision, the Appellate Court of New York ruled for the plaintiff in a food-poisoning case resulting from eating ice cream made at a drug store and sold for immediate consumption at the same drug store counter. The decision was appealed to the New York Court of Appeals where, in 1918, the court stated, in affirming the decision of the lower court:

> The consequences to the consumer resulting from consumption of articles of food sold for immediate use may be so disastrous that an obligation is placed upon the seller to see to it, at his peril, that the articles sold are fit for the purpose for which they are intended. The rule is an onerous one, but public policy, as well as the public health, demand such obligation should be imposed. The seller has an opportunity, which the purchaser does not, of determining whether the article is in the proper condition to be immediately consumed. If there be any poison in the article sold, or if its condition renders it unfit for consumption, and the customer be thereby made ill, some one must of necessity suffer, and it ought not to be the one who has had no opportunity of determining the condition of the article, but rather the one who has had at his command the means of doing so. I am of the opinon the trial court did not err in instructing the jury that, when defendant sold the cream to plaintiff, he impliedly warranted it was wholesome and fit to eat.[28]

The ruling in this case, *Race* v. *Krum*, was restricted to a certain type of food business, that is, one in which the product (ice cream) was made and sold for immediate consumption on the same premises. It was not intended to impress an implied warranty on hotel proprietors, restaurant keepers, dining car managers, and others. However, the door had been opened in one type of food establishment, and the other doors would soon follow.

In 1914, again in New York, another breakthrough occurred. The plaintiff had eaten lunch in the defendant's lunchroom, dining on toast, tea, and part of a piece of chocolate pie. The pie had an unusual color and taste. The plaintiff became violently ill about one hour later. A complaint was filed alleging an implied warranty on the food. The court's ruling was brief and to the point: "In *Race* v. *Krum*, decided by this court in March, 1913,. . . we held that there was an implied warranty in the sale of food under somewhat similar circumstances."[29]

This was an eventful decision, since it extended the ruling in *Race* v. *Krum* to a variety of eating establishments. It meant that in New York, an influential state, food served at any establishment with counter service would carry an implied warranty of wholesomeness. New York courts had led the way, as far as the consumer was concerned. The rulings made were in the interest of public health, and the customers of lunchrooms and drug store counters were the beneficiaries. New York was the lone pioneer, however, until the historic case of *Friend* v. *Child's Dining Hall Co.* in Massachusetts.

The decision that was ultimately to be followed by the majority of courts was rendered by the Massachusetts Supreme Court in 1918. The plaintiff in the case entered the defendant's restaurant in Boston and ordered New York baked beans and corned beef from a waitress. The baked beans contained several round hard stones that looked like baked beans. The plaintiff, thinking that she had, among the other baked beans, some that were simply overdone, bit hard on several of them and injured her mouth and teeth. She filed a complaint contending that there was an implied warranty of fitness on the food. The court ruled for the plaintiff, stating:

> It would be an incongruity in the law amounting to at least an inconsistency to hold with reference to many keepers of restaurants who conduct the business both of supplying food to guests and of putting up lunches to be carried elsewhere and not eaten on the premises, that, in case of want of wholesomeness, there is a liability to a purchaser of a lunch to be carried away founded on an implied condition of the contract, but that liability to the guest who eats lunch at a table on the premises rests solely on negligence. The guest of a keeper of an eating house or innkeeper is quite as helpless to protect himself against deleterious food or drink as is the purchaser of a fowl of a provision dealer. The opportunity for the innkeeper or restaurant keeper, who prepares and serves food to his guest, to discover and provide against deleterious food is at least as ample as that of the retail dealer in foodstuffs. The evil consequences in the one case are of the same general character as in the other. Both concern the health and physical comfort and safety of human beings. On principle and on authority it seems to us that the liability of the proprietor of an eating house to his guest for serving bad food rests on an implied term. . . .[30]

This decision was the inevitable result of the decision rendered in *Commonwealth* v. *Worcester* (see note 10). In that case, the dispensing of food for immediate consumption was ruled a sale in Massachusetts. If such serving of food was a sale, the seller of that food then became a merchant or dealer. In previous instances (*Van Bracklin* v. *Fonda* [22] and *Brigge* v. *Parkinson* [24]), the courts had ruled that dealers were liable for the wholesomeness of the provisions they sold. The Massachusetts Supreme Court simply stated that it was inconsistent to hold a dealer liable for food leaving the premises in a lunchsack and not liable for the same type of food eaten on the premises as a restaurant purchase.

In the preceding section, we discussed two decisions (*Merrill* v. *Hobson* [15] and *Nisky* v. *Child's Co.* [17]) that became known as the Connecticut-New Jersey rule. These two decisions strongly denied the existence of a sale in the dispensing of food in public eating places. These two decisions also rejected the existence of an implied warranty accompanying the dispensing of food in these establishments.

Opposed to the Connecticut-New Jersey rule was the New York-Massachusetts rule. The Massachusetts contribution to this rule was the case of *Friend* v. *Child's Dining Hall Co.*, just discussed. The New York contribution occurred in 1924 in the case of *Temple* v. *Keeler*. The plaintiff in this case sustained food poisoning from eating fish in the defendant's restaurant. Judge Andrews of the New York Court of Appeals wrote, in ruling for the plaintiff:

> We hold that, where a customer enters a restaurant, receives, eats, and pays for food, delivered to him on his order, the transaction is the purchase of goods. We hold also that under such circumstances the buyer does by implication make known to the vendor the particular purpose for which the article is required, and where the buyer may assume that the vendor has had an opportunity to examine the article sold, it appears conclusively that he relies upon the latter's skill or judgment. Consequently there is an implied warranty that the food is reasonably fit for consumption.[31]

The Connecticut-New Jersey rule represented the past. It was based on English law regarding victualers and innkeepers. The food industry was growing, however, and sales were increasing. The eating houses were evolving from home businesses into the now familiar merchant-type restaurant business in a building separate from the dwelling place of the owner. The number of different items on the menu was increasing, and the restaurant business was becoming more complex. Because of the increase in the number of customers being served and the increase in the variety of foods on the menu, more help was being hired to prepare and serve this food. Some foods now being served on this larger scale were readily perishable. Others were excellent media for the growth of food-poisoning bacteria. This increased handling of easily perishable and contamination-prone foods was resulting in an increasing public health hazard.

The Massachusetts-New York rule was to become the rule of the future. As the food industry grew, so did their imposed responsibilities. As early as 1898, a glimpse of the philosophy that would eventually shape the law could be seen in an Illinois decision involving a meat market that sold contaminated meat to a customer. In its ruling, the court stated:

> Where, however, articles of food are purchased from a retail dealer for immediate consumption, the consequences resulting from the purchase of an unsound article may be so serious and may prove so disastrous to the health and life of the customer that public safety demands that there should be an implied warranty on the part of the vendor that the article sold is sound and fit for the use for which it was purchased.[32]

The old law did not die quickly or quietly, however. The ensuing controversy lasted for many years. Sharply worded decisions, many times from neighboring states, rattled courtroom walls with directly opposing views. As recently as 1965, we find a decision by the Delaware Superior Court in which it is stated:

> While I recognize that the rule denying liability has been repudiated in a majority of sister jurisdictions and is factually indefensible, I am of the opinion that an overturning of the well settled rule now prevailing in this state should be done by either the legislature or the Supreme Court. It is noted that the Uniform Commercial Code specifically repudiates the rule denying the existence of the warranty. The Uniform Commercial Code has not, however, been adopted by the General Assembly of this State.[33]

The Uniform Commercial Code

As we have seen, the Uniform Commercial Code has now been almost universally adopted in the United States. Section 2–314 of the Code states: "Unless excluded or modified a warranty that the goods shall be merchantable is implied in a contract for their sale if the seller is a merchant with respect to goods of that kind." Section 2–315 states: "Where the seller at the time of contracting has reason to know any particular purpose for which the goods are required and that the buyer is relying on the seller's skill or judgment to select or furnish suitable goods, there is unless excluded or modified under the next section an implied warranty that the goods shall be fit for such purpose."

These two subsections of Section 2, which deal with sales, finally settled the issue of implied warranties and food served in public eating establishments. The issue was settled in favor of the warranty and, as a result, in favor of public health.

SUMMARY

Products liability deals with the liability of manufacturers, wholesalers, and retailers for illness caused by their products. The basis for this area of law is the responsibility placed on one who sends goods out into the channels of trade for use by others.

Food is usually defined as: (1) articles used for food or drink for man, (2) chewing gum, and (3) articles used for components of any such articles.

American foodservice establishments evolved from their English counterparts, whose origin was the old English inn and victual house. These came into existence because of the dangerous travel conditions of the times. Some private homes turned to the business of furnishing food, drink, or overnight lodging to travelers.

Because they conducted business in their own homes, the innkeeper and victualer were not considered to be merchants but rather people who provided services. They were not regulated by sales codes and did not have to guarantee the wholesomeness of their food. In case of illness, legal action could be brought only if negligence or deceit could be shown.

In 1879, the dispensing of food in public eating houses was ruled to be a sale in the state of Massachusetts. Many states followed suit over the next 40 years. Many did not, however, and conflicting court decisions resulted. Judicial decisions are now based on the Uniform Commercial Code, which states: "The serving for value of food or drink to be consumed either on the premises or elsewhere is a sale."

A warranty is a guarantee that an article, which has been sold, is fit for its intended purpose. There are two types of warranty—express and implied. An express warranty is a written or oral expression issued at the time of the sale. An implied warranty needs no oral or written expression. It occurs when the buyer has relied on the skill or judgment of the seller.

The philosophy of an implied warranty on food first appeared in the United States in 1815, but was restricted to food sold by food dealers. Attempts to extend the philosophy to restaurants, lunch counters, and so forth were unsuccessful for many years since these establishments were not considered to be selling their food.

The first successful action against a foodservice establishment, based on implied warranty, occurred in New York in 1913. The suit resulted from contaminated ice cream sold at a drug store counter. In 1914, the same court awarded the verdict to the plaintiff in a suit against a lunchroom.

The Uniform Commercial Code states: "Unless excluded or modified a warranty that the goods shall be merchantable is implied in a contract for their sale if the seller is a merchant with respect to goods of that kind."

Review Questions

1. What is meant by products liability?

2. What is a food?

3. a. How did the English inn evolve? b. What rules governed its dispensing of food?

4. What unusual position did the innkeeper and victualer occupy under English law?

5. What role did the Illinois Supreme Court play in 1896 in continuing English legal philosophy?

6. What was the significance of the 1896 ruling by the Massachusetts Supreme Court?

7. a. What is the purpose of the Uniform Commercial Code? b. What does this code state with regard to the serving of food?

8. a. What is a warranty? b. What is the difference between an express warranty and an implied warranty?

9. a. In early 1815, a case was decided in an English court that set a precedent. What was the significance of this case? b. In late 1815, a New York court extended this philosophy to foods. What was the result of the New York case?

10. What was the significance of the two related cases, *Race* v. *Krum* and *Leahy* v. *Essex Co.*?

11. What does the Uniform Commercial Code state with regard to an implied warranty on food?

The references listed here will probably not be found in your school or institution library. The Uniform Commercial Code can be found in any law office. All references listed can be found in the library of any law school in your state and in the law library of the State Supreme Court in your state capital.

The following key should prove helpful.

1. A.—*Atlantic Reporter*
2. A.2d—*Atlantic Reporter*, 2nd edition
3. Eng. Rep.—*English Reprints*
4. F.—*Federal Reporter*
5. Ill.—*Illinois Reporter*
6. Johns.—*Johnson's Reporter* (N.Y.)
7. Mass.—*Massachusetts Reports*
8. N.E.—*Northeast Reporter*

9. N.W. 2d—*Northwest Reporter*, 2nd edition
10. N.Y.S.—*New York Supplement*
11. S.W.—*Southwest Reporter*
12. S.W. 2d—*Southwest Reporter*, 2nd edition

Example: The reference, 182 N.W. 2d 800, indicates that the case can be located in volume 182, *Northwest Reporter*, 2nd edition, page 800.

References

1. Cova v. Harley Davidson Motor Co. 182 N.W. 2d 800

2. Smith-Hurd Illinois Annotated Statutes. 402.3

3. F. W. Woolworth Co. v. Garza. 390 S.W. 2d. 90

4. Beale, J. H. (1906). *The Law of Innkeepers and Hotels*. William J. Nagel. Boston.

5. Crisp v. Pratt. 79 Eng. Rep. 1072

6. Newton v. Trigg. 91 Eng. Rep. 100

7. Parker v. Flint. 88 Eng. Rep. 1303

8. Saunderson v. Rowles. 98 Eng. Rep. 77

9. Scheffer v. Willoughby. 163 Ill. 518

10. Commonwealth v. Worcester. 126 Mass. 256

11. Commonwealth v. Miller. 18 A. 938

12. State v. Lotti. 47 A. 392

13. Commonwealth v. Phoenix Hotel. 162 S.W. 823

14. People v. Clair. 116 N.E. 868

15. Merrill v. Hobson. 91 A. 533

16. Kenny v. Wong Len. 128 A. 343

17. Nisky v. Childs. 135 A. 805

18. Uniform Commercial Code. § 2–314 (1)

19. Uniform Commercial Code. § 2–314 (1)

20. Uniform Commercial Code. § 2–315

21. Roswel v. Vaughan. 79 Eng. Rep. 171

22. Gardiner v. Gray. 171 Eng. Rep. 46

23. Van Bracklin v. Fonda. 12 Johns. 467

24. Burnby v. Vollett. 153 Eng. Rep. 1348

25. Bigge v. Parkinson. 158 Eng. Rep. 758

26. Bigelow v. Maine Central Railroad. 85 A. 396

27. Valeri v. Pullman Co. 218 F. 519

28. Race v. Krum. 118 N.E. 853

29. Leahy v. Essex Co. 148 N.Y.S. 1063

30. Friend v. Child's Dining Hall Co. 120 N.E. 407

31. Temple v. Keeler. 144 N.E. 635

32. Wiedeman v. Keller. 171 Ill. 93

33. Dickens v. Horn and Hardart Baking Co. 209 A. 2d. 169

A. Privity
    1. English Decisions
    2. Early American Decisions
    3. Later American Decisions
    4. Uniform Commercial Code
B. "Foreign" or "Natural"
    1. Early American Decisions
    2. Later American Decisions
C. Summary

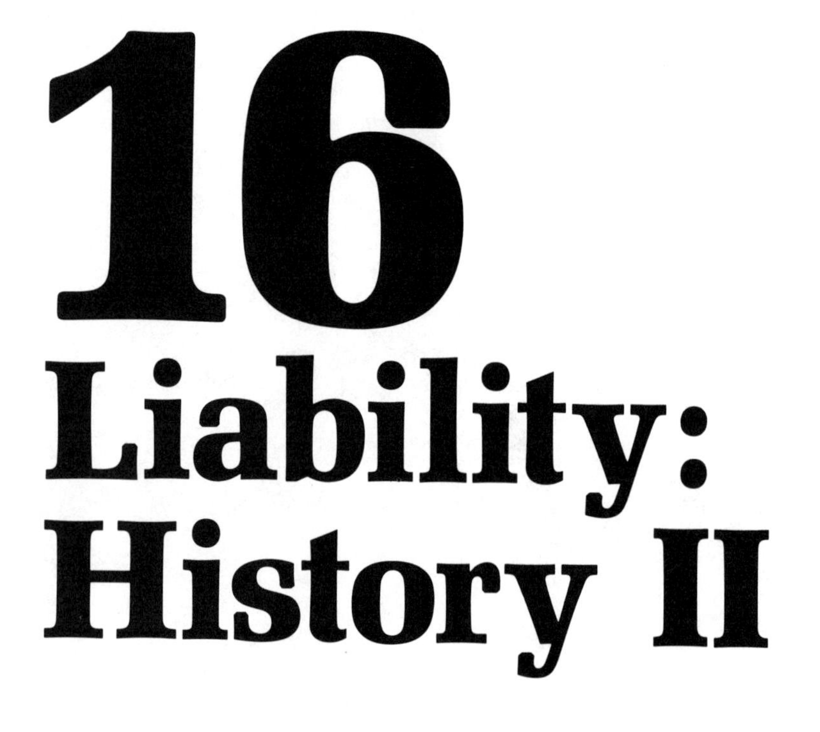

PRIVITY

Privity is the legal term used to denote the bond between two or more persons who share in a contractual relationship. For example, if during your grocery shopping, you stop at the meat counter, "window shop," and then order an appetizing roast, the purchase of that roast constitutes a sale. There is an implied sales contract between you and the store. Because of this contract, we say that privity exists between you and the store. The store purchased the meat from a meat packer. Therefore, privity exists between the store and the meat packer. When you purchased the meat, you purchased it from the grocer, not the meat packer. Therefore, there is no privity between you and the packer (Fig. 16-1).

Why is privity important? Again, consider the example of the roast. Say the roast is unwholesome. You take it home and your wife cooks it immediately for a dinner party with your neighbors. Everyone in both families comes down with food poisoning (Fig. 16-2). Who can bring action against the store? You can, because you purchased the meat, thus making a sales contract. Privity therefore exists between you and the store. Can your wife recover damages? Privity is lacking here. She neither purchased nor ordered the roast. You handled both chores yourself. Can your minor children recover damages? Can your neighbors and their children bring suit? They are far removed from the contract. Privity is certainly lacking there.

What if a canned ham had been purchased instead of the roast? You did not purchase the ham from the manufacturer. You purchased it from the grocer. It is impossible, however, for a grocer to inspect the contents of a canned ham for wholesomeness. If the ham causes food poisoning, can you recover from the manufacturer, with whom you had no sales contract and therefore no privity? These are questions we shall answer in the first part of this chapter.

English Decisions

The foundation of American law regarding privity exists in English law, particularly in one noted decision. This was *Winterbottom* v. *Wright*, in 1842, the result of a rather complex situation. For purposes of ease and clarity in the discussion, we shall denote the main characters in the case with letters. Defendant A in the case had contracted with the Postmaster-General to supply mail coaches and keep them in repair. This man, in turn, contracted with B

Figure 16-1 The concept of privity. (a) Privity exists between the customer and grocer. (b) Privity exists between the grocer and meat packer. Privity does not exist between the customer and meat packer. Based on original drawings by Cindy Paddack.

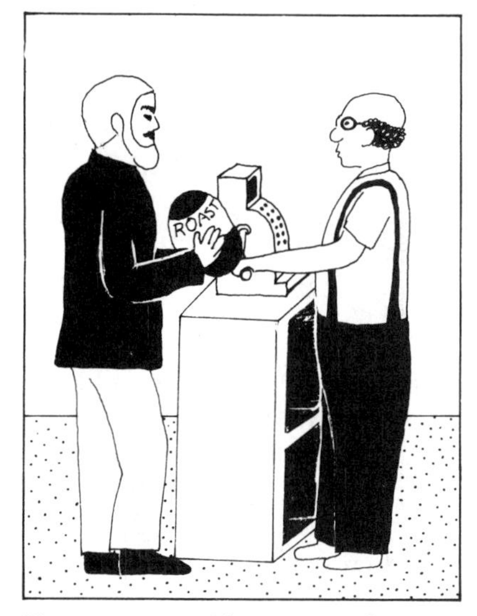

Figure 16-2 (a) The customer has privity with the grocer. (b) The customer's wife and friends do not have privity with the grocer. (c) In case of illness, only the customer could collect. Based on original drawings by Cindy Paddack.

to supply horses and hire drivers for the coaches. The drivers were to deliver the mail. On one trip, a coach collapsed, throwing the plaintiff driver C to the ground and injuring him. Driver C filed a suit against A, who was responsible for the maintenance of the coaches, contending that the defendant "improperly and negligently conducted himself" and allowed the coach to become in a "frail, weak, and infirm, and dangerous state and condition. . . ." The court ruled for the defendant, laying down this philosophy:

> I am clearly of opinion that the defendant is entitled to our judgment. We ought not to permit a doubt to rest upon this subject, for our doing so might be the means of letting in upon an infinity of actions. There is no privity of contract between these parties; and if the plaintiff can use, every passenger, or even any person passing along the road, who was injured by the upsetting of the coach, might bring a similar action. Unless we confine the operation of such contracts as this to the parties who entered into them, the most absurd and outrageous consequences, to which I can see no limit, would ensue. The contract in this case was made with the Postmaster-General alone; and the case is just the same as if he had come to the defendant and ordered a carriage, and handed it at once over to Atkinson [B]. If we were to hold that the plaintiff could sue in such a case, there is no point at which such actions would stop. The only safe rule is to confine the right to recover to those who enter into the contract: if we go one step beyond that, there is no reason why we should not go fifty.[1]

What the court has ruled here is that the plaintiff C had no recourse against the maintenance man A with whom he had no contract and with whom privity was lacking. It has stated quite firmly that where privity is lacking, no recovery is permitted. If your visit to the grocery store had taken place in 1842, neither your family, your neighbor, or his family would have been able to bring a successful action over the food-poisoning incident since privity was lacking between these persons and the store. Only in your case would recovery have been permitted against the store. If you had purchased the canned ham, no one could have brought action against the manufacturer as privity would have been lacking between all the persons involved and the manufacturer.

The effect of this ruling could be seen as late as 1936 in the commonwealth of Virginia. A father bought a container of milk from a dairy company. The milk contained undulant fever bacteria. His son contracted the disease and the father brought suit in the son's name. The Supreme Court of Virginia rejected the suit, stating:

> The fact that a seller warrants the condition or quality of a thing sold does not itself according to the better view impose any liability on him to third persons who are in no way a party to the contract. In such a case there is no privity of contract between the seller and such third person, and this precludes any right on his part to any advantage or benefit to be derived from the warranty.[2]

The court has ruled here that since the son was not a party to the contract, privity was lacking between him and the dairy company. In the absence of privity, the dairy company did not have to impliedly warrant to the son that the milk was wholesome and therefore the son could not collect for his sickness. Privity did exist between the father and the dairy company but the milk did not make the father sick. Thus, the doctrine of privity was the key factor in the rejection of the suit.

To turn to the past again, we see that the ruling philosophy from *Winterbottom* v. *Wright* was that no action in liability cases was to be successful if privity did not exist between plaintiff and defendant. At this time, then, if a wife bought food that, because of contamination at the time of purchase, made other family members sick, only the wife could recover damages from the food retailer, since privity existed only between those two. The wife could not recover damages from the manufacturer as privity was lacking between them.

Early American Decisions

The first crack in the armor of privity occurred in 1852 in the state of New York. This was the celebrated case of *Thomas* v. *Winchester*, a landmark case in the area of liability. The defendants were in the business of preparing vegetable extracts to be sold to druggists (Fig. 16-3). Their shop was in New York City. Among the extracts they prepared were extract of dandelion and extract of belladonna. Extract of belladonna is an extremely potent drug. The suit resulted from the mislabeling of a jar of belladonna with an extract of dandelion label. The mislabeled jar was sold by the defendants to a salesman who, in turn, sold it to Foord, a druggist in Cazenovia, New York. This mislabeled jar of belladonna was then sold to the husband of the plaintiff. It was then administered to the plaintiff, as a result of which, she nearly lost her life. She brought action, not against the druggist, with whom she had privity, but against the firm that prepared the extracts and with whom privity was lacking. The suit was based on negligence. The ruling of the court marked the first step in the erosion of privity as it stated:

> The defendant's negligence put human life in imminent danger. The defendant, by affixing the label to the jar, represented its contents to be dandelion; and to have been 'prepared' by his agent Gilbert. The word 'prepared' on the label, must be understood to mean that the article was manufactured by him, or that it had passed through some process under his hands, which would give him personal knowledge of its true name and quality. Gilbert, the defendant's agent, would have been punishable for manslaughter if Mrs. Thomas had died in consequence of taking the falsely labeled medicine. So highly does the law value human life, that it admits of no justification wherever life has been lost and the carelessness or negligence of one person has contributed to the death of another. Although

Figure 16-3 (a) The defendants were in the business of preparing vegetable extracts to be sold to druggists. One jar was mislabeled. (b) A druggist bought this jar from a salesman and (c) sold it to a customer. (d) The customer administered the contents to his wife (e) who became seriously ill. Recovery was allowed in spite of lack of privity. *Thomas* v. *Winchester*, 1852. Based on original drawings by Cindy Paddack.

the defendant Winchester may not be answerable criminally for the negligence of his agent, there can be no doubt of his liability in a civil action, in which the act of the agent is to be regarded as the act of the principal.[3]

Pandora's box had been opened, and an exception to the doctrine of privity had been allowed to emanate from within. One exception had been made that would eventually lead to a host of others. Lack of privity would not be allowed to prevent recovery where negligence was present in the manufacturing of dangerous drugs.

In 1882, the decision in *Thomas* v. *Winchester* was extended. This case involved a man who had constructed a scaffold for a painting contractor who was to paint the inside of the dome of the King's County Courthouse in New York. The scaffold collapsed during use, and one of the painters hired by the contractor fell and was fatally injured. His widow brought suit, not against the contractor, but against the constructor of the scaffold. Referring to *Thomas* v. *Winchester* during its decision, the New York Court of Appeals ruled:

As a general rule the builder of structure for another party, under a contract with him, or one who sells an article of his own manufacture, is not liable to an action by a third party who used the same with the consent of the owner or purchaser, for injuries resulting from a defect therein, caused by negligence. The liability of the builder or manufacturer for such defects is, in general, only to the person with whom he contracted. But, notwithstanding this rule, liability to third parties has been held to exist when the defect is such as to render the article in itself imminently dangerous, and serious injury to any person using it is a natural and probable consequence of its use.[4]

We see that the exception to the privity requirement, as set down in *Thomas* v. *Winchester*, has now been extended from negligence in preparing dangerous drugs to negligence in the manufacture of any inherently dangerous article. It only remained now for some court to classify food as an inherently dangerous article. This ruling was some years in the future, however.

It should be mentioned that the courts in the majority of states did not rush headlong to follow the decisions of the New York courts. This is normal, of course, in the case of new and sometimes revolutionary rulings that change existing laws. Privity had long been, and in some areas still is, a firm and cherished concept of law.

As an example of resistance to *Thomas* v. *Winchester*, we refer to *Curtain* v. *Somerset*, settled in 1891 in Pennsylvania. The defendant in the case was a contractor who built a hotel. At an entertainment given at this hotel, some 20 or more people were injured when a porch on which they were standing collapsed due to faulty construction. The plaintiff was one of the injured, and he brought suit against the contractor. In spite of the fact that a negligently constructed porch is certainly an inherently dangerous article, the Supreme Court of Pennsylvania ruled for the defendant, stating:

Is he [the defendant] also liable for an injury to a third person, not a party to the contract, sustained by reason of defective construction? It is very clear that he was not responsible by force of any contractual relation, for, as before observed, there was no contract between these parties, and hence there could have been no breach. If liable at all, it can only be for a violation of some duty. It may be stated, as a general proposition, that a man is not responsible for a breach of duty where he owes no duty. What duty did the defendant owe to the plaintiff? The latter was not upon the porch by the invitation of the defendant. The proprietor of the hotel, or whoever invited or procured the presence of the plaintiff there, may be said to have owed him a duty,—the duty of ascertaining that the porch was of sufficient strength to safely hold the guests who he had invited. If a contractor who erects a house, who builds a bridge, or performs any other work, the manufacturer who constructs a boiler, piece of machinery, or a steam-ship, owes a duty to the whole world, that his work or his machine or his steam-ship shall contain no hidden defect, it is difficult to measure the extent of his responsibility, and no prudent man would engage in such occupations upon such conditions. It is safer and wiser to confine such liabilities to the parties immediately concerned.[5]

In 1905, in the state of Georgia, a purchaser of a bottle of soda water had a problem with pieces of glass. The glass was in the soda water and several pieces were swallowed along with the soda water. The purchaser brought suit against the bottling company, even though the soda water had been purchased from a grocer. The suit was brought for negligence. The decision in the case, *Watson* v. *Augusta Brewing Co.*, was for the plaintiff, in spite of lack of privity, as the court stated:

When a manufacturer makes, bottles, and sells to the retail trade, to be again sold to the general public, a beverage represented to be refreshing and harmless, he is under legal duty to see to it that in the process of bottling no foreign substance shall be mixed with the beverage which, if taken into the human stomach, will be injurious. If, then, one who buys a patent medicine may rely upon the obligation of the manufacturer not to place therein ingredients which, if taken in the prescribed doses, will injure his health, certainly the purchaser of an alleged harmless and refreshing beverage should have the right to rest secure in the assumption that he will not be fed on broken glass. It does not matter that the plaintiff in the present case did not buy the soda water from the defendant, or that there was no privity of relationship between them. The duty not negligently to injure is due by the manufacturer, in a case of the particular character of the one under consideration, not merely to the dealer to whom he sells his product, but to the general public for whom his wares are intended.[6]

As you can see from these cases, the concept of privity was slowly but inexorably being relaxed. The consumer in *Watson* v. *Augusta Brewing Co.* won his case in spite of lack of privity, since he had purchased the soda water from a grocer rather than from the bottling company.

In 1908, in the state of New Jersey, another noted food case appeared on stage. The plaintiff had purchased a canned ham from a grocer. The grocer's supplier of the ham was a nationally known meat packer. The ham caused food poisoning and the consumer brought action against the packer for negligence. The New Jersey Court of Errors and Appeals upheld the claim of the plaintiff, ruling:

Upon both reason and authority we are clearly of the opinion that the declaration before us sets up a good cause of action. The fact that the defendant was the manufacturer, presumably having knowledge, or opportunity for knowledge, of the contents of the cans and of the process of manufacture; that it put the goods upon the market for sale by dealers to consumers under circumstances such that neither dealer nor consumer had opportunity for knowledge of the contents; the fact the goods were thus manufactured and marketed under circumstances that imported a representation to intending purchasers that they were fit for food and beneficial to the human body; that in the ordinary course of business there was a probability that the goods should be purchased, and used by the parties purchasing, in reliance upon the representation; and that the defendant negligently prepared the food so that it was unwholesome and unfit to be eaten, and poisonous to the human body, whereby the plaintiff was injured—make a case that renders the defendant liable for the damages sustained by the plaintiff thereby.

The effect of this decision was to place food in the category of an inherently dangerous article because of the possibility of its poisoning the human body and causing illness or death if negligently prepared. It is, indeed, a far cry from belladonna to canned ham. It required 56 years for this accomplishment to take place. It was achieved, as we have seen, in a series of small steps, each step extending *Thomas* v. *Winchester* farther afield from its original decision. It might be compared to the expansion of major league baseball. Eight teams in each league had been the hard and fast rule, comparable to the decision reached in *Winterbottom* v. *Wright*. The American League ruled that two additional teams would be added in 1961, a break with tradition comparable with *Thomas* v. *Winchester*, and triggered a series of extensions in which two more teams were added in 1969 and an additional two in 1977.

By 1914, the *Thomas* v. *Winchester* ruling had become widely accepted in the United States. Consumers were recovering damages from manufacturers in spite of lack of privity, but only if negligence could be shown or if a violation of a state law had occurred. Recovery was restricted to the purchaser of the unwholesome food, however. The doctrine of privity was still being applied to family members and guests.

In November 1914, in Kansas, the doctrine of privity was dealt yet another blow. A suit was brought by the widow of the purchaser of a contaminated pie. The husband had died within 24 hours after eating a portion of it. The widow brought the action for negligence, even though she had not purchased the pie. The Kansas Supreme Court broke with tradition and ruled for the widow, declaring:

The degree of care required of a manufacturer or dealer in human food for immediate consumption is much greater by reason of the fearful consequences which may result from what would be slight negligence in manufacturing or selling food for animals. In the latter a higher degree of care should be required than in the manufacturing or selling of ordinary articles of commerce. A manufacturer or dealer who puts human food upon the market for sale or for immediate consumption does so upon an implied representation that it is wholesome for human consumption. Practically, he must know it is fit or take the consequences, if it proves destructive.[8]

Although the privity rule was relaxed another notch in this decision to allow recovery to one not a purchaser, recovery was denied to the two minor children of the deceased. Privity was wobbly, but still alive.

Along these same lines was *Gearing* v. *Berkson*, in March, 1916, in Massachusetts. A husband and wife were the plaintiffs. At the husband's request, his wife purchased some pork chops at a grocery store. The pork chops caused food poisoning. The husband was allowed to recover damages but the wife's suit was denied on the grounds that she was merely the agent of her husband and therefore privity was lacking between her and the grocery store. The Supreme Court of Massachusetts ruled:

The difficulty with the case on this ground [implied warranty of fitness] is that there was no contractual relation, and hence no warranty, between Mrs. Gearing and the defendants. The only sale was that made to her husband through her as his agent; and no cause of action in contract accrued to her thereon, as above set forth. The implied warranty, or to speak more accurately the implied condition of the contract, to supply an article fit for the purpose required, is in the nature of a contract of personal indemnity with the original purchaser. In the absence both of an implied warranty and of negligence on the part of the defendants, the action of Mrs. Gearing fails.[9]

A resume of cases to date (1916) is shown in Table 16-1. Note that after *Thomas* v. *Winchester* (1852), third parties were winning suits in many states, but only if negligence could be

Table 16-1
Summary of cases to 1916 showing type of suit, plaintiff, bypassed party, defendant, and result of case to plaintiff.

Year	Case	Type of Suit	Plaintiff	Bypassed Party	Defendant	Result to Plaintiff
1842	*Winterbottom* v. *Wright*	Negligence	Injured party	Employer	Supplier of coaches	Lost
1852	*Thomas* v. *Winchester*	Negligence	Consumer	Retailer	Manufacturer	Won
1882	*Devlin* v. *Smith*	Negligence	Widow	Employer	Contractor	Won
1891	*Curtain* v. *Somerset*	Negligence	Injured party	Hotel owner	Contractor	Lost
1905	*Watson* v. *Augusta Brewing Company*	Negligence	Consumer	Retailer	Manufacturer	Won
1908	*Tomlinson* v. *Armour and Co.*	Negligence	Consumer	Retailer	Manufacturer	Won
1914	*Parks* v. *Yost Pie Co.*	Negligence	Widow, children	Retailer Retailer	Manufacturer Manufacturer	Won Lost
1916	*Gearing* v. *Berkson*	Implied Warranty	Agent of purchaser (wife)	—	Retailer	Lost

shown to have occurred on the part of the manufacturer or contractor.

The most eventful decision in the area of privity since *Thomas* v. *Winchester* occurred in the state of New York in 1916. The case was *MacPherson* v. *Buick Motor Co.* The suit involved an automobile, but the implications were widespread in the entire field of products liability. The defendant manufactured automobiles. One was sold to a retail dealer. He resold it in the normal manner to the plaintiff. One of the wheels was made of defective wood and its spokes collapsed, causing the entire wheel to fail. The plaintiff was thrown from the car and injured. The wheel was not made by the defendant, but the court felt that the defect should have been discovered by the defendant during proper inspection of the automobile. The suit was based on negligence.

Judge Cordozo of the New York Court of Appeals rendered the decision, a decision that caused great waves to beat against the crumbling wall of privity. In its ruling, the court stated:

> We hold, then, that the principle of *Thomas* v. *Winchester* is not limited to poisons, explosives, and things of like nature, to things which in their normal operation are implements of destruction. If the nature of a thing is such that it is reasonably certain to place life and limb in peril when negligently made, it is then a thing of danger. Its nature gives warning of the consequences to be expected. If to the element of danger there is added knowledge that the thing will be used by persons other than the purchaser, and used without new tests, then, irrespective of contract, the manufacturer of this thing of danger is under a duty to make it carefully. We have put aside the notion that the duty to safeguard life and limb, when the consequence of negligence may be foreseen, grows out of contract and nothing else. We have put the source of the obligation where it ought to be. We have put its source in the law.[10].

With just two sentences, "We have put aside the notion that the duty to safeguard life and limb, when the consequences may be foreseen, grows out of contract and nothing else. We have put its source in the law," Judge Cordozo imprinted his name in legal history.

Privity is a contractual relationship. Without that relationship, privity is lacking between the parties. At the time of this case, consumers who were the original purchasers were being allowed recovery in most states against manufacturers of defective products. However, their wives, husbands, children, and others were being denied recovery against the same manufacturers because of lack of privity, even though they may have been injured as badly or worse from the product than the purchaser was. The MacPherson decision changed this philosophy by declaring that the responsibility of a manufacturer for the health and safety of the public does not rest upon a contractual relationship, alone. The responsibility is required by law.

In addition, MacPherson describes the type of product to be covered under this new philosophy: "If the nature of a thing is such that it is reasonably certain to place life and limb in peril when negligently made, it is then a thing of danger."

We find in this case, then, a description of what was to be regarded as a "thing of danger." A manufacturer of this type of product was now liable for its fitness, not only to the purchaser, but to others using it. Suits for recovery could now be brought in action other than negligence. The dangerous drug philosophy of *Thomas* v. *Winchester* had now been extended to any article that could be foreseen to result in injury if defectively made, privity notwithstanding. It now only remained for that decision to appear that would classify food as a "thing of danger."

Later American Decisions

The 1920 case of *Davis* v. *Van Camp* was typical of what occurred after MacPherson removed the negligence necessity. This suit was brought in implied warranty, a type of action not requiring the rigorous evidence of negligence. The mother of the plaintiffs had purchased a can of pork and beans from her grocer in Gypsom, Iowa. The can appeared to be normal. The pork and beans were eaten within an hour after purchase. The mother, father, and seven children constituted the family. The parents and three of the children ate none of the beans and remained well; four children ate the beans. All four suffered food poisoning and one, who ate nothing but pork and beans for the meal, died. The action was brought on behalf of one of the children who became ill from the pork and beans. In spite of the fact that the child was a minor and was lacking in privity, the Iowa Supreme Court ruled:

> We are of the opinion that the duty of a manufacturer to see to it that food products put out by him are wholesome, and the implied warranty that such products are fit for use runs with the sale, and to the public, for the benefit of the consumer, rather than to the wholesaler or retailer, and that the question of privity of contract in sales is not controlling, and does not apply in this case.[11]

Elements of the MacPherson decision stand out in this ruling. The manufacturer is responsible for dispensing a wholesome (safe) product; he is responsible to the public rather than just to the purchaser, and privity of contract is not controlling in the case. The difference was that food was involved rather than an automobile. Food was now a "thing of danger," and MacPherson and his collapsed automobile had crashed the door of the food industry.

It has been mentioned that landmark decisions that alter current law are rarely embraced quickly by the courts in a majority of states. As an example, *Chysky* v. *Drake* occurred in 1923, three years after *Davis* v. *Van Camp* and seven years after *MacPherson* v. *Buick Motor Co.* The plaintiff in this case had been given a piece of cake by her employer, the owner of a lunch room. A nail had been baked into the cake and it injured the plaintiff's gums so seriously that three teeth required extraction. The suit was brought against the baking company. The Court of Appeals of New York ruled:

> The plaintiff received the cake from her employer. By reason of its condition it was not fit for human consumption. Her employer bought the cake from the defendant. Is it liable to the plaintiff for the injury sustained? We do not think so. If there were an implied warranty which inured to the benefit of the plaintiff it must be because there was some contractual relationship between her and the defendant, and there was no such contract. She never saw the defendant, and, so far as it appears, did not know from whom her employer purchased the cake. The general rule is that a manufacturer or seller of food, or other articles of personal property, is not liable to third persons, under an implied warranty, who have no contractual relations with him. The reason for this rule is that privity of contract does not exist between the seller and such third persons, and unless there be privity of contract, there can be no implied warranty. The benefit of a warranty, either express or implied, does not run with a chattel on its resale. . . .[12]

As you can see, this decision was in opposition in every respect to *Davis* v. *Van Camp*. The doctrine of privity was the controlling factor in the case, and because of this lack of privity, an implied warranty was denied the plaintiff. And so, in usually liberal New York, the hallowed home of MacPherson, food was not yet a "thing of danger", while in Iowa, the heart of the conservative Midwest, food was an inherently dangerous substance and for this reason its public was entitled to recover damages for defective food in a breach of warranty suit, in spite of lack of privity.

By 1939, the assault on the doctrine of privity was in full swing, however. In *Jensen* v. *Berris*, the plaintiff was a member of a ladies pinochle club. The club members met for luncheon every two weeks, with each member taking a turn in providing and paying for the entertainment and luncheon. On one of these occasions, the plaintiff suffered food poisoning. She brought suit against the restaurant. The California Court of Appeals allowed recovery, even though the plaintiff was not the host for this luncheon. Since she did not pay for her meal, privity was lacking between her and the restaurant.[13]

In 1943, in *Haut* v. *Kleene,* the Illinois Supreme Court granted the right of all family members to recover damages when it stated: "To say that in the case at bar there was an implied warranty to . . . who purchased the rabbits for food but that it did not extend to his wife and children in our opinion does not make sense."[14]

And in 1945, again in Illinois, one of the plaintiffs was a third party. Parties A, B, and C were coworkers. Party A purchased a roast beef sandwich at lunchtime for party B. The purchase was made from a restaurant and at the request of party B. Party A returned with the sandwich to the place of employment of the parties, an apparel shop. Party B, noting that the sandwich neither looked nor tasted palatable, split the sandwich with party C. The sandwich made parties B and C quite ill and, as a result, they were hospitalized for four days each. Parties B and C brought suit against the restaurant, although party C was obviously a third party and lacking in privity with the restaurant, since she neither ordered nor payed for the sandwich. During the course of its ruling, however, the Court stated: "We are of the opinion that the implied warranty that the roast beef sandwich was fit for human consumption extended to both plaintiffs, although there was no contractual relation between defendant and . . . , one of the plaintiffs.[15]

In following the downfall of the doctrine of privity with regard to lawsuits resulting from unwholesome food, we are really following the development of a philosophy regarding a manufacturer's duty of care to a remote buyer or user of his product.

Winterbottom v. *Wright* held, in 1842, that the only safe rule to follow in liability cases was to confine the right to recover to those who were parties to the contract. Most courts quickly adopted this doctrine. However, injustices began to result from the uniform application of the privity doctrine. This led to the creation of exceptions.

The first exception of real importance occurred in *Thomas* v. *Winchester* in 1852. Here it was held that the general rule of nonliability did not apply to products that were inherently dangerous to human life. This exception was later extended to a wide variety of products and eventually spawned the historic *MacPherson* v. *Buick Motor Co.* decision of Justice Cordozo. Here it was held that any article that may be foreseen to result in injury to anyone properly using it is a "thing of danger." If the article is defectively made, then the manufacturer is liable for damages, not only to the purchaser, but to the public. Privity was not to be the controlling factor.

This decision was followed by rulings such as *Davis* v. *Van Camp,* which stated that manufacturers of canned goods must impliedly warrant their products, not only to the purchaser, but also to the public. This meant that food was also to be classified as a "thing of danger" and that the public was to be just as well protected, as far as recovery for sickness or injury was concerned, as was the user of any other inherently dangerous product, providing the product was defectively made.

Uniform Commercial Code

The matter was finally settled with the adoption of the Uniform Commercial Code. The privity provision in the Code was drafted in three alternatives. Each state adopting the code was to pick one of the three alternatives to follow in its court rulings.

Alternative A states: "A seller's warranty whether express or implied extends to any natural person who is in the family or household of his buyer or who is a guest in his home if it is reasonable to expect that such person may use, consume or be affected by the goods and who is injured in person by breach of the warranty. A seller may not exclude or limit the operation of this section."[16]

The following states have adopted Alternative A, either in toto or with revisions:

Alaska	Montana
Arizona	Nebraska
Arkansas	Nevada
Connecticut	New Jersey
District of Columbia	New Mexico
Florida	North Carolina
Georgia	Ohio
Idaho	Oklahoma
Illinois	Oregon
Indiana	Pennsylvania
Kentucky	Tennessee
Maryland	Virgin Islands
Michigan	Washington
Mississippi	West Virginia
Missouri	Wisconsin

Alternative B states: "A seller's warranty whether express or implied extends to any natural person who may reasonably be expected to use, consume or be affected by the goods and who is injured in person by breach of the warranty. A seller may not exclude or limit the operation of this section."[17]

The following states have adopted Alternative B, either in toto or with revisions:

Alabama	South Carolina
Colorado	South Dakota
Delaware	Vermont
Kansas	Wyoming

Alternative C states: "A seller's warranty whether express or implied extends to any person who may reasonably be expected to use, consume or be affected by the goods and who is injured by breach of the warranty. A seller may not exclude or limit the operation of this section with respect to injury to the person of an individual to whom the warranty extends."[18]

The following states have adopted Alternative C, either in toto or with revisions:

Hawaii
Iowa
Minnesota
North Dakota

Some states have enacted legislation of their own to deal with the problem. As an example, the state of Maine has enacted the following provision:

> Lack of privity between plaintiff and defendant shall be no defense in any action brought against the manufacturer, seller or supplier of goods for breach of warranty, express or implied, although the plaintiff did not purchase the goods from the defendant, if the plaintiff was a person whom the manufacturer, seller or supplier might reasonably have expected to use, consume or be affected by the goods.[19]

All states now have legislation that declares when privity is necessary and when it is not. If your state does not appear on the list of one of the three alternatives, it means that your legislators have enacted a code unique to your state. It will be found in that section of the state statutes covering sales.

"FOREIGN" OR "NATURAL"

We shall close this chapter with a discussion of another area in which courts have been at variance with one another in cases involving food. This difference of opinion pertained to the question of what substances were to be considered foreign to food that had been purchased, either in a grocery store or a restaurant. Such substances as glass, wires, nails, buttons, insects, and mice posed no problem for the courts. They were obviously foreign to the food that had been purchased and were generally considered to be an indication of negligence. We are concerned here with following the decisions of the courts as they attempted to grope with the problem of less obvious alien substances.

Suppose that you are dining out for supper and order chicken pie. During your meal, you unwittingly swallow a small chicken bone that was in the pie; it lodges in your throat causing injury. You require the services of a doctor. Can you collect for this unfortunate experience?

Early American Decisions

A case that resulted from just such an event was *Mix* v. *Ingersoll Candy Co.*, ruled on by the California Supreme Court in 1936. The customer who swallowed the chicken bone filed a suit for negligence and breach of implied warranty against the restaurant that served the chicken pie. In deciding in favor of the restaurateur, this court set forth a line of reasoning that was soon followed by other courts:

> It is not necessary to go so far as to hold that chicken pies usually contain chicken bones. It is sufficient if it may be said that as a matter of common knowledge chicken pies occasionally contain chicken bones. We have no hesitancy in so holding, and we are of the opinion that despite the fact that a chicken bone may occasionally be encountered in the chicken pie, such chicken pie, in the absence of some further defect, is reasonably fit for human consumption. Bones which are natural to the type of meat served cannot legitimately be called a foreign substance, and a consumer who eats meat dishes ought to anticipate and be on his guard against the presence of such bones. At least he cannot hold the restaurant keeper whose representation implied by law is that the meat dish is reasonably fit for human consumption, liable for any injury occurring as a result of the presence of a chicken bone in such chicken pie.
>
> Certainly no liability would attach to a restaurant keeper for the serving of a T-bone steak, or a beef stew, which contained a bone natural to the type of meat served, or if a fish dish should contain a fish bone, or if a cherry pie should contain a cherry stone—although it be admitted that an ideal cherry pie would be stoneless. The case of a chicken bone in a chicken pie is, in our opinion, analogous to the cited examples and the facts set forth in the first count of the complaint do not state a cause of action.[20]

Most courts were following the MacPherson and Tomlinson rulings by this time. Food was considered to be an inherently dangerous substance that, in most states, carried an implied warranty of wholesomeness. Many food-poisoning cases were being won by consumers against manufacturers. Cases involving glass, wires, insects, etcetera in food were being decided in favor of the consumer. This case, however, presented the courts with a new problem; the problem of a philosophy to cover the type of injury-causing substance that appeared in *Mix* v. *Ingersoll*.

The decision in *Mix* v. *Ingersoll* gave rise to the "natural to the type of food served" doctrine that held that if a substance or object, such as a chicken bone, remained in the food of which it was originally a part, it would not be considered a foreign substance to that food. Since it was natural to that type of food, no recovery for injury sustained from eating that food would be allowed unless negligence could be shown.

In 1938, a case similar to Mix was decided, also in California. The plaintiff was served roast turkey and dressing in the defendant's restaurant. Upon eating some of the dressing, she choked (Fig. 16-4) and finally coughed up a small turkey bone, injuring her throat in the process. She brought suit against the restaurant for damages. She won her suit in a lower court but the restaurant keeper appealed to a higher court, which reversed the decision stating: "that case [*Mix* v. *Ingersoll*] is controlling, and hence the judgment must be reversed."[21]

The case of *Brown* v. *Nebiker* occurred in the state of Iowa in 1941. A death had finally resulted from one of those errant bones. In this case, a sliver of bone from a pork chop punctured the deceased's esophagus and two days later he died. The unfortunate accident occurred in a restaurant, and the widow brought suit against the establishment. Once again, the court, in this case the Supreme Court of Iowa, held that the restaurateur was not liable for the appearance in a food of a substance natural to that food. Relying heavily on the Mix Case, the Court stated:

> In the case at bar, the meat served was pork chops. There are bones in all pork chops. If a sliver of bone is not a foreign substance in a chicken pie, or a turkey bone in turkey dressing, certainly small

Figure 16-4 "Bones which are natural to the type of meat served cannot legitimately be called a foreign substance." *Mix* v. *Ingersoll Candy Co.*, 1936. Based on original drawing by Cindy Paddack.

bones in a pork chop are not a foreign substance to the pork chop.

One who eats pork chops, or the favorite dish of spareribs and sauerkraut, or the type of meat that bones are natural to, ought to anticipate and be on his guard against the presence of bones, which he knows will be there.[22]

In 1944, another death occurred, this time in Illinois. A patron of a country club in Peoria swallowed a chicken bone that was in the creamed chicken she had ordered. The bone punctured her esophagus and she died from the resultant infection. Once again, however, a recovery suit was rejected, this time by the Illinois Supreme Court, as it ruled:

It is also common knowledge that bone is as much a part of the carcass of animals, fowl and fish as the meat. The bone usually remains with the meat during the process of cooking and preparation for human consumption.

The importance of pure food to the public must not be ignored. Modern conditions require that establishments serving food shall be operated in a sanitary way and furnish food that is wholesome and fit to be eaten. However, such rule should be construed and applied in a reasonable manner, taking into consideration the common experience of life. When viewed in this light, it must be conceded that practically all meat dishes, whether they consist of beef, pork, fish or fowl, do contain bones peculiar to the food being served.[23]

In 1949, the Georgia courts received their first case of this nature. The Court of Appeals, quoting from Mix, Silva, Brown, and Goodwin, also established itself as a member of the "natural to that food" group. The case involved a man who swallowed a pork bone that was contained in a barbecued pork sandwich. His suit to recover for medical expenses was not allowed, the Court stating:

A particle of bone in a food prepared from meat is something which one might ordinarily expect to find in the food, and one should anticipate its presence and guard against possible injury from swallowing it. Under the facts of this case the defendant was not required, in the exercise of ordinary care, to discover and eliminate every single particle of bone from the barbecued pork sandwich and the mere presence of a particle of bone in the sandwich does not authorize an inference of negligence in preparing and furnishing the food to the plaintiff.[24]

Later American Decisions

In 1951, a new philosophy, which would soon spread, was cautiously introduced by the Supreme Court of Rhode Island. In *Wood* v. *Waldorf System, Inc.*, the plaintiff, in consuming chicken soup in the defendant's restaurant, also consumed a chicken bone. The medical expenses were small but the emotional damage to the plaintiff was high, the result of choking on the bone. During the course of awarding $4,000 to the plaintiff for her experience, the Court introduced a new reasoning to the "foreign substances" area as it ruled:

We are asked by defendant's argument to take judicial notice of the fact that a chicken bone of this type is natural or perhaps necessary to the preparation and serving of chicken soup. We are not disposed to speculate as to how chicken soup is or should be prepared. Assuming that chicken bones are natural to and are used in the preparation of such soup, we do not think that it is necessary, natural or customary that harmful bones be allowed to remain concealed in this type of soup as finally dispensed to a customer so as to relieve the purveyor of such food of all responsibility. In our judgment the question is not whether the substance may have been natural or proper at some time in the early stages of preparation of this kind of soup, but whether the presence of such substance, if it is harmful and makes the food unfit for human consumption, is natural and ordinarily expected to be in the final product which is impliedly represented as fit for human consumption.[25]

This philosophy, which became known as the "reasonable expectation" philosophy, was drawn on in deciding *Bryer* v. *Rath Packing Co.* in 1959. A can of boned chicken was used to prepare chow mein in a school cafeteria located in Montgomery County, Maryland. While eating the chow mein, a student swallowed a small chicken bone, which lodged in her throat. The father brought suit against the packer of the boned chicken. The Court of Appeals of Maryland followed *Wood* v. *Waldorf*, rather than the previous traditional decisions and stated:

It is our opinion that the representations of the packer, the to-be-anticipated use of the chicken expected to be boneless, and the nature of the food in which the bones were found would permit the trier of fact to find that (a) the chicken bones or slivers that were in the chow mein were 'something that should not be there,' and (b) because of their presence the chow mein was not reasonably fit or safe for human consumption.[26]

This case was followed by *Betehia* v. *Cape Cod Corp.* in 1960. It was the first case of this nature in Wisconsin. In a lengthy decision, the Supreme Court of Wisconsin developed the "reasonable expectation" doctrine to a high degree. After discussing the "natural to that food" doctrine, the Court concluded:

This reasoning is fallacious because it assumes that all substances which are natural to the food in one stage or another of preparation are, in fact, anticipated by the average consumer in the final product served. A bone natural to the meat can cause as much harm as a foreign substance such as a pebble, piece of wire, or glass. All are indigestable and likely to cause injury. Naturalness of the substance to any ingredients in the food served is important only in determining whether the consumer may reasonably expect to find such substance in the particular type of dish or style of food served.

"The 'foreign-natural' test applied as a matter of law does not recommend itself to us as being logical or desirable. It is true one can expect a T-bone in a T-bone steak, chicken bones in roast chicken, pork bone in a pork chop, pork bone in spare ribs, a rib bone in short ribs of beef, and fish bones in a whole baked or fried fish, but the expectation is based not on the naturalness of the particular bone to the meat, fowl, or fish. There is a distinction between what a consumer expects to find in a fish stick and in a baked or fried fish, or in a chicken sandwich made from sliced white meat and in roast chicken. The test should be what is reasonably expected by the consumer in the food as served, not what might be natural to the ingredients of that food prior to preparation. What is to be reasonably expected by the consumer is a jury question in most cases; at least, we cannot say as a matter of law that a patron of a restaurant must expect a bone in a chicken sandwich either because chicken bones are occasionally found there or are natural to the chicken."[27]

We have, then, two solutions to the problem that existed. One solution, as set forth in *Silva* v. *F. W. Woolworth Co.*, states that "the criterion upon which liability is determined in such cases is whether the object causing the injury is 'foreign' to the dish served." The other solution, as stated in *Betehia* v. *Cape Cod Corp.*, is that "the test should be what is reasonably expected by the consumer in the food as served, not what might be natural to the ingredients of that food prior to preparation."

Every person involved in the preparation of food should realize that this difference in philosophy is still with us. The Uniform Commercial Code did not solve the problem in this area. The "natural to that food" philosophy has been followed in recent years in Ohio (oyster shell in fried oysters, 1960)[28] and Massachusetts (fish bone in fish chowder, 1964).[29]

The "what should be reasonably expected" philosophy has been followed in Oregon (cherry pit in cherry pie, 1966),[30] Florida (walnut shell in maple walnut ice cream, 1967),[31] New York (walnut shell in "nutted cheese" sandwich, 1974),[32] and Texas (chicken bone in chicken meat, 1976).[33]

Carelessness in the preparation and dispensing of some foods can cause injury or death. In states that follow the "reasonable expectation" philosophy, carelessness can be very expensive in this day of high medical costs and large jury awards. States that adhere to the "reasonable expectation" philosophy are currently in the majority.

SUMMARY

Privity is the legal term used to denote the bond between two or more persons who share in a contractual relationship. Early philosophy was set in *Winterbottom* v. *Wright* when the plaintiff's suit was rejected because of lack of privity. In *Thomas* v. *Winchester* it was ruled that privity need not exist in the case of mishandling dangerous drugs. In *Devlin* v. *Smith* it was ruled that privity was not necessary if negligence could be shown in the case of an article that was inherently dangerous.

In *Tomlinson* v. *Armour and Co.*, a consumer recovered damages for food poisoning against a meat packer in spite of lack of privity. Privity was dealt a further blow in *Parks* v. *Yost Pie Co.* when the widow of a purchaser of a contaminated pie won her suit.

The case of *Davis* v. *Van Camp* was brought in implied warranty. The suit was filed in behalf of a child who became seriously ill with food poisoning. The decision was awarded to the plaintiff, stating that privity is not a controlling factor and that a manufacturer is responsible to the public for dispensing a wholesome product. The decision in *Haut* v. *Kleene* allowed all family members to recover damages.

The Uniform Commercial Code Alternative A states: "A seller's warranty whether express or implied extends to any natural person who is in the family or household of his buyer or who is a guest in his home if it is reasonable to expect that such person may use, consume or be affected by the goods and who is injured in person by breach of the warranty."

The development of a uniform philosophy as to what shall be considered "foreign" and what shall be considered "natural" to foods has long been a problem in the courts. In *Mix* v. *Ingersoll Candy Co.*, a suit resulting from the swallowing of a chicken bone by a customer, it was ruled that chicken bones are natural to chicken pies, and the customer's suit was rejected. This gave rise to the "natural to the type of food served" philosophy.

In *Wood* v. *Waldorf System, Inc.*, a suit resulting from the swallowing of a chicken bone in chicken soup, it was ruled that no substance, if it is harmful, should be allowed to remain in the final product. This gave rise to the "reasonable expectation" philosophy. In recent years, most courts have followed this philosophy.

Review Questions

1. What is the meaning of the legal term *privity*?

2. What important legal philosophy was set in the case of *Winterbottom* v. *Wright*?

3. How did the decision in *Thomas* v. *Winchester* change the philosophy of privity?

4. What was the significance of the decision in *Tomlinson* v. *Armour and Co.*?

5. In what way was privity relaxed in the case of *Parks* v. *Yost Pie Co.*?

6. How did the case of *Davis* v. *Van Camp* differ from other actions brought in food-poisoning cases?

7. What does the Uniform Commercial Code Alternative A state with regard to privity?

8. a. What was the early philosophy with regard to bones in food as set forth in *Mix* v. *Ingersoll*? b. How did the case of *Wood* v. *Waldorf System, Inc.* change this philosophy?

References

1. Winterbottom v. Wright. 152 Eng. Rep. 402
2. Colonna v. Rosedale Dairy. 186 S.E. 94
3. Thomas v. Winchester. 6 N.Y. 397
4. Devlin v. Smith. 89 N.Y. 470
5. Curtain v. Somerset. 21 A. 244
6. Watson v. August Brewing Co. 52 B.E. 152
7. Tomlinson v. Armour and Co. 144 P. 202
8. Parks v. Yost Pie Co. 144 P.
9. Gearing v. Berkson. 111 N.E. 785
10. MacPherson v. Buick Motor Co. 111 N.E. 1050
11. Davis v. Van Camp. 176 N.W. 382
12. Chysky v. Drake Bros. Co. 139 N.E. 576
13. Jensen v. Berris. 85 P.2d 220
14. Haut v. Kleene. 50 N.E.2d 855
15. Blarjeske v. Thompson's Restaurant Co. 59 N.E.2d 320

16. Uniform Commercial Code. § 2-318, Alternative A

17. Uniform Commercial Code. § 2-318, Alternative B

18. Uniform Commercial Code. § 2-318, Alternative C

19. Maine Annotated Statutes. § 2,318

20. Mix v. Ingersoll Candy Co. 59 P.2d 144

21. Silva v. F. W. Woolworth Co. 83 P.2d 76

22. Brown v. Nebiker. 296 N.W. 366

23. Goodwin v. Country Club of Peoria. 54 N.E.2d 612

24. Norris v. Pig 'N Whistle Sandwich Shop, Inc. 53 S.E.2d 718

25. Wood v. Waldorf System, Inc. 83 A.2d 90

26. Bryer v. Rath Packing Company. 156 A.2d 442

27. Betehia v. Cape Cod Corporation. 103 N.W.2d 64

28. Allan v. Grafton. 164 N.E.2d 167

29. Webster v. Blue Ship Tea Room, Inc. 198 N.E.2d 309

30. Hunt v. Ferguson-Paulus Enterprises, Inc. 415 P.2d 13

31. Zabner v. Howard Johnson's Inc. 201 So.2d 824

32. Stark v. Chock Full O'Nuts. 356 N.Y.S.2d 403

33. Jim Dandy Fast Foods, Inc. v. Carpenter. 535 S.W.2d 786

165 2 by 2 inch slides, illustrating the topics discussed in *Sanitation for Foodservice Workers,* are available for rental or purchase. Please write Wade R. Nicodemus, 220 S. Logan St., Mason City, Illinois 62664 for details.

A 40 page printed commentary is included in the slide package. The commentary suggests brief comments for each picture and serves to coordinate the slides with the textbook material.

Index